STEM Road Map 2.0

Featuring a team of over thirty STEM education professionals from across the United States, the updated and revised edition of this landmark book provides an integrated STEM curriculum encompassing the entire K-12 spectrum, with complete grade-level learning based on a spiraled approach to building conceptual understanding.

Taking into account the last five years of evolution in STEM education, the second edition includes an increased focus on computer science, computational thinking, mathematics, and the arts, as well as cultural relevance and addressing the needs of diverse learners and underrepresented students. Divided into three main parts – *Conceptualizing STEM*, *STEM Curriculum Maps*, and *Building Capacity for STEM* – each section is designed to build common understandings of integrated STEM, provide rich curriculum maps for implementing integrated STEM at the classroom level, and offer supports to enable systemic transformation to an integrated STEM approach.

Written for teachers, policymakers, and administrators, this second edition is fully updated to account for the needs of K-12 learners in the innovation age. *STEM Road Map 2.0* enables educators to implement integrated STEM learning into their classroom without the need for extensive resources, empowering educators and supporting students.

Carla C. Johnson is Professor of Science Education and Faculty Research Fellow at North Carolina State University.

Erin E. Peters-Burton is Professor of Science Education and Educational Psychology at George Mason University.

Tamara J. Moore is Professor of Engineering Education in the School of Engineering Education at Purdue University.

STEM Road Map 2.0
A Framework for Integrated STEM Education in the Innovation Age

2ND EDITION

Edited by Carla C. Johnson, Erin E. Peters-Burton, and Tamara J. Moore

Routledge
Taylor & Francis Group
NEW YORK AND LONDON

Second edition published 2021
by Routledge
605 Third Avenue, New York, NY 10158

and by Routledge
2 Park Square, Milton Park, Abingdon, Oxon, OX14 4RN

Routledge is an imprint of the Taylor & Francis Group, an informa business

© 2021 Taylor & Francis

The right of Carla C. Johnson, Erin E. Peters-Burton, and Tamara J. Moore to be identified as the authors of the editorial material, and of the authors for their individual chapters, has been asserted in accordance with sections 77 and 78 of the Copyright, Designs and Patents Act 1988.

All rights reserved. No part of this book may be reprinted or reproduced or utilised in any form or by any electronic, mechanical, or other means, now known or hereafter invented, including photocopying and recording, or in any information storage or retrieval system, without permission in writing from the publishers.

Trademark notice: Product or corporate names may be trademarks or registered trademarks, and are used only for identification and explanation without intent to infringe.

First edition published by Routledge 2016

Library of Congress Cataloging-in-Publication Data
Names: Johnson, Carla C., 1969– editor of compilation, author. | Peters-Burton, Erin E., editor. | Moore, Tamara J., editor.
Title: STEM road map 2.0: a framework for integrated STEM education in the innovation age / Edited by Carla C. Johnson, Erin E. Peters-Burton, and Tamara J. Moore.
Description: Second edition. | New York, NY: Routledge, 2021. | Includes bibliographical references and index.
Identifiers: LCCN 2020038946 | ISBN 9780367473303 (hardback) | ISBN 9780367467524 (paperback) | ISBN 9781003034902 (ebook)
Subjects: LCSH: Science—Study and teaching (Elementary) | Technology—Study and teaching (Elementary) | Engineering—Study and teaching (Elementary) | Mathematics—Study and teaching (Elementary) | Science—Study and teaching (Secondary) | Technology—Study and teaching (Secondary) | Engineering—Study and teaching (Secondary) | Mathematics—Study and teaching (Secondary)
Classification: LCC LB1585 .S748 2021 | DDC 372.35/044—dc23
LC record available at https://lccn.loc.gov/2020038946

ISBN: 978-0-367-47330-3 (hbk)
ISBN: 978-0-367-46752-4 (pbk)
ISBN: 978-1-003-03490-2 (ebk)

Typeset in Sabon
by codeMantra

Contents

List of Tables	vii
List of Figures	xii
About the Editors	xiii
Preface	xiv

PART I
Conceptualizing STEM   1

1. The Need for a STEM Road Map   3
   CARLA C. JOHNSON, TAMARA J. MOORE, ERIN E. PETERS-BURTON, AND SELCEN S. GUZEY

2. The Emergence of STEM   14
   CATHERINE KOEHLER, IAN C. BINNS, AND MARK A. BLOOM

3. Integrated STEM Education   25
   TAMARA J. MOORE, LYNN A. BRYAN, CARLA C. JOHNSON, AND GILLIAN H. ROEHRIG

PART II
STEM Curriculum Maps   43

4. The STEM Road Map for Grades K-2   45
   CATHERINE KOEHLER, MARK A. BLOOM, AND ANDREA R. MILNER

5. The STEM Road Map for Grades 3–5   73
   BRENDA M. CAPOBIANCO, CAROLYN PARKER, AMANDA LAURIER, AND JENNIFER RANKIN

6   The STEM Road Map for Grades 6–8 — 102
    CARLA C. JOHNSON, TAMARA J. MOORE, JULIANA UTLEY, JONATHAN BREINER, STEPHEN R. BURTON, ERIN E. PETERS-BURTON, AND JANET B. WALTON

7   The STEM Road Map for Grades 9–12 — 133
    ERIN E. PETERS-BURTON, PADMANABHAN SESHAIYER, STEPHEN R. BURTON, JENNIFER DRAKE-PATRICK, AND CARLA C. JOHNSON

PART III
Building Capacity for STEM — 175

8   Data-Driven STEM Assessment — 177
    TONI A. SONDERGELD, KRISTIN L.K. KOSKEY, GREGORY E. STONE, AND ERIN E. PETERS-BURTON

9   Sociotransformative STEM Education — 202
    ALBERTO J. RODRIGUEZ

10 Effective STEM Professional Development — 219
    CARLA C. JOHNSON AND TONI A. SONDERGELD

11 Frameworks and Advocacy for STEM — 227
    SHAUN YODER, SUSAN BODARY, AND CARLA C. JOHNSON

Appendix A: Sample STEM Module One: Grade 7 — 255
JANET B. WALTON AND JAMES M. CARUTHERS

Appendix B: Sample STEM Module Two: Grade K — 327
JENNIFER SUH

Index — 353

Tables

3.1	Distinguishing Characteristics of Integrated STEM	28
4.1	Overview of K-2 STEM Road Map Themes and Topics by Grade	47
4.2	Kindergarten STEM Road Map Themes, Topics, and Problems/Challenges	49
4.3	STEM Road Map – Kindergarten Cause and Effect Theme: Motion	50
4.4	STEM Road Map Grades K-2 – Kindergarten The Represented World Theme: Patterns on Earth and in the Sky	51
4.5	STEM Road Map Grades K-2 – Kindergarten Sustainable Systems Theme: Habitats	53
4.6	STEM Road Map K-2 – Kindergarten Optimizing the Human Condition Theme: Our Changing School Environment	55
4.7	Overview of the Grade 1 STEM Road Map Themes, Topics, and Problems/Challenges	57
4.8	STEM Road Map K-2 – First-Grade Cause and Effect Theme: Influence of the Waves	58
4.9	STEM Road Map K-2 – First-Grade Innovation and Progress Theme: Coding a Rainbow	60
4.10	STEM Road Map K-2 – First-Grade The Represented World Theme: Patterns and the Plant World	61
4.11	STEM Road Map K-2 – First-Grade Sustainable Systems Theme: Protecting Endangered Species	63
4.12	Overview of the Grade 2 STEM Road Map Themes, Topics, and Problems/Challenges	65
4.13	STEM Road Map K-2 – Second-Grade Cause and Effect Theme: The Changing Environment	66
4.14	STEM Road Map K-2 – Second-Grade Innovation and Progress Theme: Launching a Spaceship	67
4.15	STEM Road Map K-2 – Second-Grade The Represented World Theme: Change over Time – Investigating Environmental Changes	69

4.16	STEM Road Map K-2 – Second-Grade Sustainable Systems Theme: Environmentally Friendly Playscapes	70
5.1	Third-Grade STEM Road Map Themes, Topics, and Problems/Challenges	75
5.2	STEM Road Map – Third-Grade Cause and Effect Theme: Predicting the Weather	77
5.3	STEM Road Map – Third-Grade Innovation and Progress: Transportation	78
5.4	STEM Road Map – Third-Grade Represented World: Recreational STEM	79
5.5	STEM Road Map – Third-Grade Sustainable Systems: Ecosystem Preservation	81
5.6	STEM Road Map – Third-Grade Optimizing the Human Experience: Reducing our Footprint	82
5.7	Fourth-Grade STEM Road Map Themes, Topics, and Problems/Challenges	84
5.8	STEM Road Map – Fourth-Grade Cause and Effect: Field Station Mapping	86
5.9	STEM Road Map – Fourth-Grade Innovation and Progress: Harnessing Solar Energy	87
5.10	STEM Road Map – Fourth-Grade Represented World: Erosion Modeling	88
5.11	STEM Road Map – Fourth-Grade Sustainable Systems: Hydropower Efficiency	89
5.12	STEM Road Map – Fourth-Grade Optimizing the Human Experience: Water Conservation	91
5.13	Fifth-Grade STEM Road Map Themes, Topics, and Problems/Challenges	92
5.14	STEM Road Map – Fifth-Grade Cause and Effect: Schoolyard Engineering	94
5.15	STEM Road Map – Fifth-Grade Innovation and Progress: Interactions	95
5.16	STEM Road Map – Fifth-Grade Represented World: Rainwater Analysis	97
5.17	STEM Road Map – Fifth-Grade Sustainable Systems: Composting	98
5.18	STEM Road Map – Fifth-Grade Optimizing the Human Experience: Mitigating Climate Change	100
6.1	Sixth-Grade STEM Road Map Themes, Topics, and Problems/Challenges	105
6.2	STEM Road Map – Sixth-Grade Cause and Effect: Human Impacts on Climate	106
6.3	STEM Road Map – Sixth-Grade Innovation and Progress Theme: Amusement Parks of the Future	107

6.4	STEM Road Map – Sixth-Grade Represented World: Communication	109
6.5	STEM Road Map – Sixth-Grade Sustainable Systems: Global Water Quality	110
6.6	STEM Road Map – Sixth-Grade Optimizing the Human Experience: Natural Hazards	111
6.7	Seventh-Grade STEM Road Map Themes, Topics, and Problems/Challenges	113
6.8	STEM Road Map – Seventh-Grade Cause and Effect: Transportation – Motorsports	115
6.9	STEM Road Map – Seventh-Grade Innovation & Progress: Space Travel	117
6.10	STEM Road Map – Seventh-Grade Represented World: Genetic Disorders	118
6.11	STEM Road Map – Sustainable Systems: Population Density	119
6.12	STEM Road Map – Seventh-Grade Optimizing the Human Experience: Genetically Modified Organisms	120
6.13	Eighth-Grade STEM Road Map Themes, Topics, and Problems/Challenges	123
6.14	STEM Road Map – Eighth-Grade Cause and Effect: Earth on the Move	124
6.15	STEM Road Map – Eighth-Grade Innovation and Progress: Medicine	125
6.16	STEM Road Map – Eighth-Grade Represented World: Learning from the Past	127
6.17	STEM Road Map – Eighth-Grade Sustainable Systems: Minimizing our Impact	128
6.18	STEM Road Map – Eighth-Grade Optimizing the Human Experience: The Role of the Sun in Life	130
7.1	Ninth-Grade STEM Road Map Themes, Topics, and Problems/Challenges	136
7.2	STEM Road Map – Ninth-Grade Cause and Effect Theme: Formation of the Earth	138
7.3	STEM Road Map – Ninth-Grade Innovation and Progress: Erosion and Weathering	140
7.4	STEM Road Map – Ninth-Grade Represented World: Global Models and their Uses	141
7.5	STEM Road Map – Ninth-Grade Sustainable Systems: Vital Systems of the Earth	143
7.6	STEM Road Map – Ninth-Grade Optimizing the Human Experience: Evaluating Human Impact on Nature	144
7.7	Tenth-Grade STEM Road Map Themes, Topics, and Problems/Challenges	146

7.8	STEM Road Map – Tenth-Grade Cause and Effect: Healthy Living	148
7.9	STEM Road Map – Tenth-Grade Innovation and Progress: Environmental Management	149
7.10	STEM Road Map – Tenth-Grade Represented World: Modeling Ecosystems	151
7.11	STEM Road Map – Sustainable Systems: Survival and Reproduction	152
7.12	STEM Road Map – Tenth-Grade Optimizing the Human Experience: Rebuilding the Natural Environment	153
7.13	Eleventh-Grade STEM Road Map Themes, Topics, and Problems/Challenges	155
7.14	STEM Road Map – 11th-Grade Cause and Effect: Standing on the Shoulders of Giants	157
7.15	STEM Road Map – 11th-Grade Innovation and Progress: Construction Materials	158
7.16	STEM Road Map – 11th-Grade Represented World: Radioactivity	160
7.17	STEM Road Map – 11th Grade Sustainable Systems: Green Building Rooftops	161
7.18	STEM Road Map – 11th-Grade Optimizing the Human Experience: Mineral Resources	162
7.19	Twelfth-Grade STEM Road Map Themes, Topics, and Problems/Challenges	165
7.20	STEM Road Map – 12th-Grade Cause and Effect: The Business of Amusement Parks	166
7.21	STEM Road Map – 12th Grade Innovation and Progress: Creating the Next Smart Phone	167
7.22	STEM Road Map – 12th-Grade Represented World: Car Crashes	169
7.23	STEM Road Map – 12th-Grade Sustainable Systems: Creating Global Bonds	170
7.24	STEM Road Map – 12th-Grade Optimizing the Human Experience: Dealing with Natural Catastrophes	172
8.1	Bloom's Revised Taxonomy Defined with Key Words and STEM Examples	181
8.2	Analytic Rubric Example for Potential/Kinetic Energy Task	190
8.3	Holistic Rubric Example for Potential/Kinetic Energy Task	191
8.4	What to Include in Your DDDM Process	194

Appendix A

A.1.1	Content Standards Addressed in STEM Road Map Module – Transportation – Motorsports	257
A.1.2	21st Century Skills Addressed in the STEM Road Map Module	260

A.1.3	Desired Outcomes and Monitoring Success	262	
A.1.4	Assessment Plan	262	
A.1.5	STEM Road Map Module Timeline – Weeks 1–3	263	
A.1.6	STEM Road Map Module Timeline – Weeks 4 and 5	264	
A.1.7	Key Vocabulary – Lesson One	266	
A.1.8	Key Vocabulary – Lesson Two	275	
A.1.9	Key Vocabulary – Lesson Three	285	
A.1.10	Key Vocabulary – Lesson Four	293	
A.1.11	Key Vocabulary – Lesson Five	303	
A.1.12	Key Vocabulary – Lesson Six	309	

Appendix B

A.2.1	Content Standards Addressed in STEM Road Map Module – Patterns on Earth and in the Sky	329
A.2.2	21st Century Skills Addressed in the STEM Road Map Module	330
A.2.3	Prerequisite Key Knowledge	331
A.2.4	Desired Outcomes and Monitoring Success	332
A.2.5	Assessment Plan	332
A.2.6	STEM Road Map Module Schedule Weeks 1 and 2	334
A.2.7	STEM Road Map Module Schedule Weeks 3 and 4	335
A.2.8	STEM Road Map Module Schedule Week 5	336
A.2.9	Key Vocabulary	337
A.2.10	Weather Pattern Flip Book Rubric	342
A.2.11	Key Vocabulary – Lesson Two	343
A.2.12	Calendar of Events and Infomercial Rubrics	348

Figures

8.1	Ideal interaction between state standards and classroom curriculum, instruction, and assessments. State standards guide development of the classroom learning environment, and classroom learning environment components all influence each other	179
8.2	Hierarchical structure of Bloom's revised cognitive taxonomy. Lower-level skills need to be mastered before higher-level skills	180
8.3	Matrix of results for an objective-type assessment. Students are ordered from highest to lowest performing based on total score	197
8.4	Matrix of results for a self-constructed type assessment. Ratings range from 1 to 3, whereby 1 = Not Evident, 2 = Needs Improvement, and 3 = Proficient	198
11.1	Transformative STEM practices	230
11.2	Horizontally aligned STEM partnerships	236
11.3	Vertically aligned STEM partnerships	237
11.4	Horizontally and vertically aligned STEM partnerships	240
11.5	Pre-K-12 STEM drivers	242
11.6	Five supporting enablers for PK-postsecondary success	251

About the Editors

Carla C. Johnson, Ed.D Professor of Science Education and Faculty Research Fellow North Carolina State University, USA. Dr. Johnson's research is primarily focused on enabling opportunity and access in STEM for all students. A main thread of this research is examining the implementation of STEM education policy, including STEM teaching and learning, STEM professional development, STEM schools/programs, and STEM networks and/or partnerships. Her research informs policy investments, new STEM models and frameworks, and STEM curriculum development.

Erin E. Peters-Burton, Ph.D. Donna R. and David. E. Sterling Endowed Professor in Science Education George Mason University, USA. Dr. Peters-Burton investigates STEM school models, the use of educational psychology to promote diversity in STEM, supporting student science and engineering practices with self-regulated learning tactics, integration of computational thinking and data practices, and K-12 STEM curriculum.

Tamara J. Moore, Ph.D. Professor, Engineering Education and Interim Executive Director, INSPIRE Research Institute for Pre-College Engineering Purdue University, USA. Her research is centered on the integration of STEM concepts in K-16 classrooms.

Preface

The original 2015 edition of the *STEM Road Map: A Framework for Integrated STEM Education in the Innovation Age* was the first resource for educators, administrators, community stakeholders, and advocates of STEM to guide K-12 schools in the direction of integrated STEM education. A team of over 30 STEM education professionals from across the United States collaborated on the important work of mapping out the Common Core standards in mathematics and English/language arts, the Next Generation Science Standards performance objectives, and the Framework for 21st Century Learning (ww.p21.org) into a coordinated, integrated, STEM education curriculum map.

A lot in the world of STEM has changed since 2015. We have a new federal STEM strategic plan that was developed with multi-organizational representation including industry, government, and educational entities. Many states now have STEM networks and support for doing STEM in K-12. Federal and other agencies are investing heavily into STEM out-of-school experiences for students, including competitions, programs, and apprenticeships K-20. Computer science and computational thinking are now front and center in what is considered to be some of the most innovative approaches to education.

Since the publication of the first edition, our team has developed 30 of the 60 total K-12-integrated STEM units that were included, and the National Science Teaching Association (NSTA) has partnered with us to publish each of these units into full books at a very low cost – making them available to teachers.

This book is the second edition of the STEM Road Map – which we have appropriately titled it as the *STEM Road Map 2.0: A Framework for Integrated STEM Education in the Innovation Age*. In this book, you will see updated chapters that include the latest thinking on STEM. Our curriculum maps have been infused with computational thinking as a central part of instruction.

The purpose of the *STEM Road Map: A Framework for Integrated STEM Education in the Innovation Age* is to make STEM for all students a reality. Currently, the STEM Road Map is the only integrated STEM curriculum map that is available that encompasses the entire K-12 spectrum with complete grade-level learning that has a spiraled approach to building conceptual understanding. The K-12 STEM Road Map is organized around five major STEM themes that include: Cause and Effect, Innovation and Progress, The Represented World, Sustainable Systems, and Optimizing the Human Experience. At each grade level, students will engage with a topic that was derived from the academic standards (e.g. Common Core and Next Generation Science Standards) that aligns with the selected theme. The *STEM Road Map: A Framework for Integrated STEM Education in the Innovation Age* places the power into the educators' hands to implement integrated STEM learning within their classrooms without the need for extensive resources.

The book is structured into three main parts designed to build common understandings of integrated STEM, provide rich curriculum maps for implementing integrated STEM at the classroom level, and support to enable systemic transformation to an integrated STEM approach. The three corresponding parts are Conceptualizing STEM, STEM Curriculum Maps, and Building Capacity for STEM.

The first section of the book, Conceptualizing STEM, is comprised of three chapters. Chapter 1 provides an overview of the need for a STEM Road Map and presents the five STEM themes that serve as the anchor for integrated STEM curriculum. Chapter 2 discusses the historical evolution of STEM and ties to the new academic standards. Chapter 3 establishes a conceptual and practical framework for integrated STEM.

In part II, STEM Curriculum Maps, there are four chapters with each corresponding to a grade band within K-12. Chapter 4 is the K-2 STEM Road Map that presents the framework for early childhood STEM learning. Chapter 5 focuses on grades 3–5 in upper elementary and presents the STEM curriculum maps for each theme and grade level. Chapter 6 moves into middle school grade levels (6–8) and continues the spiraling curriculum maps for STEM. Chapter 7 is the final chapter in this section, which focuses on integrated STEM in the high school setting (9–12).

Part III of the book is focused on Building Capacity for STEM and the series of chapters in this section are meant to serve as resources for implementing STEM. Chapter 8 provides an overview of effective STEM assessment and using data to drive integrated STEM instruction. Chapter 9 is focused on making STEM accessible to all learners through a sociotransformative approach. Chapter 10 provides guidelines for effective STEM professional development. Chapter 11 is the final chapter in the book and presents frameworks for effective STEM programs and STEM advocacy.

In the appendix of the book, two fully developed STEM Road Map sample curriculum modules are included. These are meant to serve as a model for schools that may want to develop their own, community-based and local context curriculum for STEM using the curriculum maps provided in the book.

Part I
Conceptualizing STEM

1 The Need for a STEM Road Map

*Carla C. Johnson, Tamara J. Moore,
Erin E. Peters-Burton, and Selcen S. Guzey*

Introduction

Since the publication of the first STEM Road Map book in 2015, there have been considerable changes in the world and within the United States overall. One constant in the system has been the pervasive need for improving STEM education to meet the needs of students in their future lives – regardless of career choice – and to grow a talent base for future innovation in STEM and non-STEM fields. Policy makers and educational leaders continue to advocate that the key to future prosperity of the United States is improving STEM teaching and learning opportunities for our children (Committee on Prospering in the Global Economy of the 21st Century, 2007; Committee on STEM Education of the National Science and Technology Council, 2018; Carnevale, Smith, & Melton, 2011; Dickman, Schabe, Schmidt, & Henken, 2009). This is driven by two distinct realities: current jobs as well as the jobs of the future are integrally STEM-driven and the foundation of STEM knowledge students receive in K-12 has been directly linked to the prosperity of our country. Our reality is that there are still formidable challenges for the United States in STEM education. In 2020, students in the United States continue to score below the international average in mathematics. Science performance is only average compared to our global peers (National Science Board, 2018). Further, students in the United States are not adequately prepared for success in STEM in college (ACT2), as access to equitable STEM education opportunities still remains a great challenge.

The need for STEM talent has continued to grow and is projected to increase nearly 10% from 2014 to 2024, and individuals working in STEM fields earned nearly 30% more than non-STEM counterparts in 2015 (United States Department of Commerce, 2017). Student mastery of STEM disciplines in K-12 schools is directly connected to success in college as well as economic growth and development, national security, and global competitiveness (Business Roundtable, 2008; Committee on Science, Engineering, and Public Policy, 2007; Committee on STEM Education of the National Science and Technology Council, 2018).

There have been several reports that have provided strategies for preparing our children for the STEM-wave of change. In 2018, the Committee on STEM Education of the National Science and Technology Council released *Charting a Course for Success: America's Strategy for STEM Education* which engaged an unprecedented group of stakeholders from governmental agencies, higher education, K-12 education, and informal education agencies. For all intents and purposes, this "federal STEM strategic plan" provided a five-year vision for moving the United States into a future "where Americans will have lifelong access to high-quality STEM education and the U.S. will be a global leader in STEM literacy, innovation, and employment" (National Science and Technology Council, p. v.). There are three goals of the plan, which include focus on STEM literacy (including computational thinking and digital literacy), increasing diversity and access in STEM, and preparing individuals for success in STEM careers.

This book, STEM Road Map 2.0, leverages the foundational work in integrated STEM education that was developed and presented in the first edition and further expands the thinking and strategies to align with the goals of the new plan. Additionally, as suggested in the Carnegie Foundation report (2009), the STEM Road Map provides an innovative curriculum design for delivering STEM learning more effectively across the continuum of K-12 schooling. The STEM Road Map project started as an ambitious undertaking by 25 leaders in STEM education from the various STEM disciplines (science, technology, engineering, and mathematics) as well as English/Language Arts and stakeholders from the realm of educational policy and reform. The focus of the effort was to address the need for innovative, integrated, problem, and project-based, high-quality curriculum for K-12 that would begin to address the prevalent issues within our educational system and provide teachers with a tool that would enable them to teach the Common Core (mathematics and English/language arts) along with the Next Generation Science Standards (NGSS) while infusing the 21st Century Skills Framework (www.p21.org) in a real-world, meaningful way. An integrated STEM approach is necessary for addressing global and local challenges as well as for success in careers of today and those anticipated in the future. Roehrig, Moore, Wang, and Park (2012) argued that our daily challenges are "multidisciplinary, and many require integration of multiple STEM concepts to solve them" (p. 31). The emerging new standards have responded to the call for a more interdisciplinary approach and have infused more critical thinking and integration of other content areas (e.g. English/Language Arts inclusion of science, NGSS focus on mathematics, and engineering).

The STEM Road Map 2.0 provides a complete, K-12 mapping of academic standards (i.e. Common Core and NGSS) organized by five STEM themes that students will experience in a spiraled curriculum that will grow their content knowledge and skills through application within five-week sequences of instruction organized around a problem or a project. The STEM Road Map curriculum is designed to be delivered by teachers in a collaborative, integrated manner where explicit ties to the actual project and/or problem are made within each content area each week of instruction, while one or more of the disciplines serve as the lead for delivery of the module. As a result, students will experience the overlapping nature of integrated STEM learning, and deeper conceptual understanding will be achieved in both STEM and non-STEM disciplines.

Since our publication of the original edition of *STEM Road Map: A Framework for Integrated STEM Education* (2015), we have partnered with the National Science Teaching Association (NSTA) to develop and publish 30 additional books – each of which represents one of the STEM Road Map grade-level units that were mapped out in the original STEM Road Map book. Our approach for STEM Road Map 2.0 is to leave those 30 units as they were originally mapped for this reason. The remaining 30 units were reimagined based upon the need to include more focus on computational thinking and digital literacy (including computer science) within the curriculum. Our plan is to move forward with mapping out and writing those additional curriculum books in the near future – making the entire series available for transforming K-12 STEM education.

Integrated STEM Themes in the STEM Road Map

The foundation of the STEM Road Map 2.0 is meaningful integration of the STEM disciplines within the context of real-world challenges and problems in K-12 classrooms (e.g. Breiner, Harkness, Johnson, & Koehler, 2012; Johnson, 2013; Rennie, Venville, & Wallace, 2012; Roehrig et al., 2012) with attention to computational thinking and digital literacy. Research has demonstrated the ability of an integrated STEM curriculum focus to improve academic outcomes and college preparedness (Johnson & Sondergeld, 2020). Integrated STEM is primarily about providing opportunities for students to learn in settings that require interdisciplinary boundaries to be crossed; in particular, integrated STEM education is an effort by educators to have students participate in engineering design and engineering thinking as a means to develop and/or explore technologies in a manner that requires deep learning and application of mathematics and/or science as well as consideration of other disciplines (e.g. social studies and English/language arts).

The STEM Road Map is organized around five real-world STEM themes that serve as the focus for delivery of the spiraled curriculum in grades K-12. Each of these themes will have a focused STEM topic within each grade level that is tied to the appropriate academic content standards. An overview of each theme is presented in this chapter to provide the context for the grade-level theme-based topics that will appear in this book.

Cause and Effect

The concept of Cause and Effect is a powerful and pervasive notion in the STEM fields. It is the foundation of understanding how and why things happen as they do. Humans spend considerable effort and resources trying to understand the causes and effects of natural and designed phenomena to gain better control over events and environment and to be prepared to react appropriately. Equipped with the knowledge of a specific cause and effect relationship, one can lead better lives or contribute to the community by altering the cause leading to a different effect. For example, if a person recognizes that irresponsible energy consumption leads to global climate change, that person can act to remedy their contribution to the situation. Although cause and effect is a core idea in the STEM fields, it is actually very difficult to determine. Students should similarly be capable of understanding when evidence points to cause and effect as well as when evidence points to relationships, but not direct causality. The major goal of education is to foster students to be empowered, analytic thinkers, capable of thinking through complex processes to make important decisions. An understanding of causality, as well as understanding when causality cannot be determined, will help students become better consumers, global citizens, and community members.

Innovation and Progress

The theme of Innovation and Progress as conceptualized for the STEM Road Map consists of ideas that use established concepts to move the STEM fields forward. One of the most important factors in determining if humans will have a positive future is innovation. Innovation is the driving force behind progress, which helps to make possibilities that did not exist before. Innovation and progress are creative entities, but in the STEM fields, they are anchored by evidence and logic. In creating something new, students must consider what is already known in the STEM fields and apply this knowledge appropriately. When we innovate, we create value that was not there previously and create new conditions and possibilities for even more innovations. Students should consider how their innovations might affect progress and use their STEM thinking to change current human burdens to benefits. For example, if we develop more efficient cars that use by-products from another manufacturing industry, such as food processing, then we have used waste productively and reduced the need for the waste to be hauled away, an indirect benefit of the innovation.

The Represented World

When we communicate about the world we live in, how the world works, and how we can meet the needs of humans, sometimes we can use the actual phenomena to explain a concept. However, sometimes the concept is too big, too slow, too small, too fast, or too complex for us to explain using the actual phenomena, and we must use a representation or a model to help communicate the important features. We need representations and models such as graphs, tables, mathematical expressions, and diagrams because it makes our thinking visible. For example, when explaining about geologic time, we cannot actually observe the passage of such large chunks of time, so we create a timeline or a model that uses a proportional scale to visually illustrate how much time has passed for different eras. Another example may be something too complex for students in a particular grade level, such as explaining the p subshell orbitals of electrons to fifth graders. Instead, we use the Bohr model, which more closely represents the orbiting of planets, which is accessible to fifth graders. When we create models, they are always helpful because they point out the most important features of a phenomenon. We also create representations of the world with mathematical functions, which help us to change parameters to suit the situation. Creating representations of phenomenon engages students because they enhance the information and communicate it directly. However, models also leave out some of the details that occur with the phenomena. Because models are helpful, but are also estimates of a phenomenon, it is important for students to evaluate their usefulness as well as what they leave out because they are estimates of an occurrence.

Sustainable Systems

We encounter sustainable systems in everything we do. Looking at a garden, you will see flowers blooming, weeds sprouting, insects buzzing, and various forms of life living within its boundaries. This is an example of an ecosystem, a collection of living organisms that survive together. This happens to be one type of "system" but if you look around, systems are all around us. From an engineering perspective, the term "systems" is the use of "concepts of component need, component interaction, systems interaction, and feedback. The interaction of subcomponents to produce a functional system is a common lens used by all engineering disciplines for understanding, analysis, and design" (Koehler, Faraclas, Giblin, Moss, & Kazarounian, 2006, p. 8). Systems can either be open (as in the example of an ecosystem) or closed (as in the example of a combustion engine). Ideally, a system should be sustainable (e.g. being able to maintain equilibrium without much energy from outside the structure). In our example of an ecosystem, the interaction of the organisms within the system and the influences of the environment (e.g. water and sunlight can maintain the system for a period of time thus demonstrating its ability to endure). Being sustainable is an ideal system as it allows for existence of the entity for the long-term. In our STEM Road Map project, we identified different standards that we consider to be oriented toward "systems" that students should know and understand in the K12 setting. Included in this systems-thinking, we have identified examples such as ecosystems, the rock cycle, earth processes (such as erosion, tectonics, ocean currents, and weather phenomena), Earth–Sun–Moon cycles, heat transfer, and the interaction between the geosphere, biosphere, hydrosphere, and/or atmosphere, to give a few. Students and teachers need to understand that we live in a world of systems and they are not independent of each other but instead, intrinsically linked so that disruption in one part of the system will have reverberating effects on other parts of the system.

Optimizing the Human Experience

The theme of Optimizing the Human Experience as conceptualized for the STEM Road Map consists of the notion that science, technology, engineering, and mathematics as disciplines have the capacity to continuously improve the ways humans live, interact, and find meaning in the world. This idea has two components: being more suited to our environment and being more fully human. For example, the progression of STEM ideas can help humans live more comfortably by providing unique ways to access water sources, design energy sources that do not have as much of an impact on our environment, develop new ways of communication and expression, and build efficient shelters. STEM ideas can also help humans to be self-actualized by providing access to the secrets and wonders of nature. Learning in STEM requires students to think logically and systematically, which is a way of knowing the world that is markedly different from knowing the world as an artist. However, we feel that when students can utilize various ways of knowing and understand when it is appropriate to use a different way of knowing or integrate ways of knowing, they are fully experiencing the best of what it is to be human. Learning to think like a STEM professional via the problem-based learning scenarios provided in the STEM Road Map helps students to optimize the human experience by innovating improvements in the designed world students live in as well as helping students develop ways of thinking like a STEM professional to ask questions and design solutions.

Infusion of Technology and Engineering in the STEM Road Map

In 2009, the National Academy of Engineering produced a report, *Engineering in K-12 Education: Understanding the Status and Improving the Prospects*, which explained some of the factors that make incorporating the topic of engineering difficult. First, most teachers do not have an engineering background, and as a result, there is not a critical mass of teachers who would feel comfortable or qualified to teach a curriculum that is exclusively about engineering subject matter. Second, the curricular demands on teachers are already overwhelming, and adding another topic to teach is not productive, particularly given the high stakes testing environment. Therefore, the National Academies of Engineering suggested two different strategies for implementation of engineering education in the current K-12 curriculum: infusion and mapping. Infusion is the proactive strategy of taking engineering standards and embedding them into the science and mathematics standards. The science and engineering practices in NGSS is an example of infusion because engineering standards have been added along with the science standards, for example, asking questions (science) and identifying problems (engineering). Mapping involves integrating big ideas in engineering onto current standards in other disciplines. The big ideas suggested in this report include engineering design, systems thinking, optimization, modeling, identifying constraints, analysis, communication, and engineering habits of mind. The STEM Road Map incorporates both mapping and infusion in the designed curriculum. Since the themes in the book are aligned to the NGSS and engineering standards are mapped into the Science and Engineering Practices as well as the Disciplinary Core Ideas, mapping of engineering standards is folded into the curriculum. Similarly, the themes were designed to support engineering ideas such as the ones recommended in the National Academies report; therefore, the STEM Road Map also infuses major engineering ideas into the integrated curriculum.

The Nature of STEM

The practice of integrating STEM topics has been around for a long time. However, on closer examination, STEM education is often accelerated or enriched science and mathematics education rather than integration. In the STEM Road Map, we have embraced a truly integrated STEM approach as a response to workforce and societal needs. Learning through multiple, integrated subjects can produce deeper conceptual understandings, better development of skills, and higher achievement than learning the subjects in isolation. Similar to the philosophy of the NGSS, we feel that learning concepts to pass a test is not enough; students should also be learning what it is like to think like a STEM professional and develop the requisite STEM habits of mind. A multidisciplinary approach can help students reinforce their learning across all four subjects in STEM. Creating a STEM learning environment can be accomplished by examining the nature of each discipline and considering what is alike and what is different about the core content areas that are integrated into STEM learning experiences.

There has been a great deal of work developed on the nature of science, technology, engineering, and mathematics individually, and the intention in this section of the book is not to delve deeply into each one, but to look at common features that might enhance teachers' understandings of integrated STEM learning. The nature of science (NOS) has been defined as the inherent guidelines that scientists follow in order to cultivate valid ideas about the natural world (Lederman, 1992; McComas & Olson, 1998). The nature of technology (NOT) explains features of technological advancements that extend humans' abilities to shape the world for goals ranging from survival needs to aesthetics (American Association for the Advancement of Science, 1993). The nature of engineering (NOE) can be described as what engineers do in the cyclical design process, how engineering impacts society, and how society impacts engineering (National Research Council, 2014). The nature of mathematics (NOM) can be considered the cycle of inquiry that begins with the representation of quantities as abstract symbols, accounting for all possibilities through manipulation of the rules (although there is some flexibility), and validating the quality of solutions and models by understanding the differences between mistakes and reasonable choices that did not turn out to be successful (Schoenfeld, 1992). All of these disciplines depend on iterative cycles of inquiry that lead to the development of valid and productive ideas. In these iterative cycles of inquiry, there are no rigid steps in the processes of the development of ideas, although they are guided by reasoned arguments. Therefore, STEM can be characterized as the human endeavor of anticipating outcomes based on background knowledge, making sense of what is observed, the use of logical reasoning, approaching unknowns systematically, and the necessity of transparency for the purposes of replicability and evaluation. An important feature of the outcome of the iterative cycles is that the process is self-righting. That is, if there is an error along the way, peer review, replication, and evaluation will help straighten out issues with the process of the investigation, a model created as a tool or a product, or the design process. STEM professionals, and STEM students, should recognize that choices in the cycles of inquiry are made for a reason, and the attempts to try to account for all possibilities are central features of their discipline.

Role of Computational Thinking in the STEM Road Map

In the STEM Road Map 2.0, computational literacy has been expanded as a central component of integrated STEM education. Digital devices and the internet have permeated every facet of our daily lives, and this includes the K-12 classroom. The importance of supporting students to gain digital literacy and empowering them to use tools for research and creation of new products and ideas is key. Computational thinking is a critical skill for all students today, and it is defined within the STEM Road Map 2.0 as a skill set that enables problem-solving using digital platforms and computer science. Computational thinking, sometimes known as proto-computational thinking (Tatar, Harrison, Stewart, Frisina, & Musaeus, 2017), are a suite of ideas that aid in solving problems by drawing on concepts central to computer science. When one understands computational thinking, they have additional tools to approach learning tasks. Computational thinking practices can be described with the following four components (Wing, 2006):

- Decomposition
- Pattern recognition
- Abstraction
- Algorithmic thinking

Decomposition is the process of breaking down a complex idea into less complex ideas. Students who may be overwhelmed with a large, long-term, or complex learning task may need to use decomposition to help them chunk a problem into smaller, more approachable parts. For example, when a student wants to test the factors that influence a period of a pendulum, they must first break down the phenomena of a pendulum swinging into the different influential variables. Only then can they begin to design the investigation.

Pattern recognition helps students to see regularities and irregularities in data, whether it is observational or numerical. When students actively and mindfully pursue pattern recognition in data collection, they can make meaning of the information as they collect it, resulting in efficient analysis. For example, when measuring temperature of ice melting on a heating pad, a student may notice that the temperature was rising, but at a certain point when there is both ice and water in the container, the temperature remains constant. A student who understands pattern recognition may try to find why the pattern of increasing temperature in the phenomena suddenly stopped, rather than immediately thinking it was an error of the equipment.

Abstraction is the process of finding the essence of an idea in cluttered information. Students who can use abstraction can recognize and separate the key information from the unnecessary, extraneous, or misleading information. For example, when a student collects data on an electromagnet, they may collect the weight of the system, but then realize that it is not central to the phenomena, employing abstraction.

Algorithmic thinking is the recognition and application of steps to reach a goal. When one is proficient in algorithmic thinking, they can create efficiencies in learning. For example, if a student can recognize the steps taken in choosing a graph for a particular type of data, they can employ the steps the next time they choose a graph to represent data.

STEM Road Map Curriculum Module Planning Template

The STEM Road Map: A Framework for Integrated STEM book includes the K-12 academic content standards, and Framework for 21st Century Learning (Partnership for 21st Century Learning, 2009) mapped out in a full pathway for implementing integrated STEM. Additionally, two full STEM Road Map curriculum modules have been included in the appendix to serve as a resource for implementation. Further, the STEM Road Map Curriculum Module Planning Template is also included to guide individual teachers, schools, districts, and other educational programs in the development of their own locally contextualized curriculum.

The STEM Road Map Curriculum Module Planning Template includes a summary of the module, established goals/objectives, content standards, 21st century themes and skills, as well as the overall challenge or problem that drives instruction in the grade-level topic of study. Other included areas of the template are module launch activity, key concepts, desired outcomes, assessment plan, resources, timeline, and then individual lesson plans.

References

American Association for the Advancement of Science. (1993). *Benchmarks for scientific literacy.* New York: Oxford University Press.

American College Testing. (2018). The condition of college and career readiness 2018. Retrieved from http://www.act.org/content/dam/act/unsecured/documents/cccr2018/National-CCCR-2018.pdf

Breiner, J., Harkness, M., Johnson, C.C., & Koehler, C. (2012). What is STEM? A discussion about conceptions of STEM in education and partnerships. *School Science and Mathematics, 112*(1), 3–11.

Carnegie Foundation. (2009). *The opportunity equation: Transforming mathematics and science education for citizenship and the global economy.* New York: Institute for Advanced Study.

Carnevale, A. P., Smith, N., & Strohl, J. (2010). *Help wanted: Projections of jobs and education requirements through 2018.* Washington D.C.: The Georgetown University Center on Education and the Workforce. Retrieved from https://cew.georgetown.edu/cew-reports/help-wanted/

Committee on Prospering in the Global Economy of the 21st Century. (2007). Rising above the gathering storm: Energizing and empowering America for brighter economic future. Retrieved from http://www.nap.edu/catalog/11463.html

Committee on STEM Education of the National Science and Technology Council (2018). *Charting a course for success: America's strategy for STEM education.* Washington, DC: National Science and Technology Council.

Dickman, A., Schabe, A., Schmidt, J., & Henken, R. (2009). *Preparing the future workforce: Science, technology, engineering and math (STEM) policy in K-12 education in Wisconsin.* Milwaukee, WI: Public Policy Forum.

Johnson, C.C. (2013). Conceptualizing integrated STEM education – Editorial. *School Science and Mathematics Journal, 113*(8), 367–368.

Johnson, C.C., & Sondergeld, T.A. (2020). Outcomes of an integrated STEM high school: Enabling access and achievement for all students. *Urban Education.* https://doi.org/10.1177/0042085920914368

Koehler, C., Faraclas, E., Giblin, D., Moss, D.M., & Kazarounian, K. (2006, June). Are concepts of technical and engineering literacy included in state science curriculum standards: A regional overview of the nexus between technical & engineering literacy and state science frameworks. Paper presented at the 2006 Proceedings of the American Society for Engineering Education Conference, Chicago, IL.

Lederman, N.G. (1992). Students' and teachers' conceptions of the nature of science: A review of the research. *Journal of Research in Science Teaching, 29,* 331–359.

McComas, W.F., & Olson, J.K. (1998). The nature of science in international standards documents. In W.F. McComas (Ed.), *The nature of science in science education: Rationales and strategies* (pp. 3–39). Dordrecht, The Netherlands: Kluwer Academic Publishers.

National Science and Technology Council. (2018). Charting a course for success: America's strategy for STEM education. Washington, DC: National Science and Technology Council. Retrieved from https://www.whitehouse.gov/wp-content/uploads/2018/12/STEM-Education-Strategic-Plan-2018.pdf

National Science Board. (2007). A national action plan for addressing the critical needs of the U.S. Science, Technology, Engineering, and Mathematics System. (Rep. No. NSB-07-114), Washington, DC: National Science Foundation.

National Research Council. (2014). *STEM integration in K-12 education: Status, prospects, and an agenda for research.* Washington, DC: National Academies Press.

Rennie, L., Venville, Gr., & Wallace, J. (2012). *Integrating science, technology, engineering, and mathematics: Issues, reflections, and ways forward.* New York: Routledge.

Roehrig, G.H., Moore, T.J., Wang, H.H., & Park, M.S. (2012). Is adding the E enough? Investigating the impact of K-12 engineering standards on the implementation of STEM integration. *School Science and Mathematics, 112*(1), 31–44.

Schoenfeld, A. (1992). Learning to think mathematically: Problem solving, metacognition, and sense making in mathematics. In D. Grouws (Ed.), *Handbook of research on mathematics teaching and learning* (pp. 334–370). New York: Macmillan Publishing Company.

Tatar D., Harrison S., Stewart M., Frisina C., & Musaeus P. (2017) Proto-computational thinking: The uncomfortable underpinnings. In P. Rich & C. Hodges (Eds.), *Emerging research, practice, and policy on computational thinking* (pp. 63–81). Springer. https://doi.org/10.1007/978-3-319-52691-1_5

United States Department of Commerce. (2017). STEM jobs: 2017 update. Retrieved from: https://www.commerce.gov/news/reports/2017/03/stem-jobs-2017-update

Wing, J.M. 2006. Computational thinking. *Communications of the ACM, 49*(3), 33–35. doi: 10.1145/1118178.1118215

2 The Emergence of STEM

Catherine Koehler, Ian C. Binns, and Mark A. Bloom

In the rapidly changing technological world where there has been a fundamental shift in the composition of the workforce, America needs to compete in a global market (Business-Higher Education Forum, 2002; Friedman, 2005; National Academy of Engineering [NAE], 2005; National Academy of Science [NAS], 2007; National Science Board, 2004; National Science Foundation, 2005; Smalley, 2003). Our nation's well-being depends upon how well we educate our children in science, technology, engineering, and mathematics (STEM) and prepare them for careers within these fields. It is through proficiency in these STEM fields that our economic and national security will maintain our competitiveness in this global competition. In *Rising Above the Gathering Storm* (NAS, 2007), the National Academy of Science warns us of the danger that "…Americans may not know enough about science, technology, or mathematics to contribute significantly to, or fully benefit from, the knowledge-based economy that is already taking shape around us" (p. 121). It is estimated that only approximately 6% of American undergraduate students major in engineering, while other countries boast much higher numbers: European countries (12%), Singapore (20%), and China (40%). Other indicators in this report included: (a) the US economy, though strong, has more investments in foreign stocks than in US stocks (remember that this report was prior to the 2008 US financial collapse); (b) the United States is sending many jobs overseas; and (c) advanced research in physics (e.g. the particle accelerator) is located outside the United States (NAS, 2007). Traditional classroom lecture methods are not preparing our youth for the challenges of the coming global change, and so we need to teach differently.

The *Rising Above the Gathering Storm* report (2007) provides guidance to improve global competitiveness of the United States through engagement with STEM and STEM education:

1 Increase America's talent pool by vastly improving K-12 science and mathematics education.
2 Sustain and strengthen the nation's traditional commitment to long-term basic research that has the potential to be transformational

in order to maintain the flow of new ideas that fuel the economy, provide security, and enhance the quality of life.
3   Make the United States the most attractive setting in which to study and perform research so that we can develop, recruit, and retain the best and the brightest students, scientists, and engineers from within the United States and throughout the world.
4   Ensure that the United States (a) is the premier place in the world to innovate, (b) invests in downstream activities such as manufacturing and marketing, and (c) creates high-paying jobs based on innovations.

This chapter focuses on the first recommendation: Increase America's talent pool by vastly improving K-12 science and mathematics education.

Historical Evolution of STEM

On October 4, 1957, the Soviet Union launched Sputnik 1 and rocked the world of science and science education. This small, silver satellite orbited the Earth approximately 1, 400 times before reentry into the atmosphere on January 4, 1958, 92 days after it was launched (NASA, 2014). The launch of Sputnik reverberated fear throughout the United States and the *Race to Space* was on. This monumental occasion marked an era that would change how curriculum would be evaluated, particularly the subjects of science and mathematics. It was clear to the American public that reform was needed in science and mathematics instruction. As a rapid reaction to the launching of *Sputnik 1*, NSF began funding curriculum projects such as the Physical Science Study Committee (PSSC), Earth Science Curriculum Project (ESCP), and Biological Science Curriculum Study (BSCS) (among others) that were developed and taught in schools across the United States. Mathematics also had its share of curriculum projects, including the School Mathematics Study Group (SMSG), the University of Maryland Mathematics project (UMMaP), and the Madison Project to name a few. With the development of these curriculum projects, teachers had difficulty with implementation as they did not have the content background to support these new reform efforts. Unfortunately, without content support and professional development, teachers reverted back to teaching content that was familiar to them using familiar pedagogical strategies: not representative of these new approaches to teaching and learning (Bybee, 2013). Technology and engineering were also onboard with curriculum initiatives in the 1970s with the development of *The Man Made World*, part of the Engineering Concepts Curriculum project (ECCP), but unfortunately, there was no place in schools to teach these concepts (International Technology Education Association [ITEA], 2009).

Despite the curricular efforts of the 1960s and 1970s, the 1983 report by the National Commission on Excellence in Education (NCEE),

A Nation at Risk, revealed a distressing picture of the education system in the United States (National Commission on Excellence in Education, 1983). Among other things, this report indicated that (a) US students were behind their peers from other developed nations with regard to science and mathematics, (b) many students did not possess "higher order" thinking skills, and (c) the average achievement of high school students was even lower than when Sputnik was launched. One of the many recommendations from this report was the development of standards of learning. It was this report that led to the development of *Project 2061: Science for All Americans* (American Association for the Advancement of Science [AAAS], 1989), which provided a framework for K-12 education and established the goal that all Americans must be literate in science, technology, and mathematics by 2061, the year Halley's Comet returns. *Project 2061: Science for All Americans* led to the development of the *Benchmarks for Science Literacy* (Benchmarks) (AAAS, 1993). The *Benchmarks* served as a set of coherent learning objectives leading to the outcomes of *Science for All Americans* for K-12 education and a foundation for most state's science standards. In 1996, the National Research Council (NRC, 1996) released the *National Science Education Standards* (NSES), which has been the last attempt at publishing a set of national science standards until 2013.

The national science standards as described in *Project 2061* and *Benchmarks* are not strictly focused on science content; they include engineering and technology standards. Both reform documents included five specific chapters related to STEM areas. In *The Nature of Mathematics* (Chapter 2) and *The Mathematic World* (Chapter 9), mathematics is described as a "science of patterns and relationships" and an "applied science" (AAAS, 1989, p. 16) and used as a "modeling process" that "plays a key role in almost all human endeavors" (p. 129). The *Nature of Technology* (Chapter 3) recommends that students have knowledge about the nature of technology as a requirement for scientific literacy (p. 25). *The Designed World* (Chapter 8) recommends that students have an understanding how technology and human activity shape our environment and our lives. The technologies this chapter focuses on include agriculture, manufacturing, energy sources/use, communication, information processing, and health technology.

It is not only important to know about the concepts of science, technology, engineering, and mathematics, but it is also equally important to be able to engage in the practices of these disciplines. In a chapter that brings together these ideas about science and technology practices, *Habits of Mind* (Chapter 12) outlines the values and attitudes toward science, mathematics, and technology. This chapter focuses on thinking skills that are necessary to engage in these disciplines: computation and estimation, manipulation and observation, communication, and critical response. Although the acronym STEM was not used in the context of these reform documents, all essential elements of the disciplines were mentioned.

What Is STEM?

The term STEM (science, technology, engineering, and mathematics) has its original roots in government policy and was coined by the National Science Foundation (NSF) in the early 1990s. The original term was actually "SMET" (science, math, engineering, and technology), but due to its similarity to a vulgar term, a program officer at NSF suggested that STEM be adopted (NAS, 2009; Saunders, 2009).

Recent research has indicated that even persons who deal with STEM on a daily basis are a bit confused as to its meaning and context. Breiner, Harkness, Johnson, and Koehler (2012) conducted a survey at a major research university in the Midwest and asked faculty members two questions, "What is STEM?" and "How does STEM influence and/or impact your life?" They reported that faculty members were able to identify STEM as separate disciplines, e.g. science, technology, engineering, and mathematics, but their conceptualization of the term was based solely on their academic discipline. For example, a faculty member who studied biology or worked in medicine might answer the first question with a response such as STEM as stem cell research or the stem of a plant.

In response to the second question, "How does STEM influence and/or impact your life?," it was noted that the faculty responses fell into three main categories: societal reasons, personal reasons, and a null (no) relationship to STEM. In the societal category, responses included: "It is life" and "develops competencies about basic skills used in life." In the personal category responses included: "I teach math" and "I used a bit of technology and I truly enjoy reading about science." Some faculty were unaware of the notion of STEM (the null relationship to STEM category) and their response consisted of not knowing what STEM was or "none that I am aware of." The most interesting finding under the personal reasons of how STEM influences/impacts your life, and these responses included a faculty member who was disenfranchised about STEM stating, "It further marginalizes my field since I am in the Humanities. It makes my field seem irrelevant, which STEM programs already do. It furthers narrow-minded thinking" (pp. 8–9). There has been little further research exploring these questions and, as such, the operational definition of STEM is left up to the parties as to how they will use it for their purposes of argument (Breiner et al., 2012).

As STEM is made up of four disciplines, one concern is the perception that the "T" (technology) and "E" (engineering) are oftentimes secondary to the "S" (science) and "M" (math) (ITEA, 2009; NAE, 2005). When we refer to STEM in K-12, it does not mean that students are learning math and science with a little sprinkle of technology and engineering mixed in, but instead, it refers to integration of the disciplines. ITEA advocates that students learn about the development of technology, with a sense toward "the study of all modifications humans have made in their natural environment for their own purposes" and as a disciple that includes the "study and application of learning experiences that relate to inventions, innovations, and changes intended to meet human needs and wants" (ITEA, 2009, p. 22). Different forms of technology have been included in the school setting for many years; however, engineering education has not yet made such inroads.

Engineering has not been adopted in the K-12 setting until very recently, and in only selected schools. Engineering has been strengthened in the K-12 system by the development of technology standards by ITEA.

Engineering as a discipline in the K-12 setting is often referred to as the missing letter in STEM. Because there are no nationally adopted academic standards for engineering for the K-12 setting, there is no student assessment in engineering education; thus, policy makers and school administrators pay little attention to it in K-12 schools. However, the National Academy of Engineering recommends that engineering concepts be infused into other subjects to illustrate the nature of big ideas such as design and systems thinking. The infusion approach is practical for curriculum design because the engineering design process is an iterative decision-making process that uses the content knowledge of mathematics and science as its foundation (Koehler, Faraclas, Giblin, Moss, & Kazerounian, 2013). This leaves an opportunity for STEM educators to design and implement innovative engineering activities that integrate the STEM disciplines in meaningful learning opportunities for students. The *Next Generation Science Standards* (NGSS Lead States, 2013) provides several options of how to implement the integration of standards in novel ways throughout grades K through 12, particularly in the field of engineering. In the last section of this chapter, we will discuss NGSS in more detail, and in particular, how it will guide science education in the future.

Federal Funding for STEM Initiatives

The Federal government has been a driving force behind STEM initiatives in the United States. STEM funding has been plentiful since 2007 when the Bush Administration signed into law the *America Creating Opportunities to Meaningfully Promote Excellence in Technology, Education, and Science Act*, known as America COMPETES Act. The emphasis of this law was "to invest in innovation through research and development, and to improve the competitiveness of the United States" (Government Printing Office, 2007, p. 1) and authorized $32.7 billion dollars between 2008 and 2010 for programs and activities in STEM-related disciplines.

It also established the creation of a National Science and Technology Summit, a group of Federal agencies, that were tasked to examine pathways for the US STEM initiatives, support basic research in physical sciences, propose improved instruction in mathematics, increase access for low-income students for AP/IB coursework, and to authorize Teacher Corps programs that would bring 30,000 mathematics and science teachers into the classroom (Bush, 2007).

The Obama Administration reauthorized this Act in 2010 and as part of the reauthorization established an office under the National Science and Technology Council (NSTC) that managed the coordination of STEM education activities in Federal agencies such as the NSF, National Aeronautics and Space Administration (NASA), National Oceanic and Atmospheric Administration (NOAA), and the Department of Education, among others.

Within the 2010 America COMPETES Reauthorization Act, there was a call for the NSTC to create a five-year federal STEM education strategic plan. In the 2013 progress report, the NSTC outlined five goals to drive Federal investment in STEM education. These goals include:

1. Improve STEM instruction by preparing 100,000 excellent new K-12 STEM teachers by 2020 and support the existing STEM teacher workforce;
2. Increase and sustain youth and public engagement in STEM by supporting a 50% increase in the number of US youth who have authentic STEM experiences each year prior to completing in high school;
3. Enhance STEM experience for undergraduate students by graduating one million additional students with degrees in STEM fields over the next 10 years;
4. Better serve groups historically under represented in STEM fields by increasing the number of underrepresented in STEM fields with STEM degrees (including women) over the next 10 years;
5. Design graduate education for tomorrow's STEM workforce by providing graduate-trained STEM professionals with basic and applied research expertise to acquire specialized skills in areas of national importance (NSTC, 2013, p. 15).

In another Federal initiative, *Race to the Top*, President Obama announced a challenge to states to create comprehensive education reform by establishing state-wide strategies to turn around student achievement, adopt rigorous and high-quality student assessments, teacher evaluations and professional development, and data systems to track student performance. This reform was rolled out as a competition among states. This program was funded with $4.35 billion dollars; an unprecedented amount for any education reform initiative. Within this plan, the President advocated what we now know as the Common Core, a common set of rigorous, career ready standards for math, and reading. Some of the funds from *Race to the Top* promoted the adoption of these standards. In the first-round competition, two states, Delaware and Tennessee, were awarded *Race to the Top* funds and a total of 18 states and the District of Columbia received funds through this program (US Department of Education [USDOE], 2014a).

Fund for STEM relies on the Federal budget and, as such, based on this five-year strategic plan written by the NSTC. In 2015, then President Obama proposed to support $170 million dollars for STEM education. In this budget, President Obama proposed several initiatives designed to improve teaching and learning in STEM subject areas for teachers and students and to train the next generation of innovators. He also proposed money allocated for STEM innovation networks to support partnerships between school districts and universities that would develop streamlined pathways to STEM education and careers. Teacher training is paramount, and this budget included funding for STEM teacher pathways to recruit and train STEM educators for high-need schools as well as a national program for STEM Master Teacher Corps that will develop teacher leaders who will advocate for STEM education in their communities (USDOE, 2014b).

Simultaneously in 2015, the Every Student Succeeds Act (ESSA, 2015) was signed into law, replacing the No Child Left Behind Act of 2001. The ESSA document, aligning with the funding for the 2015 Federal budget, stresses that a well-rounded education includes components of STEM in K-12 schools. It also advocates for the need to recruit and train teachers in STEM disciplines as well as establish STEM specialty schools to increase STEM access for all learners. A concerted effort to include computer science within the STEM disciplines was promoted.

In 2018, the United States launched a five-year strategic vision for STEM education in a report titled, *Charting a Course for Success: America's Strategy for STEM Education* (NSTC, 2018). This plan, referred to as the "North Star," guides the Federal government's commitment to STEM education by using evidence-based practices that engage stakeholders in equity and diversity to build a strong STEM ecosystem. A STEM ecosystem fosters a collaboration between business and education promoting innovation to enrich students' future career opportunities. In this vision, "all Americans will have lifelong access to high-quality STEM education and the United States will be the global leader in STEM literacy, innovation and employment (p. 1)."

In order to achieve this vision, the strategic plan sets three main goals: (1) build strong foundations for STEM literacy; (2) increase diversity, equity, and inclusion in STEM; and (3) prepare a STEM workforce for the future. In addition to these goals, the strategic plan sets forth four pathways described as cross-cutting approaches for achieving the goals:

(1) develop and enrich strategic partnerships; (2) engage students where disciplines converge; (3) build computational literacy; and (4) operate with transparency and accountability. However, there is no mention of how these programs are to be financed, and a review of the 2021 Federal budget indicates that the Administration will repeat cutting funding across all federal agencies. On a positive note, the Department of Education will receive more funding (from 7.4 to 90 million dollars) to expand Career and Technical Education, particularly in relation to computer science training. It will also receive increased funding ($150 million) for STEM initiatives targeted toward Minority Serving Institutions such as Historically Black Colleges and Universities and Hispanic Serving Institutions (American Institute of Physics, 2020).

NGSS and STEM Education

Funding streams for STEM initiatives are well defined, but the question remains, *how does this funding impact education*? To create a seamless pipeline from childhood to career, science educators created *A Framework for K12 Science Education (Framework)* (NRC, 2012) and a set of accompanying science standards, *Next Generation Science Standards (NGSS)* (NGSS Lead States, 2013) to address the need for content in that will drive the K-12 science education agenda for the foreseeable future. The format of *Framework* (and later *NGSS*) is much different than the older reform documents, *Project 2061, Benchmarks,* and *NSES,* as *Framework* outlines three very distinctive areas, or dimensions that K-12 science education needs to focus on for 21st century learners. These three dimensions include (a) science and engineering practices, (b) crosscutting concepts, and (c) core ideas in four disciplinary areas: physical sciences, life sciences, earth/space sciences, and engineering, technology, and applications of science (NRC, 2012, p. 2).

Just as *Project 2061* is the framework for *Benchmarks*, so is *Framework* the foundation for *NGSS*. Consider NGSS as a road map of student performance expectations that connects areas of practices, content, and crosscutting concepts as they relate to the disciplines of science. What makes *NGSS* so unique is the design of the standards. Each grade level has specific content standards and cross-matched to these standards are science and engineering practices, disciplinary core ideas, and crosscutting concepts. Teachers may be familiar with the terminology of science practices from the older reform documents, but the new language of NGSS changes the focus of the notion of practices. These practices describe how scientists and engineers approach problems and engage in investigations to solve these problems. The language in which we refer is pervasive throughout the document and consists of iterative conceptual modeling, engaging in argument from evidence, and constructing explanations and designing solutions. No longer is the student expected to be the passive learner by "merely learning about (these concepts) secondhand" (NGSS Lead States, 2013, p. xv), but instead, they are active learners that are "engaging in scientific investigations that require not only skills, but also knowledge that is specific to each practice" (p. xv). Students are presented with a "phenomenon" (a problem/project-based theme) where they can develop their own learning trajectory based on the questions, they ask about that phenomenon. The phenomenon is rich using the three-dimensions of learning, so students' understanding is deep. It is our intent with the STEM Road Map book to foster the development of these practices in more detail throughout later chapters in this book.

Next Steps in STEM Education

The 21st century learning and teaching approaches must go beyond the traditional ways of dispensing knowledge and rote memorization to one where the students take more responsibility for learning and the teacher becomes a facilitator of activities. As recommended in the *Framework*, problem/project-based learning (PBL) scenarios (or phenomenon) are active learning strategies that contextualize science. In a PBL scenario, students engage in their lessons by considering "problems as the starting point for gaining new knowledge" (Lambros, 2002, p. 1). Although *project*-based learning and *problem*-based learning are often used interchangeably, each approaches a situation through a problem scenario, but the end result differs. Project-based learning culminates with a tangible creation of a product whereas problem-based learning results in new knowledge (Capraro & Slough, 2009). Ideally, the integration of STEM disciplines within PBL allows the learner to holistically approach a real-world problem learning the content and tools necessary to provide its answer.

In the STEM Road Map book, we utilize PBL scenarios/phenomenon as challenges that are based on the five themes outlined in Chapter 1: Cause and Effect, Innovation and Progress, The Represented World, Sustainable Systems, and Optimizing the Human Experience to which the NGSS and Common Core mathematics and language arts standards were aligned. The grade bands K-2, 3–5, 6–8, and 9–12 were divided into chapters, and each chapter will describe how the standards align to each theme.

The STEM education initiatives in conjunction with NGSS three-dimensional learning strategies provide the opportunity for teachers to integrate PBL scenarios/phenomenon in their classrooms. These strategies not only include the disciplines of science, technology, engineering, and mathematics, but also integrate Common Core. It is a win–win situation for both K-12 teachers and students, but most importantly, the students will develop the skills and knowledge that are necessary to engage as informed global citizens and be prepared to proceed to STEM fields.

References

American Association for the Advancement of Science. (1989). *Project 2061: Science for All Americans*. New York: Oxford University Press.

American Association for the Advancement of Science. (1993). *Benchmarks for science literacy*. New York: Oxford University Press.

American Institute of Physics. (2020). *The Federal science budget tracker*. Retrieved from https://www.aip.org/fyi/2020/fy21-budget-request-stem-education

Anderson, R. D. (2002). Reforming science teaching: What research says about inquiry. *Journal of Science Teacher Education, 13*(1), 1–12.

Breiner, J. M., Harkness, S. S., Johnson, C. C., & Koehler, C. M. (2012). What is STEM? A discussion about conceptions of STEM in education and partnerships. *Journal of School Science and Mathematics, 112*(1), 3–11.

Bush, G. W. (2007). *Fact sheet: America Competes Act of 2007*. Retrieved from http://georgewbush-whitehouse.archives.gov/news/releases/2007/08/20070809-6.html.

Business-Higher Education Forum. (2002). *Investing in people: Developing all of America's talent on campus and in the workplace*. Washington, DC: American Council on Education

Bybee, R. W. (2013). *A case for STEM: Challenges and opportunities*. Arlington, VA: NSTA Press.

Capraro, R. M., & Slough, S. W. (2009). *Project-based learning: An integrated science, technology, engineering, and mathematics (STEM) approach*. Rotterdam, The Netherlands: Sense Publishers.

Engineering is Elementary. Retrieved from www.eie.org

Every Student Succeeds Act 20 U.S.C. § 6301. (2015). Retrieved from https://www.congress.gov/bill/114th-congress/senate-bill/1177

Friedman, T. L. (2005). *The world is flat: A brief history of the twenty-first century*. New York: Farrar, Straus and Giroux.

Government Printing Office. (2007). *H.R. 2272 (110th): America COMPETES Act of 2007*. Retrieved from https://www.govtrack.us/congress/bills/110/hr2272/text

Government Printing Office. (2010). *H.R. 5116 (111th): America COMPETES Reauthorization Act of 2010*. Retrieved from https://www.govtrack.us/congress/bills/110/hr2272/text

International Technology Education Association. (2009). *The overlooked STEM imperatives: Technology and engineering K-12 education*. Reston, VA: Author.

Koehler, C. M., Faraclas, E. W., Giblin, D., Moss, D. M., & Kazerounian, K. (2013). The nexus between science literacy & technical literacy: A state by state analysis of engineering content in state science frameworks. *Journal of STEM Education, 14*(3), 5–12.

Lambros, A. (2002). *Problem-based learning in K-8 classrooms*. Thousand Oaks, CA: Corwin Press, Inc.

NASA. (2014). *NSSDC/COSPAR ID: 1957-001B*. Retrieved from http://nssdc.gsfc.nasa.gov/nmc/spacecraftDisplay.do?id=1957-001B

National Academy of Engineering. (2005). *Engineer of the 2020: Visions of engineering in the new century*. Washington, DC: The National Academies Press. Also available at http://www.nap.edu/books/0309091624/html

National Academy of Science: Committee on Science, Engineering, and Public Policy. (2007). *Rising above the gathering storm: Energizing and employing America for a brighter economic future.* Washington, DC: The National Academies Press. Also available at http://www.nap.edu/books/0309100399/html

National Commission on Excellence in Education. (1983). *A nation at risk.* Washington, DC: Author.

National Research Council. (1996). *National science education standards.* Washington, DC: National Academy Press.

National Research Council. (2012) *A framework for K-12 science education: Practices, crosscutting concepts, and core ideas.* Committee on Conceptual Framework for new K-12 Science Education Standards. Board on Science Education, Division of Behavioral and Social Sciences and Education. Washington, DC: The National Academies Press.

National Science and Technology Council. (2013). *Federal science, technology engineering, and mathematics (STEM) education 5-year strategic plan* (17 U.S.C. 105). Retrieved from http://www.whitehouse.gov/sites/default/files/microsites/ostp/stem_stratplan_2013.pdf

National Science and Technology Council. (2018). *Charting a course for success: America's strategy for STEM education.* A Report from the Committee on STEM Education. White House Press.

National Science Board. (2004). *Science indicators, 2004, Volume 2, Appendix Table 2–34.*

National Science Foundation. (2005). *The engineering workforce: Current state, issues, and recommendations.* Final Report to the Assistant Director of Engineering, National Science Foundation. Task Force Members: Charles E. Blue, Linda G. Blevins, Patrick Carriere, Gary Gabriele, Sue Kemnitzer (Group Leader), Vittal Roa, and Galip Ulsoy, May, 2005.

NGSS Lead States. (2013). *Next generation science standards: For states, by states.* Washington, DC: National Academy Press.

Saunders, M. (2009). STEM, STEM education, STEM mania. *The Technology Teacher, 68*(4), 20–26.

Smalley, R. E. (2003). *Nanotechnology, the S&T workforce, energy, and prosperity.* Presentation to the President's Council of Advisors on Science and Technology (PCAST), Rice University, March 3, 2003. Retrieved from http://cohesion.rice.edu/NaturalSciences/Smalley/emplibrary/PCAST%20March%203,%202003.ppt#432,8,Slide8

US Department of Education. (2014a). *Race to the top fund.* Retrieved from http://www2.ed.gov/programs/racetothetop/index.html

US Department of Education. (2014b). *Science, technology, engineering, and math: Education for global leadership.* Retrieved from http://www.ed.gov/stem

3 Integrated STEM Education

*Tamara J. Moore, Lynn A. Bryan,
Carla C. Johnson, and Gillian H. Roehrig*

In this chapter, we provide an overview of what "integrated STEM" is – from its forms and characteristics to the practices and pedagogical approaches involved. Integrated STEM instruction is not meant to add to an already full curriculum, but to enhance the existing curriculum and find synergies among disciplines so that students can understand the interdependence among science, technology, engineering, and mathematics – for example, as they develop an understanding of and learn to explain natural phenomena or design and propose solutions to a local, national, or global problem. In turn, student learning will be more contextualized, authentic, and meaningful. The STEM Road Map is a tool to assist teachers in this process, identifying realistic intersections of content in the STEM disciplines, suggesting themes for situating integrated STEM instruction in a meaningful context, and providing a model for planning integrated STEM learning experiences.

What Is "Integrated STEM" Education?

There is more to integrating STEM disciplines than simply teaching two disciplines together or using one discipline as a tool for teaching another. In fact, this already happens for many teachers – for example, teaching science often requires the use of mathematics: e.g. graphing, measuring, utilizing ratios, and working with geometric shapes. However, we are referring to something more intentional and more specific when we use the term "integrated STEM." Drawing on the work of scholars who are credited with inspiring the movement to more meaningfully integrate the STEM disciplines at the K-12 level (Childress & Sanders, 2007; Sanders, 2009; Sanders & Wells, 2010 as cited in Sanders, 2015), we define integrated STEM as *the teaching and learning of the content and practices of disciplinary knowledge which include science and/or mathematics through the integration of the practices of engineering and engineering design of relevant technologies.* We take the viewpoint that, while any discipline can also have learning goals in integrated STEM environments, mathematics, science, and engineering will be the primary goals.

With this in mind, we will describe the forms of STEM integration, the hallmark characteristics, and the pedagogical practices of any integrated STEM learning environment.

Forms of STEM Integration

There are many forms and definitions of STEM integration (Moore, Johnston, & Glancy, 2020); however, STEM integration in the classroom generally takes one of three forms: content integration, context integration, or application/tool integration. Content integration refers to units and activities that have multiple STEM (and potentially others) disciplinary learning objectives, whereas context integration puts the learning emphasis on one or more content areas while using a context from another discipline. Context integration is often implemented in one of two ways: (1) through the use of a story that situates the disciplinary content goals in another discipline's practices or (2) through teaching one content area meaningfully in service of the main content area's learning outcomes (e.g. teaching data analysis/statistics in support of engineering or science learning). Application/tool integration refers to an application or a tool of one discipline, such as a simulation technology, is used within the teaching of another discipline as a way to teach the main discipline's learning objectives. This is the most common and simplest form of STEM integration but is useful nevertheless. Meaningful content integration is the ultimate goal of the STEM Road Map project; however, the STEM Road Map includes versions of all three forms. It is recommended that instruction has a good mixture of all three but keep the emphasis on content integration whenever possible.

Characteristics of STEM Integration

The characteristics of STEM integration are important to understand as these characteristics are the foundations for the STEM Road Map curricula discussed later in this book and published as full modules separately. Moore and colleagues (Moore, Guzey, & Brown, 2014; Moore & Hughes, 2018; Moore, Stohlmann, et al., 2014) developed a STEM integration framework for the purpose of guiding and assessing STEM integration curricula. The "Framework for STEM Integration in the Classroom" has seven primary elements that will be incorporated in the STEM Road Map:

1. In order to engage students in meaningful learning and provide access to the content, integrated STEM learning environments include a motivating and engaging context. These contexts should be personally meaningful and allow for students to connect with the content.

Integrated STEM Education 27

2   In order to develop problem-solving abilities, creativity, and higher-order thinking skills, integrated STEM education should include engineering design challenges of relevant technologies for compelling purposes. This can also include engineering thinking, technological progress, and reverse engineering of technologies.
3   STEM integration should allow for students to learn from failure and to redesign based on what was learned. This is one of the hallmarks of engineering thinking and should not be overlooked.
4   In order for the learning to be meaningful and worth the time, it takes to participate in project and problem-based learning challenges, integrated STEM education should include standards-based mathematics and/or science objectives in the learning activities. In addition, real-world problems are interdisciplinary beyond just the STEM disciplines. This means that other disciplines, such as English/language arts and social studies, should be included as appropriate.
5   In order to provide students with opportunities to learn the standards-based content deeply, it is imperative that content be taught in a student-centered manner. Students need opportunities to grapple with the content and think for themselves in order to deepen their conceptual knowledge.
6   Integrated STEM learning environments should emphasize teamwork and communication abilities that are imperative for life in a 21st century workforce.
7   STEM integration environments should not be siloed into the different components, but rather they should be integrated throughout the unit varying emphasis throughout the unit. This is especially true of the engineering design aspects of STEM integration. Making sure that engineering design is threaded throughout the STEM integration activity provides the context for learning all of the other disciplines.

Each curricular module within the STEM Road Map has been designed using these seven elements. However, the STEM Road Map also provides an extensive breadth of themes that students will encounter in a given year or grade level.

Pedagogical Practices of STEM Integration

While there are different models of integrating STEM content and practices, five core pedagogical practices related to the characteristics of STEM integration distinguish integrated STEM learning experiences from activities, lessons, or courses that simply find superficial 'connections' among STEM disciplines. These characteristics include instruction in which (1) the content and practices of one or more anchor science and mathematics disciplines define some of the primary learning goals; (2) the integrator is the engineering practices and engineering design of technologies as the context and/or an intentional component of the content to be learned; (3) the engineering design or engineering practices related to relevant technologies require the scientific and mathematical concepts through design justification; (4) the development of 21st Century Skills is emphasized; (5) the context of instruction requires solving a real-world problem or task through teamwork (Bybee, 2013; Moore, Stohlmann, et al., 2014; National Academy of Engineering & National Research Council, 2014; National Research Council [NRC], 2012b; Partnership for 21st Century Skills, 2015; Sanders, 2009). In Table 3.1, we provide a description of these five cores, distinguishing characteristics of integrated STEM instruction. These areas map to the STEM integration curriculum framework described in Chapter 1.

Table 3.1 Distinguishing Characteristics of Integrated STEM

Distinguishing Characteristic	Description
The content and practices of one or more **anchor science and mathematics disciplines** define some of the primary learning goals.	Anchor disciplines are the primary disciplines from which the learning goals for instruction are derived. Learning goals (what you want students to know) provide coherence between the instructional activities (how students will come to know what you want them to know) and assessments (how you determine whether students have come to know what you want them to know) (Wiggins & McTighe, 2005). Explicit attention is given within the learning goals to the connections between disciplines. By emphasizing the relationships of content across different disciplines, students develop deep, transferable understandings, and more coherent frameworks for reasoning about interdisciplinary problems and phenomena.
The integrator is the **practices of engineering and engineering design** as the context and/or an intentional component of the content to be learned.	An 'integrator' brings together different parts in a way that requires those parts to work together for a whole. As the integrator in integrated STEM, the practices of engineering and engineering design provide real-world, problem-solving contexts for learning and applying science and mathematics as well as meaningfully bring in other disciplines. In addition, engineering practices require students to use informed judgment to make decisions and help them develop habits of mind such as troubleshooting, pulling from prior experiences, and learning from failure (Moore, Guzey, & Brown, 2014).

(Continued)

Table 3.1 (Continued)

Distinguishing Characteristic	Description
The engineering design or engineering practices related to relevant technologies require the scientific and mathematical concepts through **design justification**.	High-quality STEM integration learning experiences meaningfully integrate the engineering design/practices with the science and mathematics content. Design justification is one way to require the students to apply the mathematics and science to the engineering design. For example, students should make recommendations for the design to their clients that are supported by the background information and content and the results from their tests as data for their decisions (e.g. Mathis, Siverling, Moore, Douglas, & Guzey, 2018). Justification of design choices is parallel to the argumentation in science education, i.e. claims, evidence, and explanation (Toulmin, 2008; see also Hand, Norton-Meier, Staker, & Bintz, 2009; Llewellyn, 2014; Sampson, Enderle, & Grooms, 2013)
The development of **21st century skills** is emphasized.	The phrase, "21st century skills," refers to the knowledge, skills, and character traits that are deemed necessary to effectively function as citizens, workers, and leaders in the 21st-century workplace (Bybee, 2010; NRC, 2012b; Partnership for 21st Century Skills, 2011).
The context of instruction requires **solving a real-world problem or task** through teamwork and communication.	A real-world problem or task centers on an authentic issue or meaningful challenge. As opposed decontextualized or contrived tasks (e.g. "cook-book" labs in science or rote problem solving in mathematics), real-world problems engage students in issues that are significant in everyday life and have more personal and/or social relevance. Furthermore, the teamwork involved in solving real-world problems or tasks provide opportunities to understand the interdisciplinary nature of STEM through rich, engaging, and motivating experiences that require teams of students to solve them. Teams of students need to communicate their processes and results (Carlson & Sullivan, 2004; Dym, Agogino, Eris, Frey, & Liefer, 2005; Frykholm & Glasson, 2005; Smith, Sheppard, Johnson, & Johnson, 2005).

As you will see in the upcoming chapters, the STEM Road Map provides a guide at each grade level for teachers to be able to design integrated STEM instruction to reflect these core, defining characteristics.

Commitments to Teaching and Learning

We should note that while there are defining characteristics of integrated STEM instruction, there are also characteristics of effective teaching and learning *in general* to which we are committed. Thus, our vision of integrated STEM is grounded in the following commitments about teaching and learning that embody recommendations from time-honored and contemporary education research:

- Learning is a generative and revisionary process in which students are responsible for constructing knowledge.
- Teaching requires deep, flexible content knowledge, pedagogical content knowledge, and reflective practices.
- Instruction should be culturally inclusive, socially relevant, and situated in authentic contexts.
- Quality instruction is guided by the content, approaches, and pedagogical principles of standards that are rigorous, coherent, and research-based.

Learning is a generative and revisionary process in which students are responsible for constructing knowledge. Decades of research in learning and cognition have shown that learning entails the development of conceptual constructs, reasoning processes, and patterns of activity. Instruction that is based on a generative and revisionary view of learning takes into account students' relevant prior knowledge, experiences, and interests – i.e. students are not blank slates when they come to our classrooms. Their existing understandings, experiences, beliefs, and interests influence how they will interpret what we are trying to teach. Furthermore, constructing knowledge is a progressive and iterative process that necessarily involves revision of ideas (Osborne & Wittrock, 1983; Posner, Strike, Hewson, & Gertzog, 1982; von Glasersfeld, 1989).

For teachers of integrated STEM, this means that the design of learning experiences must take students' existing knowledge into account, provide them the opportunity to become explicitly aware of their ideas, and help them build/revise their knowledge. Such learning experiences, in turn, will help students understand concepts more deeply and will be more personally meaningful and engaging. Instructional approaches including problem-based learning and project-based learning, which are discussed in this chapter, are examples of the types of approaches conducive to teaching integrated STEM in K-12 classrooms.

Teaching requires deep, flexible content knowledge, pedagogical content knowledge, and reflective practices. It is an intuitive, but nonetheless critical notion that a deep, flexible, and coherent understanding of content is a prerequisite to the development of knowledge for how to teach the content. Content knowledge enables the integrated STEM teacher to design conceptually coherent lessons, lead dynamic and in-depth discussions about STEM constructs and phenomena, and relate STEM content to meaningful and authentic situations. However, content knowledge is not enough, as teachers must also have knowledge of and reflectively think about learners, curriculum, instructional strategies, and assessment to be engaging and effective (i.e. pedagogical content knowledge or PCK; Bryan, 2003; Geddis, 1993; Shulman, 1987; van Driel, Verloop, & de Vos, 1998).

For teachers of integrated STEM, this means that they will need to demonstrate deep, flexible subject-matter knowledge and pedagogical content knowledge related to the disciplines of STEM education; well-developed knowledge and skills to integrate cross-cutting content, processes, and practices beyond their discipline of expertise; and well-developed knowledge and skills for teaching diverse student populations. Teaching integrated STEM will require teachers to understand the nature of STEM through the study of the content and practices of scientists, technologists, engineers, and mathematicians. Thus, integrated STEM teachers will be those who exhibit attributes of educational leadership – e.g. they will take the lead in implementing innovations and breaking the boundaries of 'siloed' subject area instruction; they will lead their colleagues in professional development by sharing and disseminating what they know and know how to do; they will lead in developing collaborations that enrich the learning experiences of their students. They will be reflective practitioners who use evidence from student learning artifacts to inform and revise practices and possess the disposition of a life-long learner.

Instruction should be culturally inclusive, socially relevant, and situated in authentic contexts. An abundance of educational research indicates that students bring to the classroom ways of knowing, thinking, and communicating that are reflective of their home and community environments and comprise part of the foundation of a students' classroom/educational experience (e.g. Bryan & Atwater, 2002; Fradd & Lee, 1999; Gay, 2010; Lee, 1999; Lemke, 2001). In short, the social and cultural life of students is central to their learning. Thus, teaching approaches that acknowledge, respond to, and celebrate the culture, motivations, and interests of students are more likely to engage them and facilitate their learning (Brophy, Klein, Portsmore, & Rogers, 2008; Carlson & Sullivan, 2004; Frykholm & Glasson, 2005; Gay, 2010). Situating integrated STEM instruction in culturally inclusive, socially relevant, and situated in authentic contexts provides students with opportunities to make sense of the situation at hand based on extensions of their own personal knowledge and experiences. Engaging contexts also provide a compelling purpose to do the challenge at hand, including but not limited to global, environmental, and social contexts that involve contemporary events and issues.

Quality instruction is guided by the content, approaches, and pedagogical principles of standards and practices that are rigorous, coherent, and research-based. Standards provide the foundation to inform and provide coherence among curriculum, instruction, and assessment within each discipline as well as, in the context of integrated STEM education, across disciplines. National level standards such as *Next Generation Science Standards* (NGSS) and *Common Core State Standards in Mathematics* (CCSS-M) are designed with a progression of academically rigorous disciplinary core ideas and practices that are scientifically and mathematically coherent. In addition, these standards are designed to prepare K-12 students for college and career readiness at an internationally competitive level (NGSS Lead States, 2013).

At the core of the STEM Road Map are four sets of education standards that articulate what students should know (knowledge) and be able to do (skills/practices) in each content area at each grade level: NGSS (NGSS Lead States, 2013), CCSS-M (National Governors Association Center for Best Practices and Council of Chief State School Officers [NGA and CCSSO], 2010b), Common Core State Standards in English Language Arts (CCSS-ELA; NGA and CCSSO, 2010a), and 21st Century Skills Framework (Partnership for 21st Century Skills, 2015).

You have probably noted that there are no technology or engineering standards explicitly mapped in the STEM Road Map. The fields of technology and engineering certainly have been a central part of the integrated STEM conversation, working to increase and expand attention to these disciplines in K-12 instruction. For example, the International Technology Education Association (ITEA) developed the *Standards for Technological Literacy* that define what students should know and be able to do to be technologically literate and outline content standards for technological literacy in grades K-12 (International Technology Education Association, 2007). The National Academy of Engineering (NAE) report, *Standards for K-12 Engineering Education?*, determined that developing separate K-12 engineering standards was not an appropriate approach as "it would be extremely difficult to ensure their usefulness and effective implementation" (NRC, 2010b, p. 1). Instead, the committee offered the strategy of integration for K-12 engineering education – embedding relevant engineering learning goals into standards for another discipline (e.g. science). This integrative approach addresses the importance of engineering design, making connections between engineering and other STEM disciplines, and communication. So while the STEM Road Map does not explicitly map these standards at each grade level, the theme-inspired topics for each grade level are organized around an engineering design challenge or project that contextualize and motivate the learning of integrated STEM content, practices, and skills.

What Are STEM Practices and Skills in Integrated STEM Instruction?

The fields of science, technology, engineering, and mathematics use specific knowledge and skills that form distinct practices (NGSS Lead States, 2013; NRC, 2012a). 'Practices' are the behaviors that professionals (in this case, scientists, mathematicians, and engineers) engage in as they investigate, design, and problem solve as well as build models, theories, and systems. Practices involve the use of both discipline knowledge and skills specific to each practice (NGSS Lead States, 2013). Therefore, instruction that integrates across STEM disciplinary boundaries necessarily facilitates students' understanding, development, and use of the various *practices* of science, technology, engineering, and mathematics (Berlin & White, 1995; Frykholm & Glasson, 2005; NGSS Lead States, 2013; NRC, 2012a). We describe below the essential practices of each of the STEM fields for K-12-integrated STEM instruction: science inquiry, engineering design, and mathematical thinking and reasoning. In addition, we also describe 21st Century Skills – skills that are intertwined with the development of STEM content knowledge and are an integral part of integrated STEM instruction.

Scientific Inquiry

Scientific inquiry refers to the diverse thinking processes and practices that scientists use to examine and answer questions about the natural world (NGSS Lead States, 2013; NRC, 2000). While scientific inquiry occurs in various forms, several central characteristics of scientific inquiry, when incorporated into K-12 instruction, enable students to construct knowledge of scientific ideas and understand the work of scientists. Inquiry in the integrated STEM classroom mirrors scientific inquiry by emphasizing students' questioning, collecting evidence, developing explanations, and communicating findings.

Engineering and Engineering Design

The field of engineering is focused on the design, manufacturing, and operation of efficient and economical technologies for a specific purpose. These technologies can be structures, machines, processes, and/or systems. The design of these technologies requires creative and carefully planned applications of scientific and mathematical concepts (Moore, Glancy, et al., 2014). As reflected in recent STEM-related reform documents such as *The Framework of K-12 Science Education* (NRC, 2012a) and the *Next Generation Science Standards* (NGSS Lead States, 2013), design processes are the heart of engineering practice and, as such, are a focus of engineering at the K-12 level. Engineering design processes can be represented by iterative and reflective practices on stages of design, such as *problem scoping, learning the background, planning for a solution, implementing a solution, testing the solution, and evaluating the tests of the solutions* (Moore, Glancy, et al., 2014). An integral part of engineering design is engineering thinking or habits of mind: systems thinking, creativity, optimism, perseverance, innovation, collaboration, communication, and ethical thinking. Additionally, engineers must manage risk and uncertainty, learn from failure, consider the safety of those developing and using the technologies designed, and consider prior experience (Moore, Stohlmann, et al., 2014). Engineering design coupled with engineering thinking allows students to become independent, reflective thinkers who have learned to integrate multiple ideas together to solve problems.

Mathematical Thinking and Reasoning

Mathematics is a human-developed way of thinking and knowing that investigates ordering, operational, and structural relationships in a logical manner (Gilfeather & del Regato, 1999). As a discipline, mathematics is concerned with the development of new mathematical knowledge. Mathematicians develop new mathematics through considering all of the body of current mathematics and extending that body of knowledge through logical development of new mathematical structures. This results in a new mathematics that is still internally consistent with the previous body of mathematics. This way of thinking – that is the hallmark of mathematics – is also parallel to the mathematical thinking and reasoning students should develop. Often times school mathematics curricula are about learning to think 'inside-the-box;' however, mathematical thinking is learning to think flexibly and 'outside-the-box' (Devlin, 2012). The CCSS-M provides the mathematical practices as guidelines, which include:

- Make sense of the problem and persevere in solving it;
- Explain the meaning of a problem and look for solution entry points;
- Reason abstractly and quantitatively;
- Decontextualize – create abstractions of a situation and represent it as symbols and manipulate;
- Construct viable arguments and critique the reasoning of others;
- Model with mathematics;
- Use appropriate tools strategically;
- Attend to precision;
- Look for and make use of structure;
- Look for and express regularity in repeated reasoning (NGA and CCSSO, 2010b).

These represent the ways of mathematical thinking and reasoning both from the viewpoint of the mathematician and the K-12 student. Developing competency in mathematical thinking and reasoning is a fundamental goal of a mathematics education because it allows students to bring the all of the knowledge of mathematics to bear in different situations.

21st Century Skills

Cognitive, intrapersonal and interpersonal skills and abilities necessary to effectively function as citizens, workers, and leaders in the 21st century workplace are often referred to as "21st Century Skills" (Bybee, 2010; NRC, 2010a, 2012b; Partnership for 21st Century Skills, 2015). Research suggests that these skills are increasing in value across a wide range of jobs, whether low-skill, low-wage service-oriented positions or high-skill, high-wage professional-oriented positions (Bybee, 2013; Levy & Murnane, 2004). Integrated STEM is a promising context for offering K-12 students opportunities to develop these skills, as these skills are embedded to some degree in all of the STEM disciplines.

There exist several lists of 21st century skills (e.g. Bybee, 2010; NRC, 2010a, 2012b; Partnership for 21st Century Skills, 2015) with considerable overlap. In the STEM Road Map, authors have utilized the P21 Framework (Partnership for 21st Century Skills, 2015) as the conceptual guide for curriculum mapping and module development. The P21 Framework includes core components of 21st Century Themes, Learning and Innovation Skills, Information, Media, and Technology Skills, and Life and Career Skills. In addition to a main and important focus on core content, the P21 Framework promotes development of student understanding of 21st Century Themes that include global awareness, financial, economic, business, and entrepreneurial literacy, civic literacy, environmental literacy, and health literacy (Partnership for 21st Century Skills, 2015).

The Learning and Innovation Skills in the P21 Framework focus on creativity and innovation, critical thinking and problem solving, and communication and collaboration. Information, Media, and Technology skills are inclusive of information literacy, media literacy, and information, communications, and technology literacy. The final area of the P21 Framework is Life and Career Skills, and this area emphasizes flexibility and adaptability, initiative and self-direction, social and cross-cultural skills, productivity and accountability, and leadership and responsibility (Partnership for 21st Century Skills, 2015).

Pedagogical Approaches to Teaching Integrated STEM

Two of the primary pedagogical approaches to teaching integrated STEM include project-based learning and problem-based learning. These two terms often are used interchangeably, though there are distinct differences between them. Both of these pedagogical approaches utilize an open-ended problem, question, or challenge to begin the instructional sequence and are designed to be an authentic application of 21st Century Skills in the context of learning and mastering new integrated STEM concepts. Students work collaboratively and access multiple tools and data sources to solve the problem, question, and/or challenge. Project and problem-based learning often include partners from outside the school who are STEM experts and who provide a connection with STEM industry and careers to broaden the scope of the content and bring the learning alive for integrated STEM students.

Project-based learning specifically has, as the outcome of the work, a product that has been conceptualized, designed, and tested to determine if the product is a viable solution to the problem (e.g. Blumenfeld et al., 1991; Krauss & Boss, 2013). Most project-based learning is tied to a local need or problem where the question is student generated. This could be a school or community-based issue. Project-based learning units of instruction often require multiple weeks to complete. Student teams present their products to their classmates and the larger community, often to model the process that a STEM professional would complete.

Problem-based learning is driven by fictitious scenarios or case studies that may not fit within a local or community issue (e.g. Barell, 2006; Lambros, 2004). Students are presented with an ill-structured problem that may often be a global or persistent concern for society, or a potential problem that may need to be solved in the present or future (Johnson, 2003). Problem-based learning instructional units often have a more narrowly specified outcome, which may include a solution or point of view rather than a tangible product (Johnson, 2004). The duration of problem-based learning experiences tends to be shorter, lasting days or weeks.

It is essential with integrated STEM learning that the pedagogy that drives the learning has an integrated focus and students engage with an important topic, problem, or issue that is either teacher or student generated (NRC, 2011). Problem and project-based learning pedagogy provide the best-in-class model, which currently exists for implementing integrated STEM education at all levels K-12.

The Continuum of STEM Integration

The Framework for K-12 Science Education (NRC, 2012a) aligns with the recommendations for an integrated approach from the *Engineering in K-12 Education* (NRC, 2009) and *Standards for K-12 Engineering Education* (NRC, 2010b). The Framework authors emphasize that students should become familiar with engineering practices through increasingly sophisticated experiences with them across the grades. The Framework authors are clear that "not every such practice will occur in every context" (NRC, 2012a, p. 247), but rather that

> the curriculum should provide repeated opportunities across various contexts for students to develop their facility with these practices and use them as a support for developing deep understanding of the concepts in question and of the nature of science and of engineering.
>
> (NRC, 2012a, p. 247)

Thus, it is necessary for articulation of STEM to occur across K-12 to ensure that all students have these developing experiences with the practices of engineering. While individual teachers and grade-level teams will have important decisions to make about STEM integration within a single grade, it is important that the progression of learning related to STEM integration is delineated across all grade levels.

It is useful to consider the experiences of school systems within states that have already adopted engineering into their K-12 science standards. For example, Minnesota adopted engineering standards as part of their new state science standards in 2009. Professional development has been provided for K-12 teachers across the state to enhance the implementation of integrated STEM teaching (Guzey, Tank, Wang, Roehrig, & Moore, 2014; Roehrig, Moore, Wang, & Park, 2012). Associated research into STEM integration within K-12 school systems reveals some important considerations for school systems, schools, and teachers. Elementary schools already had an articulated scope and sequence for science standards across grades K-5. Engineering was blended into this existing scope and sequence by identifying content that naturally integrated STEM concepts or where existing engineering curriculum and/or kits existed like *Engineering is Elementary*, which could be integrated with existing science units. Elementary programs also provide a space for integration with non-STEM disciplines; indeed, research has shown that the addition of literacy approaches can improve learning in STEM (Tank, 2014).

Middle schools have taken two approaches: (1) required engineering content courses and (2) integration of engineering into existing science courses. If an engineering course is to be used to address engineering practices it is important that it is not an elective offering, as engineering practices are required for all students. It is also important that "every science unit or engineering design project must have as one of its goals the development of student understanding of at least one disciplinary core idea" (NRC, 2012a, p. 201). Our experiences with middle schools that have adopted engineering curriculum as a STEM course, such as the Project Lead the Way (PLTW) Gateway to Technology, is that there is limited integration of science and mathematics concepts in the engineering design challenges (Roehrig et al., 2012; Stohlmann, Moore, & Roehrig, 2012). Integration of engineering into existing science courses can be done successfully, and our experience is that professional development and team planning for engineering integration improve quality and ensure that all students have experiences with STEM integration (Roehrig et al., 2012).

While physical science offers more natural spaces for STEM integration, it is important that students experience integrated STEM across all science disciplines. Many high schools have relegated engineering to ninth-grade physical science teachers as this is seen as a more natural fit for engineering design challenges and a course normally required for all students. We note here that biology provides important opportunities to discuss other aspects of K-12 engineering standards, for example, ethics of cloning and genetic engineering. It is not necessary at the course or lesson level to include all aspects of engineering; as long as care is taken that all aspects of a quality engineering education (Moore, Glancy, et al., 2014) are included somewhere within the scope and sequence of K-12.

Overview of Integrated STEM in this Book

Each grade band map within the STEM Road Map suggest the content area standards that serve as anchor content for an integrated STEM instructional module. Each grade level in the STEM Road Map includes five key topics (modules) for exploration. The STEM Road Map for a given grade level includes at least one module that is led by science, mathematics, English/language arts, and social studies – promoting the integration of STEM across the curriculum. In addition, the 21st Century Skills emphasized in the module are aligned with the content standards. For example, in the sixth-grade Transportation – Motorsports module, student teams are challenged to design a prototype vehicle with a new safety aspect that will ensure drivers are protected from potential accidents. This module is driven by science with very important connections to mathematics, as students will test their prototypes and conduct various formula calculations to determine efficacy. The science content in this module focuses on motion, force, energy, speed, and Newton's laws. Students will utilize mathematics practices as they make sense of the problems, reason abstractly and quantitatively, and model with mathematics. Engineering design and engineering thinking are integral to the topics and modules included in the STEM Road Map. In this module, students utilize engineering design to develop, test, and modify their vehicle prototypes. Embedded throughout the module is 21st Century Skills development, as students enhance their understandings of economic and financial literacy through their application of learning and innovation skills in this module.

References

Barell, J. (2006). *Problem-based learning: An inquiry approach.* Thousand Oaks, CA: Corwin Press.

Berlin, D. F., & White, A. L. (1995). Connecting school science and mathematics. In P. A. House & A. F. Coxford (Eds.), *Connecting mathematics across the curriculum.* Reston, VA: National Council of Teachers of Mathematics.

Brophy, S., Klein, S., Portsmore, M., & Rogers, C. (2008). Advancing engineering education in P-12 classroom. *Journal of Engineering Education, 97*(3), 369–387. https://doi.org/10.1002/j.2168-9830.2008.tb00985.x

Bryan, L. A. (2003). Nestedness of beliefs: Examining a prospective elementary teacher's beliefs about science teaching and learning. *Journal of Research in Science Teaching, 40*(9), 835–868. https://doi.org/10.1002/tea.10113

Bryan, L. A., & Atwater, M. M. (2002). Teacher beliefs and cultural models: A challenge for teacher preparation programs. *Science Education, 86*(6), 821–839. https://doi.org/10.1002/sce.10043

Blumenfeld, P., Soloway, E., Marx, R., Krajcik, J., Guzdial, M., & Palincsar, A. (1991). Motivating project-based learning: Sustaining the doing, supporting learning. *Educational Psychologist, 26*(3), 369–398. https://doi.org/10.1080/00461520.1991.9653139

Bybee, R. W. (2010). *The teaching of science: 21st century perspectives.* Arlington, VA: NSTA Press.

Bybee, R. W. (2013). *The case for STEM education: Challenges and opportunities.* Arlington, VA: NSTA Press.

Carlson, L., & Sullivan, J. (2004). Exploiting design to inspire interest in engineering across the K-16 engineering curriculum. *International Journal of Engineering Education, 20*(3), 372–380. https://www.ijee.ie/articles/Vol20-3/IJEE2502.pdf

Childress, V., & Sanders, M. (2007). *Core engineering concepts foundational for the study of technology in grades 6–12.* http://citeseerx.ist.psu.edu/viewdoc/download?doi=10.1.1.618.6916&rep=rep1&type=pdf

Devlin, K. (2012). *Introduction to mathematical thinking.* Palo Alto, CA: Author. http://www.mat.ufrgs.br/~portosil/curso-Devlin.pdf

Dym, C., Agogino, A., Eris, O., Frey, D., & Leifer, L. (2005). Engineering design thinking, teaching, and learning. *Journal of Engineering Education, 94*(1), 103–120. https://doi.org/10.1002/j.2168-9830.2005.tb00832.x

Fradd, S. H., & Lee, O. (1999). Teachers' roles in promoting science inquiry with students from diverse language backgrounds. *Educational Researcher, 28*(6), 14–20, 42. https://doi.org/10.2307/1177292

Frykholm, J., & Glasson G. (2005). Connecting science and mathematics instruction: Pedagogical context knowledge for teachers. *School Science and Mathematics, 105,* 127–141. https://doi.org/10.1111/j.1949-8594.2005.tb18047.x

Gay, G. (2010). *Culturally responsive teaching: Theory, research, and practice.* New York: Teacher College Press.

Geddis, A. N. (1993). Transforming subject-matter knowledge: The role of pedagogical content knowledge in learning to reflect on teaching. *International Journal of Science Education, 15,* 673–683. https://doi.org/10.1080/0950069930150605

Gilfeather, M. & del Regato, J. (1999). Mathematics defined. *Mathematics experience-based approach*. Indianapolis, IN: Pentathlon Institute. Retrieved from http://www.mathpentath.org/pdf/meba/mathdefined.pdf.

Guzey, S. S., Tank, K. M., Wang, H.-H., Roehrig, G. H., & Moore, T. J. (2014). A high-quality professional development for teachers of grades 3–6 for implementing engineering into classrooms. *School Science and Mathematics, 114*(3), 139–149. https://doi.org/10.1111/ssm.12061

Hand, B., Norton-Meier, L., Staker, J., & Bintz, J. (2009). *Negotiating science: The critical role of argument in student inquiry*. Portsmouth, NH: Heinemann.

International Technology Education Association. (2007). *Standards for technological literacy: Content for the study of technology* (3rd ed.). Reston, VA: ITEA. https://www.iteea.org/File.aspx?id=67767&v=b26b7852

Johnson, C. (2003) Bioterrorism is real-world science: Inquiry-based simulation mirrors real life. *Science Scope, 27*(3), 19–23. https://www.jstor.org/stable/43180420

Johnson, C. (2004) NASA rocks: Problem based learning in Earth science. *Science Scope, 28*(1), 47–49. https://www.jstor.org/stable/43179266

Krauss, J., & Boss, S. (2013). *Thinking through project-based learning: Guiding deeper inquiry*. Thousand Oaks, CA: Corwin Press.

Lambros, A. (2004). *Problem-based learning in middle and high school classrooms: A teacher's guide to implementation*. Thousand Oaks, CA: Corwin Press.

Lee, O. (1999). Science knowledge, worldviews, and information sources in social and cultural contexts. *American Educational Research Journal, 36*(2), 187–219. https://doi.org/10.3102/00028312036002187

Lemke, J. L. (2001). Articulating communities: Sociocultural perspectives on science education. *Journal of Research in Science Teaching, 38*(3), 296–316. https://doi.org/10.1002/1098-2736(200103)38:3<296::aid-tea1007>3.0.co;2-r

Levy, F., & Murnane, R. J. (2004). Education and the changing job market. *Educational Leadership, 62*(2), 80–83.

Llewellyn, D. (2014). *Inquire within: Implementing inquiry- and argument-based science standards in grades 3–8* (3rd ed.). Thousand Oaks, CA: Corwin.

Mathis, C.A., Siverling, E. A., Moore, T. J., Douglas, K. A., & Guzey, S. S. (2018). Supporting engineering design ideas with science and mathematics: A case study of middle school life science students. *International Journal of Education in Mathematics, Science and Technology, 6*(4), 424–442. https://doi.org/10.18404/ijemst.440343

Moore, T. J., Glancy, A. W., Tank, K. M., Kersten, J. A., Smith, K. A., & Stohlmann, M. S. (2014). A framework for quality K-12 engineering education: Research and development. *Journal of Precollege Engineering Education Research, 4*(1), 1–13. https://doi.org/10.7771/2157-9288.1069

Moore, T. J., Guzey, S. S., & Brown, A. (2014). Greenhouse design to increase habitable land: An engineering unit. *Science Scope, 37*(7), 51–57. https://doi.org/10.2505/4/ss14_037_07_51

Moore, T. J., & Hughes, J. E. (2018). Chapter 12: Teaching and learning with technology in science, engineering, and mathematics. In M. D. Roblyer & J. E. Hughes (Eds.), *Integrating educational technology into teaching* (8th ed.) (pp. 386–427). New York, NY: Pearson.

Moore, T. J., Johnston, A. C., & Glancy, A. W. (2020). STEM integration: A synthesis of conceptual frameworks and definitions. In C. C. Johnson, M. J. Mohr-Schroeder, T. J. Moore, & L. D. English (Eds.), *The handbook of research on STEM education* (pp. 3–16). New York, NY: Routledge. https://doi.org/10.4324/9780429021381-2

Moore, T. J., Stohlmann, M.S., Wang, H.-H., Tank, K. M., Glancy, A. W., & Roehrig, G. H. (2014). Implementation and integration of engineering in K-12 STEM education. In Ş. Purzer, J. Strobel, & M. Cardella (Eds.), *Engineering in precollege settings: Research into practice* (pp. 35–60). West Lafayette, IN: Purdue Press. https://doi.org/10.2307/j.ctt6wq7bh.7

National Academy of Engineering and National Research Council. (2014). *STEM integration in K-12 education: Status, prospects, and an agenda for research.* Washington, DC: The National Academies Press. https://doi.org/10.17226/18612

National Governors Association Center for Best Practices and Council of Chief State School Officers. (2010a). *Common core state standards-English language arts.* Washington, DC: Author.

National Governors Association Center for Best Practices and Council of Chief State School Officers. (2010b). *Common core state standards-Mathematics.* Washington, DC: Author.

National Research Council. (2000). *Inquiry and the National Science Education Standards: A guide for teaching and learning.* Washington, DC: The National Academies Press. https://doi.org/10.17226/9596

National Research Council. (2010a). *Exploring the intersection of science education and 21st Century skills: A workshop summary.* Washington, DC: The National Academies Press. https://doi.org/10.17226/12771

National Research Council. (2010b). *Standards for K-12 engineering education?* Washington, DC: The National Academies Press. https://doi.org/10.17226/12990.

National Research Council. (2011). *Successful K-12 STEM education: Identifying effective approaches in science, technology, engineering, and mathematics.* Washington, DC: The National Academies Press. https://doi.org/10.17226/13158.

National Research Council. (2012a). *A framework for K-12 science education: Practices, cross cutting concepts, and core ideas.* Washington, DC: The National Academies Press. https://doi.org/10.17226/13165

National Research Council. (2012b). *Education for life and work: Developing transferable knowledge and skills in the 21st century.* Washington, DC: The National Academies Press. https://doi.org/10.17226/13398

NGSS Lead States. (2013). *Next generation science standards: For states, by states.* Washington, DC: The National Academies Press. https://doi.org/10.17226/18290

Osborne, R. J., & Wittrock, M. C. (1983). Learning science: A generative process. *Science Education, 67*(4), 498–508. https://doi.org/10.1002/sce.3730670406

Partnership for 21st Century Skills. (2015). *Framework for 21st century learning.* Retrieved from http://static.battelleforkids.org/documents/p21/P21_Framework_Definitions_New_Logo_2015_9pgs.pdf

Posner, G. J., Strike, K. A., Hewson, P. W., & Gertzog, W. A. (1982). Accommodation of a scientific conception: Toward a theory of conceptual change. *Science Education, 66*(2), 211–227. https://doi.org/10.1002/sce.3730660207

Roehrig, G. H., Moore, T. J., Wang, H.-H., & Park, M. S. (2012). Is adding the E enough? Investigating the impact of K-12 engineering standards on the implementation of STEM integration. *School Science and Mathematics, 112*(1), 31–44. https://doi.org/10.1111/j.1949-8594.2011.00112.x

Sampson, V., Enderle, P., & Grooms, J. (2013). Argumentation in science education. *The Science Teacher, 80*(5), 30–33. https://doi.org/10.2505/4/tst13_080_05_30

Sanders, M. (2009). STEM, STEM education, STEMmania. *The Technology Teacher, 68*(4), 20–26. http://esdstem.pbworks.com/f/TTT+STEM+Article_1.pdf

Sanders, M. E. (2015). The "integrative STEM education" definition explained. http://hdl.handle.net/10919/51624

Shulman, L. (1987). Knowledge and teaching: Foundations of the new reform. *Harvard Education Review, 57*(1), 1–22. https://doi.org/10.17763/haer.57.1.j463w79r56455411

Smith, K. A., Sheppard, S. D., Johnson, D. W., & Johnson, R. T. (2005). Pedagogies of engagement: Classroom-based practices. *Journal of Engineering Education, 94*(1), 87–101. https://doi.org/10.1002/j.2168-9830.2005.tb00831.x

Stohlmann, M., Moore, T. M., & Roehrig. (2012). Considerations for teaching integrated STEM education. *Journal of Pre-College Engineering Education Research, 2*(1), 28–34. https://doi.org/10.5703/1288284314653

Tank, K. M. (2014). *Examining the effects of integrated science, engineering, and nonfiction literature on student learning in elementary classrooms* (Doctoral dissertation). Available from ProQuest Dissertations and Theses database (Publication No: AAT 3630271).

Toulmin, S. E. (2008). *The uses of argument* (updated ed.). Cambridge: Cambridge University Press.

van Driel, J. H., Verloop, N., & de Vos, W. (1998). Developing science teachers' pedagogical content knowledge. *Journal of Research in Science Teaching, 35*(6), 673–695. https://doi.org/10.1002/(sici)1098-2736(199808)35:6<673::aid-tea5>3.0.co;2-j

von Glasersfeld, E. (1989). Cognition, construction of knowledge, and teaching. *Synthese, 80*(1), 121–140. https://doi.org/10.1007/bf00869951

Wiggins, G., & McTighe, J. (2005). *Understanding by design*. Alexandria, VA: Association for Supervision and Curriculum.

Part II
STEM Curriculum Maps

4 The STEM Road Map for Grades K-2

Catherine Koehler, Mark A. Bloom, and Andrea R. Milner

Overview of K-2 STEM Road Map

This chapter will provide a detailed overview of the integrated STEM Road Map for Kindergarten through second grade. Using the overarching themes described in Chapter 1: Cause and Effect, Innovation and Progress, The Represented World, Sustainable Systems, and Optimizing the Human Experience, the K-2 Road Map will describe innovative and integrated approaches for the teaching and learning of these themes from a problem/project-based learning (PBL)/phenomenon perspective. Each theme will be described by presenting a topic in science and a problem or challenge associated with these topics. The problem/challenge will provide the teacher with the opportunity to be creative in the instruction of the topic. We suggest that teachers also explore a library of NGSS phenomena that can help students begin the questioning and exploration process prior to instruction. These NGSS phenomena were vetted and are a free resource for teachers: https://www.ngssphenomena.com/.

In the K-2 grade band for The Next Generation Science Standards (NGSS), there are limited standards for science content as much of the focus during these grades is in English/language arts (ELA) and mathematics. As such, some of the themes have been combined to maximize the content standards as written by NGSS. In a table provided after each theme description, we have included suggested Common Core State Standards (CCSS) standards in mathematics and ELA that could align with each theme. These suggested CCSS standards were chosen because they represent the ELA and mathematics ideas that most represent the intersection between the NGSS and CC. Additional ELA and mathematics standards can be added to each scenario as appropriate. We have included a STEM Road Map standards descriptor (Elementary K-2 Road Map Excel spreadsheet) that describes each of the Common Core standards per grade. It is important to note that this STEM Road Map stresses the integration between the disciplines of science, mathematics, ELA as well as social studies, art, and music. Additionally, the National Association for the Education of Young Children (NAEYC, 2012) supports curriculum goals that focus on children's emergent knowledge and skills in all subject matter areas including language and literacy, mathematics, science, social studies, health, physical education, and the visual and performing arts. As most teachers in the K-2 grade band have self-contained classrooms, each challenge presented here can easily integrate all of these disciplines.

The inclusion of 21st century skills is an important hallmark of the STEM Road Map. It is important that *all* K-12 teachers address these very important skills, no matter how young the students are. PBL promotes responsibility, independence, discipline, as well as social learning as students practice and becomes proficient in the 21st century skills (Bell, 2010).

It is never too early to introduce students to careers in STEM fields. Each of the theme descriptions in this chapter suggests careers that align with that theme. This is an opportunity for teachers to introduce the named careers to students through a variety of sources such as a YouTube video or a short story read-aloud. Interactive videoconferencing, observing professionals at work in museums, science centers, or universities are also all effective ways primary students develop STEM career awareness (Cole, 2011). Clever Crazes for Kids is a highly interactive online resource that has a STEM career focus in a variety of educational games that will begin to build their knowledge and exposure to STEM (www.clevercrazes.com).

It is also never too early to begin teaching students about computer science concepts. In *A Framework for K-12 Science Education: Practicing, Crosscutting Concepts and Core Ideas* (National Research Council, 2012), the concept of mathematical and computational thinking is introduced to students in kindergarten. It stresses the use of mathematical and computational tools to help students make predictions about the physical world and assists them in seeing visual representations therein. Following these guidelines, NGSS (NGSS Lead States, 2013) focuses K-2 students on analyzing and interpreting data and developing and using models as a means to see visual representations and make predictions about the physical world. Although computational literacy is not directly addressed in NGSS, it is implied within the context of the science and engineering practices: mathematics and computational thinking. Computational thinking is the knowledge, skills, and disposition to solve problems or complete projects often used by computer scientists. This step-by-step process can begin as early as kindergarten. Creating activities that ask students to detail directional steps is the first concept in creating an easy algorithm, e.g. the first steps in coding (Ricketts, 2018). In this Road Map, students will be introduced to some basic coding to make colors change (in grade 1) and launch a rocket (in grade 2) using free online resources such as Scratchjr.com, Cargo-Bot.com, playcodemonkey.com, or Khan Academy tutorials. We are not endorsing any one particular coding program, but instead, suggesting that teachers explore a program that they can learn simultaneously with their students.

STEM Themes in the K-2 STEM Road Map

In Table 4.1, there is an overview of the K-2 STEM Road Map and the topics that will be covered in each grade. The topics should take up to five weeks to complete as they are designed to integrate all disciplines that a K-2 teacher will cover. The integration of these disciplines is the forefront of this STEM Road Map. For each topic, there will be a challenge or problem that will guide the capstone of the module. We will not prescribe how a teacher should instruct their students as that would take away from the creativity of the teacher. Instead, we provide teachers with the core ideas that should be covered within each STEM Road Map theme and NGSS, CCSS in ELA and mathematics, NAEYC standards and positions in science and technology for kindergartners and primary grade children, and the 21st century skills that would best integrate into that theme. Also, the use of technology is a critical component of the STEM Road Map. Effective uses of technology are active, hands-on, engaging, and empowering; give the child control; provide adaptive scaffolds to ease the accomplishment of tasks; and are used as one of many options to support children's learning (NAEYC, 2012, p. 6). It is imperative that teachers consistently scaffold the students with their learning as the challenges/problems may seem developmentally challenging for younger students. Over time, the students will become more intellectually independent; they will be able to explore the challenges/problems collaboratively.

Table 4.1 Overview of K-2 STEM Road Map Themes and Topics by Grade

STEM Road Map Theme	Kindergarten Topics	Grade 1 Topics	Grade 2 Topics
Cause and Effect	Physics in Motion	Influence of the Waves	The Changing Environment
Innovation and Progress	–	Coding a Rainbow	Launching a Spaceship
The Represented World	Patterns on Earth and in the Sky	Patterns and the Plant World	Investigating Environmental Changes
Sustainable Systems	Habitats in the United States	Protecting Endangered Species	Environmentally Friendly Playscapes
Optimizing the Human Condition	Our Changing School Environment	–	–

It is the best practice of expert teachers that students in grades K-2 keep a *STEM* journal, a collection of knowledge they have discussed in class over the entire year and should include several prescribed sections where students write down information that they learned in each activity. Sections in the STEM notebook might include, but are not limited to: (a) a list of science vocabulary; (b) KWL forms, which include K-What do I know?, W-What do I want to learn?, and L-What did I learn?; (c) tables for each exploration; (d) science drawings or illustrations with associated explanations of the students' thinking; (e) conclusions for each scenario – what did I learn?; (f) connections to STEM careers. Science is a natural part of a child's daily experiences, and they are anxious to explore it, discover answers, and build new understandings (Eliason & Jenkins, 2012). Therefore, by managing their own STEM notebooks, students can begin to work like a scientist and/or engineer, and they will begin to understand that these disciplines require careful measurement, calculation, and documentation. This notebook can be used to connect science content to math and language arts as well as art, geography, and music. Students can use their journals to illustrate their ideas about the topics being discussed and can reflect on these drawings as they learn the material. Teachers can also use this notebook as an authentic assessment tool.

The STEM Road Map for Kindergarten

Teachers who begin their instruction using the STEM Road Map concept in kindergarten have the responsibility to guide young minds on a new and exciting path of integrated learning. In Table 4.2, we have outlined the overview of the topics and challenges/problems that kindergarten students could tackle. Each of these topics can be taught over the course of five weeks as they are integrating topics such as ELA, mathematics, science, engineering, art, music, and geography. Understanding that students in kindergarten are beginning the formal learning process, the problems/challenges will be developmentally appropriate for their age (see Table 4.2).

Cause and Effect: Physics in Motion

Most amusement parks and fairs today have roller coasters that cater to the very young child (e.g. Cedar Point's *Woodstock Express*; Six Flags's *Magic Flyer*). In fact, there are curriculum units and special event days at amusement parks that focus on the "Physics of Roller Coasters" aimed at kindergarteners. Many of the children who ride such roller coasters may be left wondering how they work. This module theme introduces students to the basics of physics as it relates to motion. In this challenge, the students are presented with designing a track so that a marble can roll without jumping off the track. This is not limited to only one track, but the students can design multiple tracks to test their designs. First, they will need

Table 4.2 Kindergarten STEM Road Map Themes, Topics, and Problems/Challenges

STEM Theme	Topic	Problem/Challenge
Cause and Effect	Physics in Motion LEAD Science	Student teams will investigate different types of track designs to determine which design will help a marble go the fastest without having it jump off the track.
The Represented World	Patterns on Earth and in the Sky LEAD Mathematics	Student teams will investigate how the patterns of the sky and the animals on Earth adapt to changes over one year and create a yearlong calendar to demonstrate what they have observed throughout the year.
Sustainable Systems	Habitats in the United States LEAD Social Studies	Student teams will select various habitats in the local area and various habitats in other areas in the United States and develop a reference manual to describe these habitats' similarities and differences in relation to weather, climate, and the animals that reside there.
Optimizing the Human Condition	Our Changing School Environment LEAD English/Language Arts	Student teams will develop and print a school newspaper or blog to be distributed to other kindergarten classes in the school district and beyond to report changes in the environment around the school and community.

to research how a roller coaster works (ELA) and will have to understand about motion (science), speed (science), and the pushing and pulling effects of gravity (science). Their mathematics skills will play a major role in this challenge as they will be measuring and comparing numbers. The teacher will help develop classroom charts to collect their data. A discussion could focus on identifying the best place to situate a roller coaster from an environmental perspective and what safety precautions might be at play with their roller coaster models (see Table 4.3). This module has been published in 2019 by NSTA Press. See https://my.nsta.org/resource/118783.

Innovation and Progress: Patterns on Earth and in the Sky

In this module, students will begin to identify weather and sky patterns as they emerge during the year and the adaptability of animals, including humans, on Earth to those changing patterns. The problem/challenge is a

50 Catherine Koehler et al.

Table 4.3 STEM Road Map – Kindergarten Cause and Effect Theme: Motion

NGSS Performance Objectives	Common Core		21st Century Skills
	Mathematics	Language Arts	
K-PS2-1	CCSS Math Practices MP 1, MP2, MP3, MP4, MP6	Reading Standards CCSS.ELA. RI.K.1 RI.K.3	21st Century Themes: Economic, Business, and Entrepreneurial Literacy Health Literacy Environmental Literacy
KPS2-2	CCSS.Math. Content.K.MD.B.3	Writing Standards CCSS.ELA. W.K.2 W.K.5 W.K.7	Learning and Innovation Skills: Creativity and Innovation Critical Thinking and Problem Solving Communication and Collaboration
	CCSS.Math. Content.K.CC.C.6 CCSS.Math. Content.K.CC.C.7	Speaking and Listening Standards CCSS.ELA. SL.K.1 SL.K.3 SL.K.5	Information, Media, and Technology Skills: Information Literacy Media Literacy Information Communication and Technology Literacy
	CCSS.Math. Content.K.CC.B.4 CCSS.Math. Content.K.CC.B.4a CCSS.Math. Content.K.CC.B.4b CCSS.Math. Content.K.CC.B.4c		Life and Career Skills: Flexibility and Adaptability Initiative and Self-Direction Social and Cross Cultural Skills Productivity and Accountability Leadership and Responsibility
	CCSS.Math. Content.K.MD.A.1 CCSS.Math. Content.K.MD.A.2		

petting zoo needs your team to create a yearlong calendar to demonstrate what you have observed throughout the year. Teams will create a presentation for the petting zoo to explain to their customers the changes that animals experience over a year as a result of changing weather. Much of the observations the students will make can be recorded in their STEM notebooks and used as talking points. The lead discipline for this module is mathematics because so much of the observations will take the form of quantitative relationships backed by qualitative observations, aligning the different patterns of the sky and animals. Data can be collected using illustrations of the cycles of the Sun, the Moon, the seasons, and how animals adapt to these changing conditions. Weather observations can also be collected and analyzed based on the seasons. This module can span the entire school year so that the students can understand how patterns of the sky and the Earth change (see Table 4.4).

Table 4.4 STEM Road Map Grades K-2 – Kindergarten The Represented World Theme: Patterns on Earth and in the Sky

NGSS Performance Objectives	Common Core		21st Century Skills
	Mathematics	Language Arts	
K-ESS2-1	CCSS Math Practices MP 1, MP2, MP3, MP4, MP6 MP7	Reading Standards CCSS.ELA. RI.K.1 RI.K.3	21st Century Themes: Global Awareness Environmental Literacy Civic Literacy
K-ESS3-1	CCSS.Math. Content.K.CC.B.4	Writing Standards CCSS.ELA. W.K.2 W.K.5 W.K.7	Learning and Innovation Skills: Creativity and Innovation Critical Thinking and Problem Solving Communication and Collaboration
K-PS3-1	CCSS.Math. Content.K.CC.C.6 CCSS.Math. Content.K.CC.C.7	Speaking and Listening Standards CCSS.ELA. SL.K.1 SL.K.3 SL.K.5	Information, Media, and Technology Skills: Information Literacy Media Literacy Information Communication and Technology Literacy
K-LS1-1	CCSS.Math. Content.K.MD.B.3		Life and Career Skills: Flexibility and Adaptability Initiative and Self-Direction Social and Cross Cultural Skills Productivity and Accountability Leadership and Responsibility
	CCSS.Math. Content.K.CC.A.1 CCSS.Math. Content.K.CC.A.2 CCSS.Math. Content.K.CC.A.3 CCSS.Math. Content.K.CC.B.5		

Sustainable Systems: Habitats in the United States

Habitats are ideal to observe sustainable systems at work. With the invention of the webcam, students around the world can view habitats in distant lands such as the Serengeti in Africa or the Tundra in the Arctic. From their own classrooms, students can see how the lions of the Serengeti feed and drink around a watering hole, how the emperor penguins of Antarctica birthing their babies, and how the urban peregrine falcons nest and take care of their young on city rooftops. Comparing and contrasting habitats teach students the wonders of life in places near and far. In this module, students are challenged to choose various habitats in the local area and various habitats in other areas in the United States and develop a reference manual to describe their similarities and differences in relation to weather, climate, and the animals that reside there. In kindergarten, we introduce students to the notion of habitats, concentrate on local habitats, and compare these local habitats with those in the United States. The lead discipline in this module is geography, and students will begin to see beyond their own neighborhood, city, and state to learn about the geography and habitats found in other regions of the United States. Technology plays a huge role in this unit as it can bring different habitats to the students (via webcams) so they can make observations from afar. Later in grade 1, this topic will be expanded to habitats in other countries around the world. The NAEYC and the Fred Rogers Center (2012) recommend that early childhood educators select, use, integrate, and evaluate technology and interactive media tools in intentional and developmentally appropriate ways, giving careful attention to the appropriateness and the quality of the content, the child's experience, and the opportunities for co-engagement (p. 11). At the end of this module, students will develop a reference manual that describes and compares different habitats around the United States (see Table 4.5).

Optimizing the Human Condition: Our Changing School Environment

It is never too early to introduce the notion of the changing environment to students. In this module, students are challenged to develop a means to communicate changes in the environment that they see around their school and their community. In this module, students will become aware of the changes in their environment and record those changes for wider distribution to other area kindergarten classes as well as beyond. Their challenge is *to make observations about their surrounding school environment and introduce a bird feeder to the school playground.* This bird feeder will be observed over the year and students will make observations in their own personal science notebooks about what they see. They will be feeding the birds throughout the year and will be taking

Table 4.5 STEM Road Map Grades K-2 – Kindergarten Sustainable Systems Theme: Habitats

NGSS Performance Objectives	Common Core Mathematics	Language Arts	21st Century Skills
K-LS1-1	CCSS Math Practices MP 1, MP2, MP3, MP4, MP5, MP6, MP7, MP8	Reading Standards CCSS.ELA. RI.K.1 RI.K.3	21st Century Themes: Global Awareness Environmental Literacy
K-ESS3-1	CCSS.Math.Content.K.CC.B.4 CCSS.Math.Content.K.CC.B.4a CCSS.Math.Content.K.CC.B.4b CCSS.Math.Content.K.CC.B.4c	Writing Standards CCSS.ELA. W.K.2 W.K.5 W.K.7	Learning and Innovation Skills: Creativity and Innovation Critical Thinking and Problem Solving Communication and Collaboration
K-PS3-1	CCSS.Math.Content.K.CC.C.6 CCSS.Math.Content.K.CC.C.7	Speaking and Listening Standards CCSS.ELA. SL.K.1 SL.K.3 SL.K.5	Information, Media, and Technology Skills: Information Literacy Media Literacy Information Communication and Technology Literacy
	CCSS.Math.Content.K.MD.A.1 CCSS.Math.Content.K.MD.A.2		Life and Career Skills: Flexibility and Adaptability Initiative and Self-Direction Social and Cross Cultural Skills Productivity and Accountability Leadership and Responsibility
	CCSS.Math.Content.K.MD.B.3 CCSS.Math.Content.K.CC.A.1 CCSS.Math.Content.K.CC.A.2 CCSS.Math.Content.K.CC.A.3 CCSS.Math.Content.K.CC.B.5		

observations and reporting these observations to a wider audience. The lead discipline is ELA, and this project will allow students to have literary freedom to express their findings of environmental issues as they pertain to the topics such as localized effects of weather conditions and how birds are affected by the weather, erratic weather conditions, and the effects weather has on local animal habitats all found in and around their school and their community. Students can produce an online newspaper or blog to distribute the information to a wider audience about their observations with the birds. They may develop a school newspaper to be distributed in the school district and their community to report the changes in the environment observed around the school and community. They may want to share their findings by adopting another kindergarten class and share their information with this class (see Table 4.6).

STEM Careers in Kindergarten

It is important to introduce students to different careers through the general description of the career. In this section, we will introduce you to a broad definition of different careers. Later, in this chapter, we will discuss different careers more specifically.

An *engineer* is a person who solves problems that helps society and/or the community. They can be a man or a woman. Many times, they work with other people who are also engineers in solving the problem. Engineers always try to design products to help make life easier. The engineer, after presenting a problem, uses a design process that helps them tackle the problem. Engineers always work with constraints (parameters around the problem), whether it is financial (how much does the design cost), ethical (how will this design harm the environment, animals, or humans), or time (how much time do we have to complete this task) (Koehler, Faraclas, Giblin, Moss, & Kazarounian, 2013). There are many different types of engineers that focus on different types of designs and products and a few are listed here; mechanical (manufacturing, robotics), biomedical (works with medical designs), civil and environmental (work to help transportation, construction, water resource management, waste treatment facilities), chemical (nanotechnology, uses chemistry and physics), electrical (electricity, electronics, electromagnetic, communication), computer (software and hardware design), structural (building, bridges, dams), and aerospace (airplanes, space shuttles, "rocket" science).

A *scientist* is a person (either female or male) who studies phenomena on Earth and/or in space in an attempt to explain the natural world. A scientist uses a methodology to study these phenomena oftentimes referred to as a scientific method. Science begins with an observation followed by a question. Scientists explore these questions by collecting empirical data, analyzing the data, and drawing conclusions based on

The STEM Road Map for Grades K-2   55

Table 4.6 STEM Road Map K-2 – Kindergarten Optimizing the Human Condition Theme: Our Changing School Environment

NGSS Performance Objectives	Common Core		21st Century Skills
	Mathematics	Language Arts	
K-ESS2-1	CCSS Math Practices MP1, MP2, MP3, MP4, MP6	Reading Standards CCSS.ELA. RI.K.1 RI.K.3	21st Century Themes: Economic, Business, and Entrepreneurial Literacy Environmental Literacy Civic Literacy
K-ESS3-1	CCSS.Math. Content.K.CC.B.4	Writing Standards CCSS.ELA. W.K.2 W.K.5 W.K.7	Learning and Innovation Skills: Creativity and Innovation Critical Thinking and Problem Solving Communication and Collaboration
K-PS3-1	CCSS.Math. Content.K.CC.C.6 CCSS.Math.Content. K.CC.C.7	Speaking and Listening Standards CCSS.ELA. SL.K.1 SL.K.3 SL.K.5	Information, Media, and Technology Skills: Information Literacy Media Literacy Information Communication and Technology Literacy
K-LS1-1	CCSS.Math. Content.K.MD.A.1 CCSS.Math. Content.K.MD.A.2		Life and Career Skills: Flexibility and Adaptability Initiative and Self-Direction Social and Cross Cultural Skills Productivity and Accountability Leadership and Responsibility
	CCSS.Math. Content.K.MD.B.3 CCSS.Math. Content.K.CC.A.1 CCSS.Math. Content.K.CC.A.2 CCSS.Math. Content.K.CC.A.3 CCSS.Math. Content.K.CC.B.5		

evidence that they have collected. They can use an experimental procedure to explore their research questions as most people perceive how science is conducted, or they can use their observations to explore the questions the way astronomers and some field biologists do. Scientists use their creativity in all aspects of the scientific endeavor from making the observation to drawing conclusions based on evidence. A scientist is not restricted to a laboratory in which to work, but instead can work out in the field, e.g. outdoors or in space or on computers (Koehler, Bloom, & Binns, 2013). Fields of science include life sciences (biology, medicine, environmental science) and physical science (chemistry, physics, geology, astronomy, meteorology).

A *mathematician* is a person (female or male) that studies phenomena related to numbers, models, and structures related to numbers and patterns (American Mathematical Society, 2014). The disciplines of engineering and science often use mathematics to explain the data collected and used. The person who studies mathematics can pursue careers in statistics (study of the collection, analysis, interpretation, presentation, and organization of data [Dodge, 2006]), actuaries (studies financial risk), and work with scientists in the fields of climate change and astronomy.

A *journalist* writes about a variety of topics for publication. They have an excellent command of the English language and are able to discuss issues that are related to current events. They often work in the field exploring stories that are newsworthy. They can sometimes live in foreign countries reporting on the events that are happening there.

A *meteorologist* is a person who studies the weather and different atmospheric changes that occur short term. They have a strong background in science, geography, and mathematics. Many times they are on television broadcasting the weather. Many work with maps and study how precipitation and pressure changes affect geographic areas.

An *astronomer* is a person who studies the stars, planets, the Sun, solar system, and the universe. Much of their study is done using computers and telescopes. They try to answer questions about the origins of the universe or is there life on distant planets? An astronomer needs a strong background in physics, computers, mathematics, and mapping.

An *ecologist* is a person who studies biomes, habitats, ecosystems, organisms, and their relationship to the environment. They have a strong background in biology, the environment, and climate. They often work with policy makers on environmental issues.

A *geographer* is a person who studies the land and why people settle in the area that they do. They have a strong background in mapping, geology, and anthropology (study of where people live in the past and the present), and meteorology.

The STEM Road Map for First Grade

In the first grade, students will explore themes that connect to the ideas learned in kindergarten. Each of the STEM themes has topics associated with it and a problem/challenge for the students to address that center around our environment. Although the STEM themes are the same, the topics vary depending on the NGSS standards that align with that

Table 4.7 Overview of the Grade 1 STEM Road Map Themes, Topics, and Problems/Challenges

STEM Theme	Topic	Problem/Challenge
Cause and Effect	Influence of Waves LEAD Science	Student teams will develop a model to demonstrate how humans experience and interact with light and sound waves.
Innovation and Progress	Coding a Rainbow LEAD Science	Student teams will design and create a rainbow using basic coding skills.
The Represented world	Patterns and the Plant World LEAD Mathematics	Student teams will design a window-box garden and follow their products over an extended period of time.
Sustainable Systems	Protecting Endangered Species LEAD Language Arts/ Social Studies	Student teams will develop a plan to save their selected endangered species through mitigating weather, climate, and other factors that contribute to their vitality.

theme. Each topic will be described in detail and a map of the content standards for NGSS, CCSS for English/language arts (ELA), CCSS for mathematics, NAEYC standards and positions in science and technology for kindergartners and primary grade children, and 21st century skills will accompany each description. As previously stated, each module is to be approached in an interdisciplinary way, which allows the teacher to use the lead discipline as a framework for the module while integrating other disciplines as appropriate. We provide the CCSS for ELA and mathematics as well as 21st century skills as important ingredients for the development and implementation of the module (see Table 4.7).

Cause and Effect – Influence of Waves

In this unit, students will begin to explore the notion of waves. As we know, waves can present themselves in different forms, including light and sound waves. Light waves come to the Earth in the form of visible light from the Sun while mechanical waves are waves that can produce sound. The understanding of waves is fundamental to more advanced understanding of communication and various other phenomena in science. Both types of waves are all around us at all times, constantly bombarding us, and bouncing off of us although we often do not notice this. Students will begin to understand that there are different forms of waves and that body organs (e.g. eyes, ears, and skin) respond to the waves

Table 4.8 STEM Road Map K-2 – First-Grade Cause and Effect Theme: Influence of the Waves

NGSS Performance Objectives	Common Core		21st Century Skills
	Mathematics	Language Arts	
1-PS4-1	CCSS Math Practices MP1, MP2, MP4, MP6	Reading Standards CCSS.ELA. RI.1.1 RI.1.3 RI.1.7	21st Century Themes: Economic, Business, and Entrepreneurial Literacy
1-PS4-2	CCSS.Math. Content.1. NBT.A.1c	Writing Standards CCSS.ELA. W.1.2 W.1.6 W.1.7 W.1.8	Learning and Innovation Skills: Creativity and Innovation Critical Thinking and Problem Solving Communication and Collaboration
1-PS4-3	CCSS.Math. Content.1. NBT.B.3	Speaking and Listening Standards CCSS.ELA. SL.1.1 SL.1.3 SL.1.5	Information, Media, and Technology Skills: Information Literacy Media Literacy Information Communication and Technology Literacy
	CCSS.Math. Content.1. MD.C.4		Life and Career Skills: Flexibility and Adaptability Initiative and Self-Direction Social and Cross Cultural Skills Productivity and Accountability Leadership and Responsibility
	CCSS.Math. Content.1. OA.A		

differently. This module will also explore, at a basic level, the human anatomy of hearing and sight. Students will learn about various sources of sound and light and determine how the sound and light waves reach them (see Table 4.8). This module has been published in 2019 by NSTA Press. See https://my.nsta.org/resource/118779.

Innovation and Progress: Coding a Rainbow

Introducing students in early elementary grades to the notion of computer programming is an important introduction to a career in computer science. There are free basic computer programs that students can learn to understand how coding works. They can manipulate variables to make avatars move or make wheels turn. In this module, students will be introduced to basic coding by developing a program that will make colors change in a design of their choice. The overarching science concept that will be introduced in this unit is how visible light is part of the electromagnetic waves that come from the Sun. Visible light is the wavelength that the human eye can see. Our eyes are designed to see a combination of colors that come from the Sun. In this module, the students will understand that white light from the Sun is made up of a combination of colors that can be split apart by a prism producing the colors of the rainbow. They will be using this knowledge to create a computer program that splits white light into the various colors of a rainbow (ROYGBIV); red, orange, yellow, green, blue, indigo, and violet. Computer programs that we suggest include, but are not limited to, Scratchjr.com, Cargo-Bot.com, playcodemonkey.com, or Khan Academy tutorials. It is ultimately important that teachers learn to code alongside their students so they can help them when they hit a roadblock, or vice versa. During the coding process, it is important for students to write each step they take for their final product in their science notebooks. The reason for this extra step is to capture the program in words as many of the computer-generated software is based on symbols, not coding language. The connection between the coding symbols in the computer software and written language will be useful as students' progress with their knowledge about computer programming for later projects (Table 4.9).

The Represented World: Patterns and the Plant World

Children are curious about how living things grow and change over the course of their lives. In this unit, we explore the concept of changes in the plant world and how changes on Earth (the seasons) and in the sky (daylight hours) affect these plants. In kindergarten, the students will have studied the patterns of the Sun and Moon (daylight and darkness) and how these patterns affect the Earth. In this unit, students will review these concepts again and relate them to how plants grow. The students will study how plants change over time due to the changing seasons and learn that certain plants grow in different regions. They will make observations in a real world setting about changing plant life and make notes for a design of a window-box garden. The lead discipline for this unit is mathematics, so the emphasis is on how students make their observations, measure the changes, and thus quantify the results. The challenge/problem for this unit is: *Design a window-box garden, plant several different plants, and follow their life cycle over an extended period of time (several months)*. Students

Table 4.9 STEM Road Map K-2 – First-Grade Innovation and Progress Theme: Coding a Rainbow

NGSS Performance Objectives	Common Core		21st Century Skills
	Mathematics	Language Arts	
1-PS4-4	CCSS Math Practices MP1, MP2, MP4, MP6, MP7	Reading Standards CCSS.ELA. RI.1.1 RI.1.3 RI.1.7	21st Century Themes: Economic, Business, and Entrepreneurial Literacy Global Awareness Civic Literacy
1-PS4-1	CCSS.Math. Content.1. NBT.B.3	Writing Standards CCSS.ELA. W.1.2 W.1.6 W.1.7 W.1.8	Learning and Innovation Skills: Creativity and Innovation Critical Thinking and Problem Solving Communication and Collaboration
	CCSS.Math. Content.1. MD.A.1 CCSS.Math. Content.1. MD.A.2	Speaking and Listening Standards CCSS.ELA. SL.1.1 SL.1.3 SL.1.5	Information, Media, and Technology Skills: Information Literacy Media Literacy Information Communication and Technology Literacy
	CCSS.Math. Content.1. MD.C.4		Life and Career Skills: Flexibility and Adaptability Initiative and Self-Direction Social and Cross Cultural Skills Productivity and Accountability Leadership and Responsibility
	CCSS.Math. Content.1. OA.A.1		

will begin this unit by researching which plants grow in their area and which plants do not grow in their area, and then students will explore what types of containers best serve as a foundation for the plants. Students will design a notebook to make observations of the plant and collaboratively decide what observations to make and how to organize these observations. Based on the observations the students make throughout this unit about the window-box environment, they analyze how plants develop throughout the year (see Table 4.10). This module has been published in 2018 by NSTA Press. See https://my.nsta.org/resource/114573.

Table 4.10 STEM Road Map K-2 – First-Grade The Represented World Theme: Patterns and the Plant World

NGSS Performance Objectives	Common Core		21st Century Skills
	Mathematics	Language Arts	
1-ESS1-1	CCSS Math Practices MP1, MP2, MP3, MP4, MP5, MP6, MP7, MP8	Reading Standards CCSS.ELA. RI.1.1 RI.1.3 RI.1.7	21st Century Themes: Economic, Business, and Entrepreneurial Literacy Global Awareness Environmental Literacy
1-ESS1-2	CCSS.Math.Content.1. NBT.C.4 CCSS.Math.Content.1. NBT.C.5 CCSS.Math.Content.1. NBT.C.6	Writing Standards CCSS.ELA. W.1.2 W.1.6 W.1.7 W.1.8	Learning and Innovation Skills: Creativity and Innovation Critical Thinking and Problem Solving Communication and Collaboration
1-LS3-1	CCSS.Math.Content.1. NBT.B.3	Speaking and Listening Standards CCSS.ELA. SL.1.1 SL.1.3 SL.1.5	Information, Media, and Technology Skills: Information Literacy Media Literacy Information Communication and Technology Literacy
1-LS-1–2	CCSS.Math.Content.1. MD.A.1		Life and Career Skills: Flexibility and Adaptability Initiative and Self-Direction Social and Cross Cultural Skills Productivity and Accountability Leadership and Responsibility
	CCSS.Math.Content.1. MD.C.4 CCSS.Math.Content.1. OA.A.1 CCSS.Math.Content.1. OA.A.2		

Sustainable Systems – Protecting Endangered Species

In kindergarten, the student explored habitats of animals in their own community as well as across the United States. Now, they will explore habitats on a global scale. As mentioned in the kindergarten unit, habitats are the ideal location to visualize sustainable systems at work. With the help of webcams, students can experience habitats in distant lands. By comparing and contrasting habitats, students learn about the wonders of life in places near and far. In this challenge/problem, students are asked to *choose an endangered species in habitats in other areas around the world and develop a plan to save the endangered species by describing these habitats' characteristics as it relates to weather, climate, and the animals that reside there and how humans may influence this habitat.* The lead discipline in this unit is geography, and students will begin to see beyond their own town and state to other parts of the world. Technology plays a huge role in this unit as it can bring different habitats to the students so they can make observations afar. Students will begin this unit by researching the habitat of an endangered species and developing an organizational tool to describe the characteristics of the habitat. Students will also look at campaigns to save endangered species such as those done with the World Wildlife Foundation. At the end of this unit, students will present their plan to save the endangered animal by describing the habitat characteristics, what is happening to endanger the species, and how they might be able to help (see Table 4.11).

STEM Careers in First Grade

A *horticulturalist* is a person who works with the land and how it produces fruits, vegetables, mushrooms, and other plants. They can design gardens and work on farms growing food for consumption. They can also work as gardeners, landscape designers, and farmers. They need a strong background in soil science, plant pathology, geology, chemistry, and architecture (for designing).

A *landscaper* is a person who works with the land to keep it looking neat. This person is often seen cutting grass or shoveling snow (in northern regions of the United States), but this is not their only task. They will often prune trees that means keep them healthy by cutting away dead branches or trim bushes to keep them shapely.

A *landscape architect* is a person who works planning how the outside of a building will look. They use the natural surroundings to plan the different types of trees or plants that might look perfect for that building. They can also work with people to design porches and patios at their homes.

An *ecologist* is a person who studies biomes, habitats, ecosystems, organisms, and their relationship to the environment. They have a strong background in biology, the environment, and climate. They often work with policy makers on environmental issues.

An *urban planner* is a person who works to develop plans for the use of land in communities and cities. They are particularly interested in

Table 4.11 STEM Road Map K-2 – First-Grade Sustainable Systems Theme: Protecting Endangered Species

NGSS Performance Objectives	Common Core		21st Century Skills
	Mathematics	Language Arts	
1-LS1-1	CCSS Math Practices MP1, MP2, MP3, MP4, MP6, MP7	Reading Standards CCSS.ELA. RI.1.1 RI.1.3 RI.1.7	21st Century Themes: Economic, Business, and Entrepreneurial Literacy Global Awareness Environmental Literacy Health Literacy Civic Literacy
1-LS-1–2	CCSS.Math. Content.1. NBT.C.5	Writing Standards CCSS.ELA. W.1.2 W.1.6 W.1.7 W.1.8	Learning and Innovation Skills: Creativity and Innovation Critical Thinking and Problem Solving Communication and Collaboration
	CCSS.Math. Content.1. NBT.C.6	Speaking and Listening Standards CCSS.ELA. SL.1.1 SL.1.3 SL.1.5	Information, Media, and Technology Skills: Information Literacy Media Literacy Information Communication and Technology Literacy
	CCSS.Math. Content.1. NBT.B.3		Life and Career Skills: Flexibility and Adaptability Initiative and Self-Direction Social and Cross Cultural Skills Productivity and Accountability Leadership and Responsibility
	CCSS.Math. Content.1. MD.C.4		
	CCSS.Math. Content.1. OA.A.1		

how people move into communities. They have a background in geography, anthropology, and policy.

An *environmental engineer* is a person who is interested in how the environment is affected by humans. They work with groups to ensure that water quality and waste management are maintained in communities. They also work with CIVIL Engineers in the development of roads and infrastructures in communities. Where the civil engineer is responsible for construction and design of infrastructure as it relates to transportation, an environmental engineer will work to insure that the environment is maintained as naturally as possible. These engineers have a background in environmental science, ecology, mathematics, transportation, construction, and materials.

A *climatologist* is a person who studies climate in a region or worldwide. These individuals study how weather over the long term affects plants and animals in a specific region. With the onset of Global Climate Change (GCC), it is important for climatologists to understand how GCC is affecting us. They need a strong background in meteorology, geology, chemistry, physics, botany, and mathematics.

An *environmentalist* is a person who studies all different parts of the environment and helps educate other people about it. They can work for non-profit organizations or the government. Their job is to educate people about the health and welfare of our world.

A *computer scientist* is a person who works on computers to make them do different tasks. They usually write software to be used on a computer or cell phone like Google Docs, Google Slides, or Tick Tock. The software is to allow people to use these devices such as cell phones or computers to communicate. They can also work on hardware, which means the computer machine. If you broke something on your cell phone or computer, a computer scientist might fix it if it were broken.

The STEM Road Map for Second Grade

The STEM Road Map builds on the knowledge discussed from the previous two grades. As students develop intellectually, the problems/challenges become more complex, and students should be given more independence in their thinking and problem solving. As discussed earlier, it is extremely important for students to use their science notebooks as a tool to assist in this learning. Students are encouraged to draw pictures and use labels to explain their thinking in their notebooks. Illustrating their observations requires young scientists to make close observations of the world around them. Teachers should begin very early in the year to teach students the difference between cartoon-like drawings/illustrations and scientific illustrations. Through modeling of drawing and labeling (with labels and details), students develop an understanding that their scientific illustrations give information and explain ideas. Students are encouraged to make their illustrations as realistic as possible. The teacher explains that students should think about their scientific illustrations like this: *If I weren't here to explain what is in my picture, could other scientists make sense of it?* The overarching emphasis in this section will focus on students' illustrations of their ideas and the changing conceptions as they learn the material being presented (see Table 4.12).

Table 4.12 Overview of the Grade 2 STEM Road Map Themes, Topics, and Problems/Challenges

STEM Theme	Topic	Problem/Challenge
Cause and Effect	Our Changing Environment LEAD Science and Social Studies	Student teams will develop a communication plan to inform their community in the event of a natural disaster such as a flood, tornado, earthquake, or dust storm in the area.
Innovation and Progress	Launching a Spaceship LEAD Science and Social Studies	Student teams will design a space suit using material that will protect a person from dangerous elements in space.
The Represented World	Investigating Environmental Changes LEAD Mathematics and English/Language Arts	Student teams will adopt a plot of land in your schoolyard and investigate how it changes over a school year.
Sustainable Systems	Environmentally Friendly Playscapes LEAD Science and Mathematics	Student teams will develop a schoolyard garden and explore the interaction between the Earth, plants, humans, animals, weather, and seasons.

Cause and Effect – The Changing Environment

In kindergarten and first grade, students were introduced to different habitats within the United States and the World. During these lessons and as they explore diverse habitats, they are introduced to major Earth features, such as mountain ranges, coastal regions, the Great Plains, and river basins. This lesson explores what would happen to various habitats if there were a natural hazard. In this module, students pull together the knowledge they learned from previous units on habitats local and global, research new factors, and problem-solve about how the impact of a natural hazard on the environment, the people, and the animals can be minimized (Table 4.13). The problem/challenge is *investigating your home and different regions of the United States, develop and communicate a plan to have people prepare for a natural hazard such as a flood, tornado, earthquake, or dust storm to minimize the impact of the damage on the environment*. In devising a plan for natural hazards, students can produce an infomercial about how to prepare for one of these disasters. In science, students learn about the conditions for natural hazards; in technology, students utilize technology to gather information and communicate; in engineering, students learn about how shelters are constructed and how water sources are controlled; and in mathematics, students learn about models for calculating how many people are involved and chances of weather occurrences (see Table 4.14).

Table 4.13 STEM Road Map K-2 – Second-Grade Cause and Effect Theme: The Changing Environment

NGSS Performance Objectives	Common Core		21st Century Skills
	Mathematics	Language Arts	
1-PS1–4	CCSS Math Practices MP1, MP2, MP3, MP4, MP5, MP6, MP7, MP8	Reading Standards CCSS.ELA. RI.2.1 RI.2.3 RI.2.7 RI.2.8 RI.2.9	21st Century Themes: Economic, Business, and Entrepreneurial Literacy Global Awareness Environmental Literacy Civic Literacy
2-LS2-1	CCSS.Math.Content.2. NBT.A.1 CCSS.Math.Content.2. NBT.A.2 CCSS.Math.Content.2. NBT.A.3	Writing Standards CCSS.ELA. W.2.1 W.2.2 W.2.6 W.2.7 W.2.8	Learning and Innovation Skills: Creativity and Innovation Critical Thinking and Problem Solving Communication and Collaboration
2-LS2-2		Speaking and Listening Standards CCSS.ELA. SL.2.2 SL.2.1 SL.2.3 SL.2.5	Information, Media, and Technology Skills: Information Literacy Media Literacy Information Communication and Technology Literacy
2-ESS1-1			Life and Career Skills: Flexibility and Adaptability Initiative and Self-Direction Social and Cross Cultural Skills Productivity and Accountability Leadership and Responsibility
2-ESS2-1			

Innovation and Progress – Launching a Spaceship

In the first-grade unit, students began their coding experience by challenging them to create a computer program that made a rainbow. In the second grade, the next coding challenge for students is to make a spaceship take off from the Earth's surface. Students will learn about how launching a spaceship takes very accurate timing so that there is

Table 4.14 STEM Road Map K-2 – Second-Grade Innovation and Progress Theme: Launching a Spaceship

NGSS Performance Objectives	Common Core Mathematics	Common Core Language Arts	21st Century Skills
1-PS1-3	CCSS Math Practices MP1, MP2, MP3, MP4, MP5, MP6, MP7	Reading Standards CCSS.ELA. RI.2.1 RI.2.3 RI.2.7 RI.2.8 RI.2.9	21st Century Themes: Economic, Business, and Entrepreneurial Literacy Global Awareness Health Literacy
2-PS1-2	CCSS.Math.Content.2.NBT.A.1 CCSS.Math.Content.2.NBT.A.2 CCSS.Math.Content.2.NBT.A.3	Writing Standards CCSS.ELA. W.2.1 W.2.2 W.2.6 W.2.7 W.2.8	Learning and Innovation Skills: Creativity and Innovation Critical Thinking and Problem Solving Communication and Collaboration
2-PS1-1	CCSS.Math.Content.2.NBT.A.4	Speaking and Listening Standards CCSS.ELA. SL.2.2 SL.2.1 SL.2.3 SL.2.5	Information, Media, and Technology Skills: Information Literacy Media Literacy Information Communication and Technology Literacy
	CCSS.Math.Content.2.MD.A.1 CCSS.Math.Content.2.MD.A.2 CCSS.Math.Content.2.MD.A.3 CCSS.Math.Content.2.MD.A.4		Life and Career Skills: Flexibility and Adaptability Initiative and Self-Direction Social and Cross Cultural Skills Productivity and Accountability Leadership and Responsibility
	CCSS.Math.Content.2.MD.D.10		

no wasted material or fuel. Planets are aligned during certain periods, so it is essential to launch the spaceship at just the right time. Teachers can determine constraints as to the size of the spaceship and how long it takes for it to travel to Mars or the Moon. These are real world problems that engineers and scientists face whenever there is a rocket launch from Earth. The challenge for this unit is to *create a computer program to launch a rocket from Earth to the Moon AND Earth to Mars*. During the coding process, it is important for students to write each step they take for their final product in their science notebooks.

The Represented World: Investigating Environmental Changes

In this module, students explore changes in the natural environment, focusing on ways to observe and measure these changes. Students are challenged to design and build an outdoor STEM classroom that they and other students at their school will use to observe phenomena such as plant and animal life cycles and the movement of the Earth around the sun. Students create a proposal for the outdoor classroom and use the engineering design process (EDP) to create the space. Students also devise a data collection plan to observe and analyze changes in the outdoor classroom over time (Table 4.15). This module has been published in 2019 by NSTA Press. See https://my.nsta.org/resource/116282.

Sustainable Systems: Environmentally Friendly Playscapes

In other units, we have discussed changing environments and habitats in the United States and the world. It is important for students to understand that even small areas, such as their schoolyard, undergo changes. If we act locally and think globally, we can make positive changes right here at home. The problem/challenge for this unit is to adopt a plot of land in your schoolyard for a project of your design, plan for optimal land use, responsible water consumption, and investigate and make predictions how it changes over a school year (see Table 4.16). In teams of two to three students, they will create a design proposal and present it to the class to be voted on by the other students. Student teams will need to present a proposal for this land use such as creating a playground, planting a vegetable or butterfly garden, or creating an athletic field. They will be responsible for planning for optimal land use, responsible water consumption, making observations and recording measurements throughout the school year, and documenting this data in the science notebooks. Student teams will need to make an environmental assessment about where the playscape will be located, whom will it serve, what safety hazards exist, and how it will be maintained. They are to infer what conditions change and how the playscape will change due to these conditions. Students will learn about the tools used for measurement, use technology to make observations, and realize that science is conducted not only in the classroom but in the field as well.

Table 4.15 STEM Road Map K-2 – Second-Grade The Represented World Theme: Change over Time – Investigating Environmental Changes

NGSS Performance Objectives	Common Core		21st Century Skills
	Mathematics	Language Arts	
2-PS1-1	CCSS Math Practices MP1, MP2, MP3, MP4, MP5, MP6, MP7, MP8	Reading Standards CCSS.ELA. RI.2.1 RI.2.3 RI.2.7 RI.2.8 RI.2.9	21st Century Themes: Environmental Awareness Civic Literacy
2-ESS1-1	CCSS.Math.Content.2. MD.D.10	Writing Standards CCSS.ELA. W.2.1 W.2.2 W.2.7 W.2.8	Learning and Innovation Skills: Creativity and Innovation Critical Thinking and Problem Solving Communication and Collaboration
2-ESS2-2	CCSS.Math.Content.2. NB.T.A.1 CCSS.Math.Content.2. NB.T.A.2 CCSS.Math.Content.2. NB.T.A.3	Speaking and Listening Standards CCSS.ELA. SL.2.2 SL.2.1 SL.2.3 SL.2.5	Information, Media, and Technology Skills: Information Literacy Media Literacy Information Communication and Technology Literacy
2-LS-4-1			Life and Career Skills: Flexibility and Adaptability Initiative and Self-Direction Social and Cross Cultural Skills Productivity and Accountability Leadership and Responsibility
2-LS2-2			

STEM Careers in Second Grade

An *audio engineer* is an engineer who specializes in sound. They can design different speakers to project sound from a radio or stereo. They can design earbuds to listen to your iDevices. They have a strong background in physics-waves, mathematics, computers, the human brain, and hearing.

Table 4.16 STEM Road Map K-2 – Second-Grade Sustainable Systems Theme: Environmentally Friendly Playscapes

NGSS Performance Objectives	Common Core Mathematics	Common Core Language Arts	21st Century Skills
2-LS4-1	CCSS Math Practices MP1, MP2, MP3, MP4, MP5, MP6, MP7, MP8	Reading Standards CCSS.ELA. RI.2.1 RI.2.3 RI.2.7 RI.2.8 RI.2.9	21st Century Themes: Economic, Business, and Entrepreneurial Literacy Global Awareness Environmental Literacy Civic Literacy Health Literacy
2-LS2-2	CCSS.Math. Content.2.MD.B.5	Writing Standards CCSS.ELA. W.2.1 W.2.2 W.2.7 W.2.8	Learning and Innovation Skills: Creativity and Innovation Critical Thinking and Problem Solving Communication and Collaboration
2-ESS2-2	CCSS.Math. Content.2.MD.D.10	Speaking and Listening Standards CCSS.ELA. SL.2.2 SL.2.1 SL.2.3 SL.2.5	Information, Media, and Technology Skills: Information Literacy Media Literacy Information Communication and Technology Literacy
2-ESS2-1	CCSS.Math. Content.2.NBT.A.3		Life and Career Skills: Flexibility and Adaptability Initiative and Self-Direction Social and Cross Cultural Skills Productivity and Accountability Leadership and Responsibility

A *hearing specialist* is a person who studies human hearing. They study how people hear and try to determine if a person has a hearing deficiency. They can prescribe hearing aids to help people with damaged hearing. They have a background in human anatomy and physiology, the human brain and hearing, computers, and mathematics.

A *material engineer* is a person who develops, processes, and tests materials that enhance the structure of products. A material engineer can make computer chips that run computers, design new materials for plastics or ceramics that can be used in manufacturing, develop a tissue to help burn victims, and create a new material for sports equipment like skis or golf clubs, or even airplanes. They have a background in physics, engineering, mathematics, and design (Bureau of Labor Statistics, 2014).

An *astronaut* is a person who travels in space. They are highly trained, usually engineers or scientists, to work and travel in the depths of space. They can work on the International Space Station (ISS) as a scientist to conduct experiments in space or as an engineer that works to keep the ISS working properly. They can work for NASA or for NOAA. Once you are an astronaut, you are always an astronaut, even if you do not travel in space again. Their skills are very useful to helping other astronauts learn about space travel.

An *aerospace engineer* is a person who works in the space industry. They are highly skilled people who work in all areas of flying or space travel. They can design spacecraft to fly to Mars, the Moon, or the ISS. They can work on developing the type of spacesuit that astronauts wear while in space, and they can design how the rocket will launch from Earth. They have many different skills sets, one of them usually is computer programming.

A *town planner* works for a town and plans different buildings and structures within the town. They are responsible for making sure that buildings are able to be occupied, that houses are well kept, that bridges and roads are passable for people, that stores and town greens are maintained. They are essential in the town to make sure that the physical appearance works for every citizen in the community.

A *safety engineer* is a person who maintains the safety in the community. This may take the form of road signs or stop lights. They study traffic flow patterns and intersections to make sure that they are safe for both pedestrians and automobiles. They make sure that schools are safe for students and that their playgrounds are safe for visitors.

A *computer programmer* is someone who works on a computer by writing computer software. Computer software is used to run programs such as Google Docs, Tic Tok, Instagram, and many other programs you use on your phone. They are responsible if there is a security problem with a piece of software. They can write software to prevent computers from being hacked.

K-2 Road Map Summary

This chapter outlines the STEM Road Map for grades K-2. The intent in this chapter is to provide learning modules for teachers that will integrate the NGSS with Common Core ELA and mathematics and the NAEYC standards and positions into five STEM themes. The beauty of this integration is to include other disciplines such as social studies, art, and music that are often forgotten in the classroom. The addition of 21st century skills is woven into the themes and will help students develop the skills necessary for their future learning and understanding. With each theme, we have suggested careers for teachers and students to explore together and with this exploration; they will recognize that they, too, can consider STEM in their future aspirations. It is our hope that after an integrated curriculum over three years as we have outlined here, students will begin to appreciate that they have the opportunity to contribute to the ever-changing world in which we live.

STEM Road Map Module

A complete STEM Road Map kindergarten Patterns on Earth and In the Sky module is included in the Appendix of this book.

To learn more about the STEM Road Map Curriculum Series and the available titles, visit www.nsta.org.

References

American Mathematical Society. (2014). What do mathematicians do? Retrieved from http://www.ams.org/profession/career-info/math-work/math-work.

Bell, S. (2010). Project-based learning for the 21st century: Skills for the future. *The Clearing House, 83*, 39–43.

Bureau of Labor Statistics. (2014). Occupational outlook handbook: Material engineers. Retrieved from http://www.bls.gov/ooh/architecture-and-engineering/materials-engineers.htm.

Cole, C. (2011). *Connecting students to STEM careers: Social networking strategies*. International Society for Technology in Education.

Dodge, Y. (2006) *The Oxford dictionary of statistical terms*. New York: Oxford University Press.

Eliason, C. F., & Jenkins, L. T. (2012). *A practical guide to early childhood curriculum* (9th ed.). Merrill.

Koehler, C. M., Bloom, M., & Binns, I. C. (2013). Lights, camera, action! Developing a methodology to document mainstream films' portrayal of nature of science and scientific inquiry. *Electronic Journal of Science Education, 17*(2). Retrieved from http://ejse.southwestern.edu.

Koehler, C. M., Faraclas, E. W., Giblin, D., Moss, D. M., & Kazerounian, K. (2013). The nexus between science literacy & technical literacy: A state by state analysis of engineering content in state science frameworks. *Journal of STEM Education, 14*(3), 5–11.

National Association for the Education of Young Children. (2012). Technology and interactive media as tools in early childhood programs serving children from birth through age 8. Retrieved from http://www.naeyc.org/content/technology-and-young-children.

National Research Council. (2012) A framework for K-12 science education: Practices, crosscutting concepts, and core ideas. Committee on Conceptual Framework for new K-12 Science Education Standards. Board on Science Education, Division of Behavioral and Social Sciences and Education. Washington, DC: The National Academies Press.

NGSS Lead States. (2013). *Next generation science standards: For states, by states*. Washington, DC: National Academy Press.

Ricketts, R. (2018). Computational thinking for kindergarteners. Retrieved at https://www.edutopia.org/article/computational-thinking-kindergartners.

5   The STEM Road Map for Grades 3–5

Brenda M. Capobianco, Carolyn Parker, Amanda Laurier, and Jennifer Rankin

Overview of the Grades 3–5 STEM Road Map

Learning and teaching STEM at the elementary school level means providing students multiple opportunities to develop scientific understandings and related practices necessary to function productively as problem-solvers in a scientific and technological world. Allowing elementary school students to explore, experiment, or investigate while modeling, reasoning, and communicating affords students the opportunity to build curiosity, increase interest, and moreover, construct and apply new scientific knowledge to real world problems (National Research Council, 2005).

This is critically important at the upper elementary school level (defined here as grades 3–5) because students' thought processes become more mature and they start solving problems in a more logical fashion as well as incorporating inductive reasoning (Piaget & Inhelder, 1973). From a STEM perspective, this means that teachers must consider innovative ways to engage grades 3–5 students in a more student-centered, collaborative, hands-on, problem-based approach to learning while integrating disciplinary core ideas, scientific and engineering practices, and critical thinking across multiple subject areas. In this chapter, we provide an overview of an integrated approach using our grades 3–5 STEM Road Map. Like other grade-band chapters, the STEM Road Map for grades 3–5 is anchored in the five STEM themes: Cause and Effect, Innovation and Progress, The Represented World, Sustainable Systems, and Optimizing the Human Experience. Each STEM Road Map theme is intended for a five-week sequence of integrated instruction where the theme and associated problem or project is enacted through a core content area. For the upper elementary STEM Road Map chapter, we provide an example of a five-week module for each grade level (see links to download), including the complete unit with all instructional and assessment materials.

The STEM Road Map for grades 3–5 is aligned to Common Core Mathematics, Common Core English/Language Arts, Next Generation Science Standards, and the 21st Century Skills Framework. The enactment of the curriculum should be student-centered, be facilitated in an integrated fashion, and taught by making explicit connections across multiple content areas.

STEM Themes in the 3–5 STEM Road Map

The five overarching STEM themes continue to be reinforced and spiraled from the early, elementary grades. *Cause and Effect*, the dynamic relationship between various phenomena in the world, provides a real-world context for the study of weather, seismic activity, and the changes of seasons.

Human ingenuity and its important contributions to society are described by the theme, *Innovation and Progress*. In grades 3–5, students explore the design of maglev trains and solar ovens as well as the use of multimedia resources to display the influence of Earth's systems on one another.

Different models that humans have developed to help make sense of the world are included in the theme, *Represented World*. In grades 3–5, suggested topics include the phenomenon of bungee jumping, the process of erosion, and the development of a rainwater harvesting system.

The theme, *Sustainable Systems*, challenges students to investigate the interaction of different components of a larger system and explore ways the system can be sustained over time. Students in grade 3–5 can do so by examining the interactions among living and non-livings things in an aquarium/terrarium, the study of renewable energy, and the process of making compost.

Lastly, the theme, *Optimizing the Human Experience*, encourages students to utilize STEM ideas, concepts, and principles to improve the human condition. In grades 3–5, students focus on the development of levees and their impact on humans, the history of volcanic eruptions in conjunction with the development of a mechanical device to detect vibrations, and alternative ideas for conserving energy and promoting ecological sustainability.

The STEM Road Map for Third Grade

Before mapping out an integrated approach to learning STEM, it is important to consider what students have learned and experienced prior to entering third grade. In the second grade, students are expected to develop a more informed understanding of plants, different habitats, properties of materials, Earth events, and factors that contribute to Earth events. In addition, students in second grade develop and use models, plan and carry out investigations, analyze and interpret data, construct explanations, and design solutions. Using this newly acquired knowledge, students in third grade extend their existing ideas and conceptions in life, physical and earth and space sciences by engaging in one or more problem-based challenges. Each challenge is organized around one central topic inspired by one or more of the STEM Road Map Themes. These topics align not only with the theme but also with the grade-level academic content standards (e.g. Common Core and Next Generation

Science Standards). The topics for third grade include the following: weather, transportation, motion, ecosystems, and environmental science. Each of these topics is organized around a standards-based challenge, problem, or project that student teams are assigned to tackle in the course of learning necessary content and skills in the various disciplines (see Table 5.1).

Table 5.1 Third-Grade STEM Road Map Themes, Topics, and Problems/Challenges

STEM Theme	Topic	Problem/Challenge
Cause and Effect	Storm of the Year LEAD Mathematics	Student teams will analyze data using pattern recognition to identify predictable patterns such as daily temperature variation. Teams will recognize these cause-and-effect relationships and will use this knowledge to create a local weather forecast using digital tools, which can be in the form of an in-class, live presentation, recorded video, or video blog. Finally, teams will develop preparedness plans for their school and towns for weathering the storm.
Innovation and Progress	Transportation in the Future LEAD Social Studies	Student teams will design a model of a high-speed train that will safely transport passengers.
Represented World	Swing Set Makeover LEAD Science	Student teams will conduct a survey of their school playground or a nearby park or playground and develop a proposal for design of a new swing set that is both more entertaining, yet safer environment for play.
Sustainable Systems	Community Ecosystems LEAD Science	Student teams will develop a plan to preserve local ecosystem. Teams will investigate a local ecosystem, such as a nearby stream or park to better understand the interaction between living creatures, energy, and the non-living. Next, students work together to build a model of an aquatic ecosystem and observe the relationships between aquatic plants, algae, fish (mosquito fish or guppies), and snails.
Optimizing the Human Experience	Reducing Our Footprint LEAD Language Arts	Student teams will develop a plan for more environmentally friendly transportation methods at their local school.

Cause and Effect: Storm of the Year

Weather is easily observable and impacts our daily lives in various ways. Extreme weather events and conditions including blizzards, tornadoes, and hurricanes are sources of engagement and high interest among young learners. In this project, third-grade students are engaged with the challenge of an impending storm that will be one for the decade. Some states will encounter blizzard conditions and those living further south will endure tornadoes and thunderstorms. Student teams will investigate the many factors that influence weather including temperature, air pressure, clouds, and wind direction and speed in science. Students will assume the role of the weatherperson for Channel 10 and will use tools they have created to collect daily weather data, utilizing measurement skills in mathematics, by using various instruments including thermometers, barometers, wind vanes, anemometers, and hygrometers. After taking measurements and making direct observations of the sky and outdoor conditions, students will compare their findings to current weather data provided through research in language arts on reputable sites such as that of the National Weather Service (see: www.weather.gov). Students will analyze data utilizing pattern recognition to identify predictable patterns such as daily temperature variation. Students will recognize these cause-and-effect relationships and will use this knowledge to create a local weather forecast using digital tools, which can be in the form of an in-class, live presentation, recorded video, or video blog. Student audience members will view the presentation and state whether they agree or disagree with the forecast, providing supporting evidence. Finally, teams will develop preparedness plans for their school and towns for weathering the storm (see Table 5.2).

Innovation and Progress: Transportation in the Future

The notion of innovation for third-grade students means that students can be creative, resourceful, and imaginative. In the transportation design task led by social studies, third-grade students learn the history of train technology from wagon tramways of the 1500s to the invention of the first modern day train in 1804 by Richard Tevithick to today's modern-day technology that includes improved engine efficiency and enhanced aerodynamics. In science, students examine magnetic interactions and use magnets to solve a simple design problem. Using their new knowledge, students work in small teams to design a train that can safely carry passengers (weights) down an eight-foot track. The goal is to devise a vehicle that can easily glide when pushed. Final designs are assessed on the following criteria: (a) how well the car stays on the track; (b) how well the car glides to the end of the track; (c) appearance and

Table 5.2 STEM Road Map – Third-Grade Cause and Effect Theme: Predicting the Weather

NGSS Performance Objectives	Common Core Mathematics	Common Core Language Arts	21st Century Skills
3-ESS2-1	CCSS.Math.Practices MP1, MP2, MP3, MP4, MP5, MP6	Reading Standards CCSS.ELA. RI.3.1 RI.3.3 RI.3.5 RF.3.4	21st Century Themes: Global Awareness, Environmental Literacy, and Health Literacy
3-ESS2-2	CCSS.Math. Content.3.MD.A.1 CCSS.Math. Content.3.MD.A.2	Writing Standards CCSS.ELA W.3.1. W.3.1a, W.3.1b, W.3.1c, W.3.1d W.3.2, W.3.2b W.3.3 W.3.7 W.3.8	Learning and Innovation Skills: Creativity and Innovation Critical Thinking and Problem Solving Communication and Collaboration
	CCSS.Math. Content.3.NBT.A.1 CCSS.Math. Content.3.NBT.A.2 CCSS.Math. Content.3.NBT.A.3	Speaking and Listening Standards CCSS.ELA. SL.3.1, SL.3.1d SL.3.3 SL.3.4 SL.3.5 SL.3.6	Information, Media and Technology Skills: Information Literacy Media Literacy ICT Literacy
	CCSS.Math. Content.3.MD.B.3 CCSS.Math. Content.3.MD.B.4		Life and Career Skills: Flexibility and Adaptability Initiative and Self-Direction Social and Cross Cultural Skills Productivity and Accountability Leadership and Responsibility

overall design; and (d) speed. Further, students will conduct research on design ideas in language arts and will use mathematical practices to support solving this challenge. Teams will present their designs in a group presentation (see Table 5.3). This module has been published in 2017 by NSTA Press. See https://my.nsta.org/resource/110048.

Table 5.3 STEM Road Map – Third-Grade Innovation and Progress: Transportation

NGSS Performance Objectives	Common Core		21st Century Skills
	Mathematics	Language Arts	
3-PS2-3	CCSS.Math.Practices MP1, MP2, MP4, MP5, MP6	Reading Standards CCSS.ELA. RI.3.1 RI.3.3 RI.3.8	Learning and Innovation Skills: Creativity and Innovation Critical Thinking and Problem Solving Communication and Collaboration
3-PS2-4	CCSS.Math. Content.NBT.A.2	Writing Standards CCSS.ELA. W.3.1. W.3.1a, W.3.1b, W.3.1c, W.3.2, W.3.2b W.3.3 W.3.7 W.3.8	Information, Media, and Technology Skills: Information Literacy Media Literacy ICT Literacy
	CCSS.Math. Content.MD.A.1	Speaking and Listening Standards CCSS.ELA. SL.3.1, SL.3.1d SL.3.3 SL.3.4 SL.3.5 SL.3.6	Life and Career Skills: Flexibility and Adaptability Initiative and Self-Direction Social and Cross Cultural Skills Productivity and Accountability Leadership and Responsibility
	CCSS.Math. Content.MD.B.4		

Represented World: Swing Set Makeover

This investigation requires third-grade students to examine the STEM aspects involved in constructing a swing set to propose a prototype for a new and improved swing set. In science class, students will learn about motion and forces and conduct research on available swing sets in and around their schools. If swing sets are not readily available, a teacher could provide films depicting swing sets from the Internet. As the students examine different swing sets, they respond to the following questions: "What are the best conditions when creating a fun but safe swing set?" Students may need to be prompted to look at the length of the rope or chain, the type of seat, or how a child "gives power" to the swing to create the ride. Once the student teams examine, compare, and contrast different swing sets, they will develop a sketch and small-scale model of their proposed design, using geometric shapes and precise measurements (mathematics). Finally, individual students will draft a short essay or blog, which details the key components of how their design is an improvement upon existing swing sets (see Table 5.4). This module has been published in 2018 by NSTA Press. See https://my.nsta.org/resource/114575.

Table 5.4 STEM Road Map – Third-Grade Represented World: Recreational STEM

NGSS Performance Objectives	Common Core		21st Century Skills
	Mathematics	*Language Arts*	
3-PS2-1 3-PS2-2	CCSS.Math.Practices MP1, MP2, MP4, MP5, MP7	Reading Standards CCSS.ELA. W.3.7 W.3.8	21st Century Themes: Health Literacy Learning and Innovation Skills: Creativity and Innovation Critical Thinking and Problem Solving Communication and Collaboration
	CCSS.Math.Content.3.MD.A.2	Writing Standards CCSS.ELA. W.3.1. W.3.1a, W.3.1b, W.3.1c, W.3.1d W.3.2, W.3.2b W.3.3 W.3.7 W.3.8	Information, Media, and Technology Skills: Information Literacy Media Literacy ICT Literacy
	CCSS.Math.Content.3.MD.B.4	Speaking and Listening Standards CCSS.ELA. SL.3.1, SL.3.1d SL.3.3 SL.3.4 SL.3.5 SL.3.6	
	CCSS.Math.Content.3.OA.D.8 CCSS.Math.Content.3.OA.D.9		

Sustainable Systems: Community Ecosystems

Devising, building, and maintaining models of terrestrial and aquatic ecosystems provide third-grade students the opportunity to explore factors necessary to sustain an ecosystem as well as observe the diverse and unique life cycles of living organisms. In small teams, students investigate a local ecosystem, such as a nearby stream or park, to better understand the interaction between living creatures, energy, and the non-living. Next, students work together to build a model of an aquatic ecosystem and observe the relationships between aquatic plants, algae, fish (mosquito fish or guppies), and snails. Students begin to discuss the roles of organisms in the ecosystem as well as observe first-hand the life cycles of different aquatic plants and animals. As students observe events in the aquatic ecosystems, the students review the concepts introduced earlier in the life science sequence (biotic and abiotic factors; needs and characteristics of organisms and habitats). The term "ecosystem" is then introduced to refer to the system composed of a community of organisms interacting with its environment. The concept of "sustainable" is further explored by instructing students to find out different ways to maintain their ecosystems over time using what they know about the conditions necessary for the living organisms to survive. Finally, students apply what they have learned when they constructed their own ecosystems and apply their knowledge to their community's ecosystem. For the language arts component, students will write an essay or blog on how to protect, appreciate, and take care of a local natural pond, creek, or park. In social studies, students will learn about the various types and locations of biomes locally in the United States (see Table 5.5).

Optimizing the Human Experience: Reducing Our Footprint

Third graders will assume the role of activists in their community for this unit, as they are challenged in language arts class to reduce their footprint on the environment by identifying a local challenge and working in teams to write a persuasive story that will be shared on a blog for the local community. Student teams will do research, including collecting new data that will be analyzed to inform and provide support for their story. Students may be interested in addressing issues with pollution, recycling, deforestation, or other concerns that are localized in nature. Student teams will construct their blog and respond to questions, comments from readers that will further inform their learning on this important topic (Table 5.6).

Sample STEM Careers in the Third-Grade STEM Road Map

Career development and exploration in the elementary grades are critically important in facilitating students' interest, attitude, and persistence

Table 5.5 STEM Road Map – Third-Grade Sustainable Systems: Ecosystem Preservation

NGSS Performance Objectives	Common Core		21st Century Skills
	Mathematics	Language Arts	
3-LS1-1	CCSS.Math.Practices MP1, MP2, MP4	Reading Standards CCSS.ELA RI.3.7 RI.3.1 RI.3.2 RI.3.3	21st Century Themes: Environmental Literacy
3-LS4-3	CCSS.Math. Content.3.MD.B.3	Writing Standards CCSS.ELA W.3.1. W.3.1a, W.3.1b, W.3.1c, W.3.1d W.3.2, W.3.2b W.3.3 W.3.7 W.3.8	Learning and Innovation Skills: Creativity and Innovation Critical Thinking and Problem Solving Communication and Collaboration
	CCSS.Math. Content.3.NBT.A.1 CCSS.Math. Content.3.NBT.A.2 CCSS.Math. Content.3.NBT.A.3	Speaking and Listening Standards CCSS.ELA. SL.3.5	Information, Media, and Technology skills: Information Literacy Media Literacy ICT Literacy
	CCSS.Math. Content.3.NF.A.1		Life and Career Skills: Flexibility and Adaptability Initiative and Self-Direction Social and Cross Cultural Skills Productivity and Accountability Leadership and Responsibility

in STEM. There are many online resources that teachers and parents can utilize to begin supporting children's exploration of STEM careers including Clever Crazes for Kids (www.clevercrazes.com). Careers that complement the third-grade STEM Road Map and related activities include a variety of professions that reinforce opportunities for students to pursue their interests in fields such as meteorology, climatologist, field biology, environmental science, careers in law, and rail engineer.

Table 5.6 STEM Road Map – Third-Grade Optimizing the Human Experience: Reducing our Footprint

NGSS Performance Objectives	Common Core		21st Century Skills
	Mathematics	Language Arts	
3-LS4-2 3-LS4-3	CCSS.Math.Practices. MP1, MP2, MP4, MP5	Reading Standards CCSS.ELA. RI.3.7 RI.3.3	21st Century Themes: Global Awareness Environmental Literacy
	CCSS.Math. Content.3.MD.B.4	Writing Standards CCSS.ELA. W.3.1. W.3.1a, W.3.1b, W.3.1c, W.3.1d W.3.2, W.3.2b W.3.3 W.3.7 W.3.8	Learning and Innovation Skills: Creativity and Innovation Critical Thinking and Problem Solving Communication and Collaboration
	CCSS.Math. Content.3.MD.D.8	Speaking and Listening Standards CCSS.ELA SL.3.1 SL.3.1.a SL.3.1.b SL.3.1.d SL.3.5	Information, Media, and Technology Skills: Information Literacy Media Literacy ICT Literacy
	CCSS.Math. Content.3.OA.B.5		Life Career Skills: Flexibility and Adaptability Initiative and Self-Direction Social and Cross Cultural Skills Productivity and Accountability Leadership and Responsibility

The work of professionals, such as meteorologists and climatologists, allows students to explore different skills, practices, and knowledge associated with weather forecasting by interpreting and reporting the weather patterns; predicting future climate trends; and researching, verifying, and reporting on storms of the past. Essentially, a meteorologist is a specialized scientist that focuses on some aspect of the atmosphere. There are many different types of meteorologists ranging from broadcast to research meteorologists and forensic to archive meteorologists. A climatologist is a scientist who studies the climate. In short, climatology is related to meteorology, the study of weather, except that it looks at long-term trends and the history of the climate rather than examining weather systems in the short term like meteorologists do.

Additional professions in areas, such as field biology and environmental science, provide students with the opportunity to plan and carry out investigations in the field, analyze and interpret data from the field, and communicate results from their field studies to a larger audience. Environmental scientists and specialists use their knowledge of the natural sciences to protect the environment and human health. Environmental scientists monitor the quality of the environment (air, water, and soil), interpret the impact of human activities on terrestrial and aquatic ecosystems, and develop strategies for restoring ecosystems. In addition, environmental scientists help planners develop and construct buildings, transportation corridors, and utilities in ways that protect water resources and reflect efficient and beneficial land use. Advocates of environmental science include environmental attorneys, policy makers, and state councilman who serve local, national, and international communities. Their work is to ensure the development and enactment of environmental laws, policies, and guidelines that address issues including climate change, conservation, water quality, groundwater and soil contamination, use of natural resources, waste management, disaster reduction, and air and noise pollution.

Large corporations, such as The Walt Disney Company, employ mathematicians to develop models to predict the movement and flow of visitors throughout their theme parks and resorts and to and from different sites within the parks. The mathematicians use different software applications to make their work easier and more efficient while providing information about how to minimize the wait time at each park site and enhance visitors' overall experience. The company also employs technology developers and managers to design, implement, lead, and deliver different applications that support the company's overarching mission to create and deliver unforgettable experiences for the audience.

The STEM Road Map for Fourth Grade

As students progress from third to fourth grade, they become more informed problem-solvers. Using what they learned in third grade about weather, the interaction between forces, and changes within an ecosystem, fourth-grade students now apply their newly acquired knowledge to explore the properties of waves and energy transfer, the effects of weathering, and the role of renewable energy within sustainable systems. Fourth-grade students explore topics inspired by the different STEM Road Map Themes. These topics also align with grade-level academic

Table 5.7 Fourth-Grade STEM Road Map Themes, Topics, and Problems/Challenges

STEM Theme	Topic	Problem/Challenge
Cause and Effect	Field Station Mapping LEAD Social Studies	Student teams will create a plan for the construction of a safe and accessible station to conduct research on predicted volcano activity.
Innovation and Progress	Harnessing Solar Energy LEAD Science	Student teams will design, construct, and test a system that removes salt from saltwater using solar energy that could be used for their selected region of the world.
Represented World	Living on the Edge LEAD Mathematics	Student teams will examine the challenges residents face who choose to live near nature's beauty and develop a potential plan for mitigating erosion in their teams' selected location.
Sustainable Systems	Hydropower Efficiency LEAD Science	Student teams will develop a three-dimensional model or a computer-assisted image that demonstrates how an engineer may optimize the efficiency of a dam.
Optimizing the Human Experience	Water Conversation LEAD Language Arts	Student teams will develop informational materials for their school and community focused on water conservation generally and decreasing the use of bottled water specifically by use of filtration methods for tap water.

content standards (e.g. Common Core and Next Generation Science Standards). The topics for fourth grade include the following: mapping, solar energy, soil erosion, energy, and water consumption and conservation. Each of these topics is organized around a challenge/problem or project that student teams are assigned to tackle in the course of learning necessary content and skills in the various disciplines (see Table 5.7).

Cause and Effect: Field Station Mapping

In social studies, fourth-grade students learn about different kinds of maps. Fourth-grade students discover what all maps have in common, as well as some of the features they can expect to encounter while map reading. In this project, fourth-grade students blend their mapping skills with their understanding of science principles by analyzing world maps that show the locations of volcanoes and recent seismic activity (earthquakes) and learn about types of plates (oceanic and continental) and plate boundaries (divergent, convergent, and transform). In science, students will also learn about change over time evidenced in rock formations and fossils. In small teams, students generate different volcanic activity maps and make volcano predictions. Student teams explore patterns of volcano activity on different landmasses and identify a location and design for a research station. Using their understanding from the map analysis and related volcano activity data, student teams select and present their recommendations for a research station site as well as a design that would remain safe from geological hazards and would be easily accessible (see Table 5.8).

Innovation and Progress: Harnessing Solar Energy

This project allows fourth-grade students to design, test, and refine their ideas for a device that separates salt from water. Students apply what they know about science concepts including electromagnetic radiation and solar energy to plan, construct, and test a passive solar powered desalination apparatus (a desal Adora in Spanish). Students test the performance of their desal Adora by separating saltwater that has the same concentration of the average sample of ocean water. Students will learn that ocean water contains about 35,000 ppm of salt and will use multiplication and division to determine concentrations of solutions in word problems. This task encourages students to learn more about topics through research in language arts such as the greenhouse effect and, furthermore, innovate different ways of harnessing the sun's light energy. In social studies, students will learn about how populations have used solar energy for a variety of ways to move their region forward (see Table 5.9). This module has been published in 2017 by NSTA Press. See https://my.nsta.org/resource/110049.

Represented World: Living on the Edge

Some of the most beautiful places to live in the world are adjacent to the water, often on beaches or on cliffs overlooking the majestic oceans and rivers of the world. In this unit, fourth-grade students will examine the challenges residents face who choose to live near nature's beauty and develop a potential plan for mitigating erosion in their teams' selected location. Teams will be introduced to recent news on landslides along highway 1 in southern California and other more localized examples the teacher will provide. Erosion can be a serious problem, as it naturally occurs, primarily through water and wind processes. Teams will examine rainfall and tropical storm data to examine mathematical trends associated with rainfall and flooding which lead to erosion complications

Table 5.8 STEM Road Map – Fourth-Grade Cause and Effect: Field Station Mapping

NGSS Performance Objectives	Common Core		21st Century Skills
	Mathematics	Language Arts	
4-ESS2-1 4-ESS2-2	CCSS.Math.Practices MP1, MP2, MP5, MP6	Reading Standards CCSS.ELA. RI.4.3 RI.4.4 RI.4.6 RI.4.7 RF.4.4a	21st Century Themes: Global Awareness Environmental Literacy
	CCSS.Math.Content.4.MD.A.1 CCSS.Math.Content.4.MD.A.2 CCSS.Math.Content.4.MD.A.3	Writing Standards CCSS.ELA. W.4.1, W.4.1a, W.4.1b, W.4.1c, W.4.1d W.4.2, W.4.2a, W.4.2b, W.4.2c, W.4.2.d, W.4.2.e W.4.4 W.4.5 W.4.6 W.4.7 W.4.8 W.4.9	Learning and Innovation Skills: Creativity and Innovation Critical Thinking and Problem Solving Communication and Collaboration
		Speaking and Listening Standards CCSS.ELA. SL.4.1 SL.4.4 SL.4.5	Information, Media, and Technology Skills: Information Literacy ICT Literacy
			Life Career Skills: Flexibility and Adaptability Initiative and Self-Direction Social and Cross Cultural Skills Productivity and Accountability Leadership and Responsibility

for residents. Students will learn more about erosion through research, investigation, and virtually interviewing constituents who choose to live in areas which are in a potential hazard zone. In social studies, students will explore regions that have been historically impacted by landslides

Table 5.9 STEM Road Map – Fourth-Grade Innovation and Progress: Harnessing Solar Energy

NGSS Performance Objectives	Common Core		21st Century Skills
	Mathematics	Language Arts	
4-PS3-2 4-PS3-4	CCSS.Math.Practices. MP1, MP2, MP3, MP5	Reading Standards CCSS.ELA. RI.3.7 RI.3.3	21st Century Themes: Global Awareness Environmental Literacy
4-ESS3-1 4-ESS3-2	CCSS.Math. Content.4.OA.A.2 CCSS.Math. Content.4.OA.A.3	Writing Standards CCSS.ELA. W.4.6 W.4.7 W.4.8	Learning and Innovation Skills: Creativity and Innovation Critical Thinking and Problem Solving Communication and Collaboration
	CCSS.Math.Content. MD.A.1 CCSS.Math.Content. MD.A.2	Speaking and Listening Standards CCSS.ELA. SL.4.1, SL.4.1a, SL.4.1b, SL.4.1c, SL.4.1.d SL.4.4	
	CCSS.Math. Content.4.NBT.B.4		

and sinkholes and develop a plan for informing the public of the hazards of living in areas that are more prone to erosion (Table 5.10).

Sustainable Systems: Hydropower Efficiency

In this science-led project, students will learn about the natural resources that provide our energy and fuels for everyday life, specifically hydroelectric power. The challenge for this module is focused on the development of a three-dimensional model or computer-assisted image that will demonstrate how to optimize the efficiency of a dam. In language arts, student teams will research how water has historically been used to produce energy, with an emphasis on sustainability. In social studies, using sources students explore the historical development and use of hydroelectric dams, wave power, and tidal power (see Table 5.11). In science, student teams explore how a hydroelectric dam operates through online simulations (e.g. Oregon Museum of Science).

Table 5.10 STEM Road Map – Fourth-Grade Represented World: Erosion Modeling

NGSS Performance Objectives	Common Core		21st Century Skills
	Mathematics	Language Arts	
4-ESS2-1	CCSS.Math.Practices MP1, MP2, MP3, MP4, MP5	Reading Standards CCSS.ELA. RI.3.7 RI.3.3	21st Century Themes: Environmental Literacy
4-ESS3-2	CCSS.Math.Content.4.MD.A.1 CCSS.Math.Content.4.MD.A.2 CCSS.Math.Content.4.MD.A.3	Writing Standards CCSS.ELA. W.4.1, W.4.1a, W.4.1b, W.4.1c, W.4.1d W.4.2, W.4.2a, W.4.2b, W.4.2c, W.4.2.d, W.4.2.e W.4.4 W.4.5 W.4.6 W.4.7 W.4.8 W.4.9	Learning and Innovation Skills: Creativity and Innovation Critical Thinking and Problem Solving Communication and Collaboration
		Speaking and Listening Standards SL.4.1, SL.4.1a, SL.4.1b, SL.4.1c, SL.4.1.d SL.4.4	Information, Media, and Technology Skills: Information Literacy ICT Literacy
			Life Career Skills: Flexibility and Adaptability Initiative and Self-Direction Social and Cross Cultural Skills Productivity and Accountability Leadership and Responsibility

Table 5.11 STEM Road Map – Fourth-Grade Sustainable Systems: Hydropower Efficiency

NGSS Performance Objectives	Common Core		21st Century Skills
	Mathematics	Language Arts	
4-ESS3-1	CCSS.Math.Practices MP1, MP2, MP3, MP5, MP6	Reading Standards CCSS.ELA. RI.4.1 RI.4.2 RI.4.5 RI.4.7 RI.4.9	21st Century Themes: Environmental Literacy Global Awareness
4-PS3-4		Writing Standards CCSS.ELA. W.4.1 W.4.2 W.4.7 W.4.9	Learning and Innovation Skills: Creativity and Innovation Critical Thinking and Problem Solving Communication and Collaboration
		Speaking and Listening Standards CCSS.ELA. SL.4.1 SL.4.4 SL.4.5	Information, Media, and Technology Skills: Information Literacy Media Literacy ICT Literacy
			Life Career Skills: Flexibility and Adaptability Initiative and Self-Direction Social and Cross Cultural Skills Productivity and Accountability Leadership and Responsibility

Optimizing the Human Experience: Water Conservation

The theme, Optimizing the Human Experience, asks students to apply STEM ideas, concepts, and principles to improve the human condition. In this language arts-driven module, student teams will develop informational materials for their school and community focused on water conservation. On a planet where only 1% of the water is useful for humans, yet presently sustains a growing population, helping students understand the importance of water reuse through filtration is an imperative.

Through their research and connections made in science class, students explore the ever-more-scarce natural resource, water, by investigating all of the various ways water is used and wasted where they live. Students generate, distribute, and analyze water consumption surveys; interview local residents; and meet with county engineers. Using a writing journal, students record what they learn and reflect on ways water is wasted. In social studies, student teams will learn in detail about the global water quality and access issues while also emphasizing the geography and economic vitality of the countries they study.

Using notes from their writing journals, students prepare a persuasive essay that convinces the local town council to start a water conservation campaign for the town. Students apply what they have learned to design a personal water conservation plan for their home. For example, if students identified a family member who took unusually long showers, the students could propose and design a water capture system that would allow the reuse of the water for something like flushing a toilet, watering the lawn, or washing the car. Students would then present their capture and purification systems to their peers (see Table 5.12).

Sample STEM Careers in the Fourth-Grade STEM Road Map

As students' curiosity and enthusiasm for STEM builds, it is equally important to make fourth-grade students aware of different related careers. Professions such as civil engineer, seismologist, urban planner, journalist, and topographer align well with the fourth-grade STEM Road Map and related activities.

Civil engineers design and oversee the construction and maintenance of buildings and infrastructure such as highways, tunnels, rail systems, airports, and water supply and sewage systems. The job includes analysis – especially in the planning stage – studying survey reports and maps, breaking down construction costs, and considering government regulations and potential environmental hazards. Civil engineers also may test soils and building materials, provide cost estimates for equipment and labor, and use software to plan and design systems and structures.

A seismologist is a scientist who specializes in earth science. The work of a seismologist varies depending on where the work is needed. Some of this work may include monitoring, maintaining, testing and operating seismological equipment, documenting data, supervising preparation of test sites, managing inventory on equipment, and maintaining safety standards. Most seismologists work for petroleum or geophysical companies, and data processing centers.

An urban planner combines skills in land planning, along with transportation planning, to design a community or region that is easy to live in and attractive to look at. Urban planners are typically trained as engineers or architects. They must also have an understanding of many other fields including the environment, transportation, and psychology.

A journalist is an individual who investigates, collects, and presents information in the form of a news story. This story can be presented

Table 5.12 STEM Road Map – Fourth-Grade Optimizing the Human Experience: Water Conservation

NGSS Performance Objectives	Common Core Mathematics	Language Arts	21st Century Skills
4-ESS3-2	CCSS.Math.Practices MP1, MP2, MP3, MP4, MP5, MP6	Reading Standards CCSS.ELA. RI.4.3 RI.4.4 RI.4.5 RI.4.6 RI.4.7 RI.4.8 RI.4.9	21st Century Themes: Environmental Literacy
	CCSS.Math.Content.4.OA.A.2	Writing Standards CCSS.ELA. W.4.1, W.4.1a, W.4.1b, W.4.1c, W.4.1d W.4.2, W.4.2a, W.4.2b, W.4.2c, W.4.2.d, W.4.2.e W.4.4 W.4.5 W.4.6 W.4.7 W.4.8 W.4.9	Learning and Innovation Skills: Creativity and Innovation Critical Thinking and Problem Solving Communication and Collaboration
	CCSS.Math.Content.4.NBT.A.3	Speaking and Listening Standards CCSS.ELA. SL.4.1, SL.4.1a, SL.4.1b, SL.4.1c, SL.4.1.d SL.4.4	Information, Media, and Technology Skills: Information Literacy Media Literacy ICT Literacy
	CCSS.Math.Content.MD.A.2		Life Career Skills: Flexibility and Adaptability Initiative and Self-Direction Social and Cross Cultural Skills Productivity and Accountability Leadership and Responsibility

through newspapers, magazines, radio, television, and the Internet. A journalist writes in an objective manner, stating the facts and getting multiple perspective of the story.

A topographer is an expert in geology or geography who surveys lands and creates highly accurate representations through models and maps. Topographers often use computer equipment to take precise measurements of the elevation, location, shape, and contours of a particular area. Topographers created many of the maps we use today.

The STEM Road Map for Fifth Grade

At the fifth-grade level, students are able to develop and use models, plan and carry out investigations, and analyze and interpret data. More specifically, fifth-grade students delve further into the properties of matter and the conversation of matter, the movement of matter among plants and animals within an environment, and the representation of data used to reveal the daily changes in the length and direction of shadows, day and night, and the four seasons. Fifth-grade students explore topics inspired by the various STEM Road Map Themes that also align with grade-level academic content standards (e.g. Common Core and Next Generation Science Standards) which include the following: the interpretation and representation of data on the length and direction of shadows over time; the development of a protocol for making compost; the analysis of rainwater; and the design of a technological innovation that incorporates students' understanding of the interactions between Earth's systems. Each of these topics is organized around a challenge/problem or project that student teams are to address in their course of learning necessary content and skills in the various disciplines (see Table 5.13).

Table 5.13 Fifth-Grade STEM Road Map Themes, Topics, and Problems/Challenges

STEM Theme	Topic	Problem/Challenge
Cause and Effect	Schoolyard Engineering LEAD Mathematics	Student teams will design a movable awning for a picnic table located on the schoolyard that provides enough shade throughout recess for students and adults.
Innovation and Progress	Wind Energy LEAD Social Studies	Student teams will develop a proposal (using multimedia visual display) for the location of a wind turbine in their assigned region.
Represented World	Rainwater Analysis LEAD Mathematics	Student teams will design a rainwater harvesting system for their school.

(*Continued*)

Table 5.13 (Continued)

STEM Theme	Topic	Problem/Challenge
Sustainable Systems	Composting LEAD Science	Student teams will design a compost system for their school's cafeteria that makes use of excess food and food waste that is disposed of each day.
Optimizing the Human Experience	Mitigating Climate Change LEAD Social Studies	Student teams will design a solution that will mitigate the effects of global climate change in their selected region of the world.

Cause and Effect: Schoolyard Engineering

In this project, fifth-grade students are challenged to design an awning for the schoolyard picnic table that will provide enough shade across the day for students and adults. In science class, students will explore trends and patterns in data gathered from making observations of the length and direction of shadows from day to night and from one season to the next season. Different positions of the sun, moon, and stars at different times of the day, month, and year afford students the opportunity to observe and record patterns. Over an extended period, while on the playground each day, students measure and calculate the length of their shadows while facing different directions using mathematics. Students then pool their data to identify trends and patterns, relating the length of their shadows to the position of the sun at recess. Students develop graphs to represent their data and analysis. Using data from their analysis, students work in teams to plan, design, and test a movable awning for a picnic table located in the school grounds that follows the path of the sun and creates a large enough shadow to provide shade during recess throughout different times of the day. The student teams will present their prototypes to a panel of teachers and community members who will judge their innovativeness and presentation quality (see Table 5.14).

Innovation and Progress: Wind Energy

In this social studies-led fifth-grade challenge, students are challenged to develop a proposal for the location of a wind turbine off the east coast of the United States. Students will investigate US geography, as well as economic factors and feasibility of potential locations. In science, students will learn about the interaction of the Earth's systems (e.g. geosphere, hydrosphere, atmosphere, and biosphere) as well as how the interaction of landforms and long-term weather patterns influence one another to support energy production. In mathematics, students will use multiplication of whole numbers to determine the potential wind energy generated

Table 5.14 STEM Road Map – Fifth-Grade Cause and Effect: Schoolyard Engineering

NGSS Performance Objectives	Common Core		21st Century Skills
	Mathematics	Language Arts	
5-ESS1-1 5-ESS1-2	CCSS.Math.Practices MP1, MP2, MP3, MP6	Reading Standards CCSS.ELA. RI.5.1 RI.5.4 RI.5.9 RF.5.3 RF.5.4a	21st Century Themes: Global Awareness
	CCSS.Math.Content.5.MD.A.1	Writing Standards CCSS.ELA. W.5.1 W.5.2 W.5.4 W.5.6 W.5.7 W.5.8	Learning and Innovation Skills: Creativity and Innovation Critical Thinking and Problem Solving Communication And Collaboration
	CCSS.Math.Content.5.NBT.A.3 CCSS.Math.Content.5.NBT.A.4	Speaking and Listening Standards CCSS.ELA. SL.5.1, SL.5.1d SL.5.4 SL.5.5 SL.5.6	Information, Media, and Technology Skills: Information Literacy Media Literacy ICT Literacy
	CCSS.Math.Content.5.NBT.B.5		Life Career Skills: Flexibility and Adaptability Initiative and Self-Direction Social and Cross Cultural Skills Productivity and Accountability Leadership and Responsibility
	CCSS.Math.Content.5.NF.B.4b		
	CCSS.Math.Content.5.G.A.2		
	CCSS.Math.Content.5.G.B.3		

based upon the speed of wind and duration of wind and will provide an analysis by month of the projected wind energy production. Student teams will develop a written report based upon their selection of site and criteria for the specified location. In addition, students will deliver

Table 5.15 STEM Road Map – Fifth-Grade Innovation and Progress: Interactions

NGSS Performance Objectives	Common Core		21st Century Skills
	Mathematics	Language Arts	
5-ESS2-1	CCSS.Math.Practices MP1, MP2, MP3, MP4, MP5	Reading Standards CCSS.ELA RI.5.1 RI.5.4 RI.5.9 RF.5.3 RF.5.4a RI.5.7	21st Century Themes: Global Awareness
	CCSS.Math.Content.5.NBT.B.5	Writing Standards CCSS.ELA W.5.1 W.5.2 W.5.4 W.5.6 W.5.7 W.5.8	Learning and Innovation Skills: Creativity and Innovation Critical Thinking and Problem Solving Communication and Collaboration
	CCSS.Math.Content.5.MD.A.1	Speaking and Listening Standards CCSS.ELA SL.5.1, SL.5.1d SL.5.4 SL.5.5 SL.5.6	Information, Media, and Technology Skills: Information Literacy Media Literacy ICT Literacy
	CCSS.Math.Content.5.G.A.2		Life Career Skills: Flexibility and Adaptability Initiative and Self-Direction Social and Cross Cultural Skills Productivity and Accountability Leadership and Responsibility

a multimedia presentation using PowerPoint, Prezi, or a video to present their findings to the class (see Table 5.15). This module has been published in 2017 by NSTA Press. See https://my.nsta.org/resource/110050.

Represented World: Rainwater Analysis

In this mathematics-led challenge, student teams are challenged to devise a method to capture and reuse rainwater around their school building.

96 Brenda M. Capobianco et al.

To provide context for the module, students will learn in social studies about the importance of water for agriculture and will visit a local farm or garden to learn about how they reclaim water. In science class, students will engage in study of plants and the resources that sustain and support plant life (e.g. air and water). In addition, students will learn about the hydrosphere and distribution of water on Earth. Student teams will gather data related to the amount of rainfall in various locations around the school to determine the best placement for their teams' capture system. Using self-constructed rain gauges made out of canning jars, students measure the amount of rain that falls in different areas of the playground over a one-month period. From this data, students estimate the actual amount of rainfall that falls over the entire playground. Using the engineering design process, teams of students plan, construct, and test a method to collect and reuse excess rainwater. This challenge builds upon the knowledge that the students gained from the fourth grade Represented World challenge focused on soil erosion. The fifth-grade challenge could be differentiated to include the concept of rainwater capture as a way to decrease soil erosion in the school's playground. The language arts connection will include reading a variety of children's literature that are focused on water. In addition, students will collect data in their site for one month and compile their data to share with the class. The class data set will serve as a means to base a proposal to the building principal for the location of a rainwater collection system. Finally, students will also present their findings orally to a panel of local stakeholders, including members of the town council, civil engineers, and interested citizens (see Table 5.16). This module has been published in 2019 by NSTA Press. See https://my.nsta.org/resource/116367.

Sustainable Systems: Composting

The Composting design task affords students the opportunity to devise a protocol for making compost (Dankenbring, Capobianco, & Eichinger, 2014) from the excess of water food from their school's cafeteria. Underpinning the practices associated with the engineering design process is the production of either an artifact or a process. In this challenge, students innovate and create a process for making good compost and will develop a marketing campaign to encourage students and staff to take part in the program. In doing so, students utilize what they learn in science class regarding biotic and abiotic factors, conditions for decomposition to take place, and the role of decomposers to generate a form of compost that is usable and nutrient-rich. Further, in language arts, students will learn how to develop materials from their research and experiences for the purpose of relaying a position. Over several weeks, small teams of students will monitor the progress of their compost by recording measurements such as soil temperature, pH, odor, and level of moisture

Table 5.16 STEM Road Map – Fifth-Grade Represented World: Rainwater Analysis

NGSS Performance Objectives	Common Core Mathematics	Common Core Language Arts	21st Century Skills
5-ESS2-2 5-ESS2-1	CCSS.Math.Practices MP1, MP2, MP3, MP4, MP5, MP7	Reading Standards CCSS.ELA. RI.5.1 RI.5.4 RI.5.9 RF.5.3 RF.5.4a RI.5.7	21st Century Themes: Global Awareness
5-LS1-1 5-LS1-2	CCSS.Math.Content.5.G.A.1	Writing Standards CCSS.ELA W.5.1 W.5.2 W.5.4 W.5.6 W.5.7 W.5.8 W.5.9	Learning and Innovation Skills: Creativity and Innovation Critical Thinking and Problem Solving Communication and Collaboration
	CCSS.Math.Content.5.MD.C.5 CCSS.Math.Content.5.MD.C.5A CCSS.Math.Content.5.MD.C.5B	Speaking and Listening Standards CCSS.ELA. SL.5.1, SL.5.1d SL.5.4 SL.5.5 SL.5.6	Information, Media, and Technology Skills: Information Literacy Media Literacy ICT Literacy
	CCSS.Math.Content.5.NBT.A.3 CCSS.Math.Content.5.NBT.A.4		Life Career Skills: Flexibility and Adaptability Initiative and Self-Direction Social and Cross Cultural Skills Productivity and Accountability Leadership and Responsibility
	CCSS.Math.Content.5.NBT.B.5		

while also finding ways to aerate, weed, and water. At the end of the first month, students will compile their data into a technical report that will be summarized and shared with the school community. In social studies, students will learn about landfills and other areas in the United States and world that garbage is dumped and the implications for human vitality.

In mathematics, students will calculate the savings in disposal costs as well as in the repurposing of the compost to fertilize future gardens and replenish the soil. In the end, students develop an informed understanding of the interdependent relationships in ecosystems and significant role decomposers play within these respective ecosystems (see Table 5.17).

Table 5.17 STEM Road Map – Fifth-Grade Sustainable Systems: Composting

NGSS Performance Objectives	Common Core		21st Century Skills
	Mathematics	Language Arts	
5-ESS3-1	CCSS.Math.Practice MP1, MP2, MP3, MP4, MP5, MP6, MP7, MP8	Reading Standards CCSS.ELA. RI.5.1 RI.5.4 RI.5.9 RF.5.3 RF.5.4a RI.5.7 RI.5.9	21st Century Themes: Global Awareness
5-ETS1-2 5-ETS1-3	CCSS.Math. Content.MD.A.1	Writing Standards CCSS.ELA. W.5.1 W.5.2 W.5.4 W.5.6 W.5.7 W.5.8	Learning and Innovation Skills: Creativity and Innovation Critical Thinking and Problem Solving Communication and Collaboration
	CCSS.Math. Content.MD.B2	Speaking and Listening Standards CCSS.ELA SL..5.1 SL..5.4 SL..5.5 SL..5.6	Information, Media, and Technology Skills: Information Literacy Media Literacy ICT Literacy
	CCSS.Math. Content.MD.C.5		Life Career Skills: Flexibility and Adaptability Initiative and Self-Direction Social and Cross Cultural Skills Productivity and Accountability Leadership and Responsibility

Optimizing the Human Experience: Mitigating Climate Change

By leveraging their sense of curiosity and creativity, fifth-grade students, with social studies as the lead discipline, work to learn about predicted effects of global climate change on specific Third World countries. These deleterious effects are region specific. In science class, students will explore many factors that have been found to contribute to climate change both directly and indirectly in the context of using science ideas to protect the Earth's resources and environment. Student teams will be asked to research the effects of global climate change on their assigned county of the world. Student teams will plan and develop the ideas for a prototype of a technological innovation that is designed to minimize the influence of global climate change on their selected effect. For example, students may have learned that an increase in rain will increase the erosion of different landforms. In mathematics, students will develop graphs to represent the data on climate change for their selected region and develop a model for how they project their innovation may influence these statistics. Students could develop a prototype of a device that could minimize the impact of erosion and protect resources along a local riverbank or beach. In language arts, students will construct a technologically enhanced mode to present their innovation with a broad audience (e.g. blog and webpage) and will share their products with the school community (see Table 5.18).

Sample STEM Careers in the Fifth-Grade STEM Road Map

Maintaining interest in STEM careers remains important as students complete the fifth grade. Careers that align with our fifth-grade STEM learning activities are statistician, graphic artist, hydrologist, and a geotechnical engineer. In our ever-developing complex world, statisticians are experts in data. They provide expertise research design and trustworthy data production. They analyze the data to determine practical and useful conclusions. Statisticians draw on expertise from the fields of mathematics, science, and technology. They must have an understanding of research design and how data can be generated from a broad range of scientific fields. In addition, they must use complex computer programs to efficiently analyze large data sets.

A graphic artist uses a variety of medium to convey a message of emotion. Graphic designers are often employed in advertising, working side-by-side with a client to promote a product or idea. Graphic designers must have an understanding of computerized media and common design programs.

A hydrologist applies STEM knowledge and principles to solve water-related problems of quantity, quality, and availability. They may work in environmental protection, concerned with problems of flooding or soil

Table 5.18 STEM Road Map – Fifth-Grade Optimizing the Human Experience: Mitigating Climate Change

NGSS Performance Objectives	Common Core Mathematics	Common Core Language Arts	21st Century Skills
5-ESS3-1	CCSS.Math.Practices MP1, MP2, MP3, MP4, MP5, MP6	Reading Standards CCSS.ELA. RI.5.1 RI.5.4 RI.5.9 RF.5.3 RF.5.4a RI.5.7 RI.5.9	21st Century Themes: Global Awareness
	CCSS.Math.Content. NBT.A.3	Writing Standards CCSS.ELA. W.5.1 W.5.2 W.5.4 W.5.6 W.5.7 W.5.8	Learning and Innovation Skills: Creativity and Innovation Critical Thinking and Problem Solving Communication and Collaboration
	CCSS.Math.Content. MD.A.1	Speaking and Listening Standards CCSS.ELA. SL.5.1 SL.5.4 SL.5.5 SL.5.6	Information, Media, and Technology Skills: Information Literacy Media Literacy ICT Literacy
	CCSS.Math. Content.5.G.A.2		Life Career Skills: Flexibility and Adaptability Initiative and Self-Direction Social and Cross Cultural Skills Productivity and Accountability Leadership and Responsibility

erosion. Hydrologists must have an understanding of a broad range of STEM fields, including mathematics, physics, and Earth science.

A geotechnical engineer studies Earth's material and applies the knowledge to fields such as mining and fossil fuel production. Like most STEM professionals, a geotechnical engineer must have a broad understanding of STEM fields.

Summary

This chapter presented the STEM Road Map for grades 3–5 as an approach that engages upper-elementary students in authentic, team-based problems across content areas. Using the content and processes included in this chapter, instruction can be enacted in an integrated and coordinated manner, challenging students to confront real-world scenarios. The spiral approach, building on the knowledge and processes developed in the early elementary grades, supports students in the development of the skills and dispositions necessary to succeed in middle school and in later STEM careers. Several of the grades 3–5 modules have been published as new books through the National Science Teaching Association (NSTA) Press. Those books include:

- *Transportation in the Future* (3rd)
- *Swing Set Makeover* (3rd)
- *Harnessing Solar Energy* (4th)
- *Hydropower Efficiency* (4th)
- *Water Conservation* (4th)
- *Wind Energy* (5th)
- *Rainwater Analysis* (5th)
- *Composting* (5th)

To learn more about the STEM Road Map Curriculum Series and the available titles, visit https://www.nsta.org/publications/press/stemroadmap.aspx.

References

Dankenbring, C., Capobianco, B., & Eichinger, D. (2014). How to develop an engineering design task. *Science and Children, 53*(2), 4–9.

Green Schools Initiative. (2004, September). Retrieved from http://www.greenschools.net/article.php?list=type&type=4

National Research Council. (2005). In M. S. Donovan & J. D. Bransford (Eds.), *How students learn: History, mathematics, and science in the classroom.* Washington, DC. National Academies Press.

Oregon Museum of Science and Industry. *Best dam simulation ever.* Retrieved from http://www.omsi.edu/exhibits/damsimulation/.

Piaget, J., & Inhelder, B. (1973). *Memory and intelligence.* London: Routledge and Kegan Paul.

6 The STEM Road Map for Grades 6–8

Carla C. Johnson, Tamara J. Moore, Juliana Utley, Jonathan Breiner, Stephen R. Burton, Erin E. Peters-Burton, and Janet B. Walton

Overview of 6–8 STEM Road Map

This chapter will provide a detailed overview of the integrated STEM Road Map for the middle school grade levels 6–8. The STEM Road Map for grades 6–8 is anchored in the overarching five STEM themes that comprise the continuum of the STEM Road Map from K-12, which include Cause and Effect, Innovation and Progress, The Represented World, Sustainable Systems, and Optimizing the Human Experience. Each STEM Road Map theme is designed to be a five-week sequence of integrated instruction where the theme and associated problem or project is implemented across core content areas.

The STEM Road Map for grades 6–8 is designed to be delivered in an integrated fashion, meaning this is not a curriculum that should be taught by one teacher (e.g. science, social studies, mathematics, and language arts) in isolation. Rather, the STEM Road Map and associated STEM Road Map modules reflect an integration of Common Core Mathematics, Common Core English/Language Arts, Next Generation Science Standards, and the 21st Century Skills Framework and should be delivered by one or more lead teachers with other content areas making distinct ties to the project within their own curriculums as suggested in the maps and associated modules. The middle school level provides for the most authentic and facilitative setting for implementing the STEM Road Map. Students will quickly begin to see the connections across the disciplines and will also experience greater conceptual understanding of content taught in the various areas as they begin to apply their learning within the context of the real-world STEM projects in which they are engaged. Therefore, in the middle grades (6–8), there is a clear and distinctive role for all content areas (including art and music) in the inclusive, integrated STEM approach.

STEM Themes in the 6–8 STEM Road Map

The five overarching STEM themes continue to be reinforced and spiraled within the 6–8 STEM Road Map. *Cause and Effect* is the real-world STEM theme that consists of the dynamic relationships between various phenomena in the world. Students in grades 6–8 will explore motorsports, transportation, and Earth on the move within this STEM theme. *Innovation and Progress* relate to the various landmark developments driven by human ingenuity that have moved our society and understandings forward across generations. At the middle school level, topics in the STEM Road Map within Innovation and Progress include the effects of human impacts on climate, space travel, and medicine. *The Represented World* will take a look at the various models that humans have developed to make sense of the world around them. Students will explore topics including communication, genetic disorders, and learning from the past. In the *Sustainable Systems* STEM theme, students will be engaged in challenges including global water quality, populations, and minimizing human impact on the environment. The STEM Road Map theme of *Optimizing the Human Experience* focuses on innovations that have improved the quality of life. Students in grades 6–8 will investigate natural hazards, genetically modified organisms (GMOs) and the role of the sun in life on Earth.

Each of these topics will provide middle school students with an opportunity to be immersed in an authentic, problem and project-based curriculum that spans across traditional content lines to bring engineering and technological design, scientific inquiry, and mathematical reasoning to life in the process of developing potential prototypes for innovations of the future. Further, 21st Century Skills will be part of the fabric of day-to-day instruction within the STEM Road Map at the middle level as students will further refine their abilities to leverage critical thinking, creativity, communication, collaboration, information, and media literacy, all while they continue to grow their talents in leadership and taking responsibility for their own learning. The STEM Road Map 6–8 provides teachers with an engaging focus for delivery of the curriculum through the motivating topics that are personal to student interests and experiences in middle school, while also challenging adolescents to consider some of our greatest challenges and propose potential innovative solutions for society.

The STEM Road Map for Sixth Grade

This chapter is designed to build upon the experiences that students gained in the grades 3–5 STEM Road Map but could also be used with students who have not yet received any component of this curriculum. In grades 3–5, students were presented with challenges such as developing a weather forecast, designing the transportation of the future, and conserving water, one of our most precious resources on Earth.

In the sixth grade, students will explore STEM Road Map theme-inspired topics that align with grade-level academic content standards (e.g. Common Core and Next Generation Science Standards). The topics for sixth grade include *Amusement Parks, Human Impacts on Our Climate, Communication, Global Water Quality, and Natural Hazards*. Each of these topics is organized around a challenge/problem or project that student teams are assigned to tackle in the course of learning necessary content and skills in the various disciplines (see Table 6.1).

Cause and Effect: Human Impacts on Our Climate

In sixth grade, students will begin to grapple with some of the biggest challenges and often debates, within and outside of the scientific community. In the Cause and Effect STEM Road Map theme for sixth grade, the focus is on human impacts on climate overall, and the project asks students to specifically address global warming. In this project, students in science and mathematics class will investigate aspects of climate change driven by the rise in global temperatures over the past century and develop potential solutions that might address one aspect of human activity that has contributed to global climate change. This project will require students to conduct research on the potential causes of climate change, interview experts and others with understandings of this topic, use mathematical modeling and statistics to determine what steps have been taken to mitigate climate change, and develop their own prototype or solution using existing resources to address this global challenge. Table 6.2 provides a mapping of the content standards included in the Human Impacts on Our Climate PBL.

Innovation and Progress: Amusement Park of the Future

Without a doubt, most adolescents have had some type of interaction or experience with amusement parks or local carnivals in their childhood. Therefore, the sixth-grade topic of Amusement of the Future will serve as a motivating focus for instruction across this five-week sequence that is co-led by science and social studies disciplines in the STEM Road Map. The problem that students will be presented in this PBL module is to work in teams to design a prototype of the amusement park of the future. Mathematics and English/language arts components of this project will include research on the historical origins and designs of amusement parks, development of a blueprint of the model (either on paper or using technology), building and testing a small-scale prototype, developing a cost-benefit analysis for building, and maintaining the park. This will include examining the potential impact on the local

Table 6.1 Sixth-Grade STEM Road Map Themes, Topics, and Problems/Challenges

STEM Theme	Topic	Problem/Challenge
Cause and Effect	Human Impacts on Our Climate LEAD Mathematics/Science	Student teams are challenged to develop a potential solution to one aspect of human activity that may contribute to global warming. In solving this PBL challenge, students will investigate the aspects of climate change driven by the rise in global temperatures over the past century.
Innovation and Progress	Amusement Park of the Future LEAD Science/Social Studies	Given the technological capabilities of today, student teams will be challenged to produce a prototype of the amusement park of the future. Student teams will conduct research on the advancements in amusement parks from the first World's Fair to present including rides and games, as well as function in society, to inform their prototype.
Represented World	Packaging Design LEAD ELA/Mathematics	Student teams will design nested packages – small packages within large packages – for the purpose of repurposing a product or marketing it to a new user. Students will research the functions of packages, such as protect, contain, identify, transport, stack and store, and provide information.
Sustainable Systems	Global Water Quality LEAD Science	Student teams will devise a potential product/solution to address challenges related to poor water quality and/or access to clean water for their assigned county. Students will learn about the computational thinking practice of decomposition in this challenge when they break down the historical events in global water quality to see how progress has been made step by step and will develop documentaries in English/language arts that will bring to light the daily struggle for access to water around the globe
Optimizing the Human Experience	Natural Hazards LEAD Social Studies	Student teams will develop a natural hazard awareness and emergency preparedness plan specific to their selected country of interest. Second, teams will propose a potential new innovation that may enable people to either prepare for or deal with the aftermath of the event. Students will use the computational thinking practice of pattern recognition to observe recurring events and their causes.

Table 6.2 STEM Road Map – Sixth-Grade Cause and Effect: Human Impacts on Climate

NGSS Performance Objectives	Common Core		21st Century Skills
	Mathematics	*Language Arts*	
MS-ESS2-5 MS-ESS2-6	CCSS.Math.Practices MP1, MP2, MP3, MP4, MP5	Reading Standards CCSS.ELA RI.6.1 RI.6.4 RI.6.7	21st Century Themes: Global Awareness Environmental Literacy
MS-ESS3-5	CCSS.M. Content.6.NS.C.8	Writing Standards CCSS.ELA W.6.1, W.6.1a, W.6.1b, W.6.1c, W.6.1e, W.6.2, W.6.2a, W.6.2b, W.6.2d, W.6.2f	Learning and Innovation Skills: Creativity and Innovation Critical Thinking and Problem Solving Communication and Collaboration
	CCSS.M. Content.6.EE.C.9	Speaking and Listening Standards CCSS.ELA SL.6.1, SL.6.1a, SL.6.1b, SL.6.1c, SL.6.2, SL.6.5, L.6.1	Information, Media, and Technology Skills: Information Literacy Media Literacy ICT Literacy
	CCSS.M. Content.6.SP.B.5a CCSS.M. Content.6.SP.B.5b		Life and Career Skills: Flexibility and Adaptability Initiative and Self-Direction Social and Cross Cultural Skills Productivity and Accountability Leadership and Responsibility

community where the amusement park will be situated. Finally, students will develop a marketing plan and an infomercial promoting their model with script and demonstration. The mapping of content standards associated within this theme/topic can be found in Table 6.3. This module has been published in 2017 by NSTA Press. See https://my.nsta.org/resource/110051.

Table 6.3 STEM Road Map – Sixth-Grade Innovation and Progress Theme: Amusement Parks of the Future

NGSS Performance Objectives	Common Core		21st Century Skills
	Mathematics	Language Arts	
MS-PS3-1 MS-PS3-2 MS-PS3-4 MS-PS3–5	CCSS.Math.Practices MP1, MP2, MP3, MP4, MP5, MP6	Reading Standards CCSS.ELA. RI.6.1 RI.6.4 RI.6.7	21st Century Themes: Economic, Business, and Entrepreneurial Literacy
	CCSS.M. Content.6.RP.A.3	Writing Standards CCSS.ELA. W.6.1, W.6.1a, W.6.1b, W.6.1c, W.6.1e, W.6.2, W.6.2a, W.6.2b, W.6.2d, W.6.2f	Learning and Innovation Skills: Creativity and Innovation Critical Thinking and Problem Solving Communication and Collaboration
	CCSS.M. Content.6.G.A.1 CCSS.M. Content.6.G.A.3	Speaking and Listening Standards CCSS.ELA. SL.6.1, SL.6.1a, SL.6.1b, SL.6.1c, SL.6.2, SL.6.5, L.6.1	Information, Media, and Technology Skills: Information Literacy Media Literacy ICT Literacy
	CCSS.M. Content.6.SP.B.5b		Life and Career Skills: Flexibility and Adaptability Initiative and Self-Direction Social and Cross Cultural Skills Productivity and Accountability Leadership and Responsibility

Represented World: Packaging Design

In the last decade, the ability to communicate through the use of technology has grown exponentially – from Facebook to texting and Twitter to Instagram – adolescents are engaged in communicating every day and sometimes without one spoken word. In the Represented World, students will explore the realm of communication in sixth-grade English/language arts and mathematics class. They will explore packaging (in particular nested packages) with the purpose of repurposing a product or marketing the product to a new user. Either of these will require high levels of communication through the packaging. Through this, they will also learn about the importance of gaining strong personal written and verbal communication skills. Persuasive writing will be one form of communication that will be emphasized in this module, as the students will have to convince their client that their new product is marketable. As the students are required to think about nested packages (i.e. packages within packages), this module will require students to develop deep understandings of geometrical properties of 3-D shapes and engineering design, which is the focus of the science classroom component of this module. Success in the 21st century workplace and beyond hinges upon the ability to meld communication skills with their content skills. The mapping of content standards associated with this theme/topic can be found in Table 6.4. This module has been published in 2018 by NSTA Press. See https://my.nsta.org/resource/113531.

Sustainable Systems: Global Water Quality

Despite the numerous advances that have been made on Earth to move our society forward, humans still grapple with many challenges around the globe, including access to both an adequate supply and clean water overall. In this sixth-grade science-led unit, students will learn more about this international dilemma that civilizations face each and every day and the lengths to which some go in order to get access to water. As students learn about the historical context (social studies) of progress in global water quality, they will also be challenged to use their innovative thinking to devise potential future solutions to this issue. Students will learn about the computational thinking practice of decomposition in this challenge when they break down the historical events in global water quality to see how progress has been made step by step. This will require considering materials, prototypes, cost-benefit analyses, and transportation methods that may provide much needed life resources to communities in various locations around the globe. Further, student teams will develop documentaries in English/language arts that will bring to light the daily struggle for access to water around the globe (see Table 6.5).

Optimizing the Human Experience: Natural Hazards

Students in sixth grade will take a proactive stance to addressing natural hazards that our society faces on a regular basis through an exploration of the realized impact of hazards such as hurricanes, tornados,

Table 6.4 STEM Road Map – Sixth-Grade Represented World: Communication

NGSS Performance Objectives	Common Core		21st Century Skills
	Mathematics	Language Arts	
MS-ETS1-1 MS-ETS1-2 MS-ETS1-3	CCSS.Math.Practices MP1, MP4, MP5, MP6	Reading Standards CCSS.ELA. RI.6.4, RI.6.7	21st Century Themes: Environmental Literacy and Health Literacy
	CCSS.M.Content.6.G.1 CCSS.M.Content.6.G.2 CCSS.M.Content.6.G.3 CCSS.M.Content.6.G.4	Writing Standards CCSS.ELA. W.6.1, W.6.1a W.6.1b, W.6.1e W.6.2, W.6.2a, W.6.3d, W.6.4	Learning and Innovation Skills: Creativity and Innovation Critical Thinking and Problem Solving Communication and Collaboration
		Speaking and Listening Standards CCSS.ELA. SL.6.1, SL.6.1a, SL.6.1b, SL.6.1c, SL.6.2, SL.6.5, L.6.1	Information, Media, and Technology Skills: Information Literacy Media Literacy ICT Literacy
			Life and Career Skills: Flexibility and Adaptability Initiative and Self-Direction Social and Cross Cultural Skills Productivity and Accountability Leadership and Responsibility

earthquakes, tsunamis, volcanic eruptions, and flooding. Students will learn about the culture of populations that live in historically natural hazard zones and will develop an understanding of the benefits and risks that communities experience. Sixth graders will be challenged to conduct research on a selected country and learn more about the natural hazards that occur in that region with connections to science, mathematics, language arts, and the lead subject, social studies. Students will

Table 6.5 STEM Road Map – Sixth-Grade Sustainable Systems: Global Water Quality

NGSS Performance Objectives	Common Core Mathematics	Language Arts	21st Century Skills
MS-ESS2-4	CCSS.Math.Practices MP1, MP2, MP4	Reading Standards CCSS.ELA. RI.6.4 RI.6.7	21st Century Themes: Global Awareness Environmental Literacy Health Literacy
MS-ESS3-1	CCSS.M.Content.6.RP.A.1	Writing Standards CCSS.ELA. W.6.1, W.6.1a W.6.1b, W.6.1e W.6.2, W.6.2a, W.6.3d, W.6.4	Learning and Innovation Skills: Creativity and Innovation Critical Thinking and Problem Solving Communication and Collaboration
	CCSS.M.Content.6.RP.A.2	Speaking and Listening Standards CCSS.ELA. SL.6.1, SL.6.1a, SL.6.1b, SL.6.1c, SL.6.2, SL.6.5, L.6.1	Information, Media, and Technology Skills: Information Literacy Media Literacy ICT Literacy
	CCSS.M.Content.6.RP.A.3.b CCSS.M.Content.6.RP.A.3.c		Life and Career Skills: Flexibility and Adaptability Initiative and Self-Direction Social and Cross Cultural Skills Productivity and Accountability Leadership and Responsibility

use the computational thinking practice of pattern recognition to observe recurring events and their causes. Students will work in teams to develop emergency awareness and preparedness plans for their assigned setting. Teams will also develop a potential new innovation that may inform the population of an upcoming event and/or help a society deal with the aftermath of a natural hazard (see Table 6.6).

Table 6.6 STEM Road Map – Sixth-Grade Optimizing the Human Experience: Natural Hazards

NGSS Performance Objectives	Common Core		21st Century Skills
	Mathematics	Language Arts	
MS-ESS3-2	CCSS.Math.Practices MP1, MP2, MP4	Reading Standards CCSS.ELA. RI.6.1 RI.6.4 RI.6.7	21st Century Themes: Global Awareness Environmental Literacy Health Literacy Economic and Financial Literacy
	CCSS.M. Content.6.EE.B.6	Writing Standards CCSS.ELA. W.6.1,W.6.1a W.6.1b, W.6.1e W.6.2,W.6.2a, W.6.3d, W.6.4	Learning and Innovation Skills: Creativity and Innovation Critical Thinking and Problem Solving Communication and Collaboration
	CCSS.M. Content.6.EE.B.7	Speaking and Listening Standards CCSS.ELA. CCSS.ELA. SL.6.1, SL.6.1a, SL.6.1b, SL.6.1c, SL.6.2, SL.6.5, L.6.1	Information, Media, and Technology Skills: Information Literacy Media Literacy ICT Literacy
	CCSS.M. Content.6.EE.C.9		Life and Career Skills: Flexibility and Adaptability Initiative and Self-Direction Social and Cross Cultural Skills Productivity and Accountability Leadership and Responsibility

Sample STEM Careers in Sixth-Grade STEM Road Map

Environmental scientists use their knowledge of natural sciences to protect the environment. They identify problems and find solutions that protect the health of the environment and the people living in it. Environmental scientists often work in laboratories and offices, but also spend time in the environment they're protecting. Environmental scientists need at least a bachelor's degree in natural science.

Environmental engineering technicians carry out the plans that environmental engineers develop. They test, operate, and if necessary, modify equipment for preventing or cleaning up environmental pollution. They may collect samples for testing or work to identify the sources of environmental pollution. They typically work indoors, usually in laboratories. Employers in this field prefer that environmental engineering technicians have earned an associate degree.

Architects plan and design buildings and other structures. Architects spend most of their time in offices, where they consult with clients, develop reports and drawings, and work with other architects and engineers. However, architects often visit construction sites to review the progress of projects. There are three main steps in becoming a licensed architect: earning a professional degree in architecture, gaining work experience through an internship, and passing the Architect Registration Exam.

Civil engineers design and supervise large construction projects including roads, buildings, airports, tunnels, dams, bridges, and systems for water supply and sewage treatment. Civil engineers generally work indoors in offices. However, they sometimes spend time outdoors at construction sites so they can monitor operations or solve problems at the site. Civil engineers need a bachelor's degree and must be licensed in all states and the District of Columbia.

Actuaries analyze the financial costs of risk and uncertainty. They use mathematics, statistics, and financial theory to assess the risk that an event will occur and to help businesses and clients minimize the cost of that risk. Most actuaries work in an office setting. Actuaries need a bachelor's degree and must pass a series of exams to become certified professionals. They must have a strong background in mathematics, statistics, and business.

Microbiologists study the growth, development, and other characteristics of microscopic organisms like bacteria. Microbiologists work in laboratories and offices where they conduct experiments. A bachelor's degree in microbiology or a closely related field is needed for entry-level positions.

Registered nurses take care of people with injury and illness as well as teach the public about health conditions and provide emotional support to patients and their families. Nurses work in hospitals, doctors' offices, home healthcare, nursing homes, summer camps, schools, and also in the military. To become a registered nurse, an associate's or bachelor's degree is required as well as passing a national licensing exam.

Statisticians use mathematical techniques to analyze and interpret data and draw conclusions. Although statisticians work mostly in offices, they may travel in order to supervise surveys or gather data. Some statisticians work for the government; many others work for private businesses. Most statisticians enter the occupation with a master's degree in statistics, mathematics, or survey methodology, although a bachelor's degree is sufficient for some entry-level jobs. Research and academic positions generally require an advanced degree (e.g. Ph.D. or Ed.D.).

Advertising, promotions, and marketing managers plan programs to generate interest in a product or service. They work with art directors, sales agents, and financial staff members. About 24% of advertising and promotions managers worked for advertising agencies in 2012. About 16% of marketing managers worked in the management of companies and enterprises industry. A bachelor's degree is required of most advertising promotions and marketing management positions.

The STEM Road Map for Seventh Grade

In the seventh grade, students will explore STEM Road Map theme-inspired topics that align with grade-level academic content standards (e.g. Common Core and Next Generation Science Standards). The topics for sixth grade include *Transportation – Motorsports, Space Travel, Genetic Disorders, Populations, and Genetically Modified Organisms (GMOs)*. Each of these topics is organized around a challenge/problem or project that student teams are assigned to tackle in the course of learning necessary content and skills in the various disciplines (see Table 6.7).

Table 6.7 Seventh-Grade STEM Road Map Themes, Topics, and Problems/Challenges

STEM Theme	Topic	Problem/Challenge
Cause and Effect	Transportation – Motorsports LEAD Science	Student teams are challenged to design a motorsports prototype vehicle that includes one new safety aspect from existing technologies and is powered by energy transformations.
Innovation and Progress	Life in Space LEAD Science	Student teams will research, design, and build a prototype of a human colony that could enable life in space on a selected planet or moon. Students will use computational thinking in the form of abstraction to sift through the unnecessary information and find the key ideas for their project. In mathematics, modeling will be used to determine the feasibility of models in regard to space travel, light years, and the timeline for inhabiting the colony

(Continued)

Table 6.7 (Continued)

STEM Theme	Topic	Problem/Challenge
Represented World	Mapping Genetics LEAD ELA	Student teams will develop a proposed course of intervention for a selected genetic disorder, based on research that may provide some relief from symptoms associated with the disorder. Students will use the computational thinking practice of pattern recognition to seek commonalities across the research. Teams will develop informational materials (e.g. blogs and printed media) to disseminate to the public regarding their findings.
Sustainable Systems	Population Density LEAD Mathematics	Student teams will devise a model for counting populations of a given species on Earth and develop a formal presentation of model for consideration by a panel of experts.
Optimizing Human Experience	Genetically Modified Organisms (GMO) LEAD Social Studies	Student teams will develop a documentary on the pros and cons of the use of genetically modified organisms as the main source of food for humans and other living things.

Cause and Effect: Transportation – Motorsports

The seventh-grade transportation – motorsports module is led by science. Students will take on the role of design engineers as they work in teams to design, within a set of design constraints, an innovative prototype vehicle powered by energy transformations. As they move through the module, students will investigate types of energy, energy transformations, the law of conservation of energy, the concepts of speed, friction, aerodynamic drag, and the engineering design process. Students will learn about the history of the motorsports industry, safety standards, and how it has transformed the economy of the United States through Nascar, IndyCar, and other racing associations. Mathematics is embedded throughout this module, which will culminate in the design project, The Automotive X-Challenge. Engineering, manufacturing, and motorsports careers are emphasized throughout the unit via videos, activities, and visits from industry professionals. Student teams will participate in a race day event in which cars will compete for speed and will present their design to industry professionals to be judged upon design, innovation, teamwork, and presentation quality (see Table 6.8).

Innovation and Progress: Life in Space

Advancements in space travel have taken place at a very rapid pace since the first astronaut landed on the moon. Within the last decade, we have

Table 6.8 STEM Road Map – Seventh-Grade Cause and Effect: Transportation – Motorsports

NGSS Performance Objectives	Common Core		21st Century Skills
	Mathematics	Language Arts	
MS-PS2-1 MS-PS2-2 MS-PS2-3 MS-PS2-5	CCSS.Math.Practices MP1, MP2, MP3, MP4, MP5 MP6, MP7, MP8	Reading Standards CCSS.ELA. RI.7.1 RI.7.7	21st Century Themes: Global Awareness Financial Economic Business Entrepreneurial Literacy
	CCSS.M. Content.7.RP.A.1 CCSS.M. Content.7.RP.A.2 CCSS.M. Content.7.NS.A.3	Writing Standards W.7.1, W.7.1a W.7.2, W.7.2a, W.7.2.b W.7.6 W.7.7 W.7.8 W.7.9	Learning and Innovation Skills: Creativity and Innovation Critical Thinking and Problem Solving Communication and Collaboration
	CCSS.M. Content.7.EE.B.3 CCSS.M. Content.7.EE.B.4	Speaking and Listening Standards SL.7.1, SL.7.1a, SL.7.1b, SL.7.1c, SL.7.1d, SL.7.3 SL.7.4 SL.7.5	Information, Media, and Technology Skills: Information Literacy Media Literacy ICT Literacy
			Life and Career Skills: Flexibility and Adaptability Initiative and Self-Direction Social and Cross Cultural Skills Productivity and Accountability Leadership and Responsibility

seen the closure of the United States space shuttle program, and the National Aeronautics and Space Association (NASA) has focused their work more toward exploration of Mars and other aspects of our galaxy. In this sixth-grade science-led module, students will gain an understanding of some historical aspects of space travel (social studies) and will also research current advances to design and create a prototype of a habitat that could be created on another viable planet or moon in our solar system that would support human colonization. Teams will investigate light and sound, chemical properties, and the scale of the universe as they consider design possibilities for their colony. Students will read a variety of texts in English/language arts focused on space exploration and gather information from a variety of online sources to support the development of their research for this project. Students will use computational thinking in the form of abstraction to sift through the unnecessary information and find the key ideas for their project. In mathematics, modeling will be used to determine the feasibility of models in regard to space travel, light years, and the timeline for inhabiting the colony (see Table 6.9).

Represented World: Mapping Genetics

Traditionally, students in middle school have not had the opportunity to explore genetics beyond learning about Punnett Squares and learning about genetic traits at a surface level. The reality is that our genetics today can tell us many things about our history, our potential for disease in the future, and can help determine the identity of a criminal. In this seventh-grade module, students will work in teams to select a genetic disorder based upon their own interests and engage in research to learn about historical, homeopathic, and proposed treatments and remedies for symptoms of the disorder. Students will use the computational thinking practice of pattern recognition to seek commonalities across the research. The knowledge that each team gains from their work will be communicated to the public through the development of technology-based communication tools. English/language arts class is the lead discipline for this module in the STEM Road Map, where students will conduct important research and learn how to analyze sources to gather information that will serve as the basis for their course of intervention. In science, students will learn about genetic traits and disorders. In mathematics, students will use a variety of ways to model genetic traits including the mathematically based Punnett Squares (see Table 6.10).

Sustainable Systems: Population Density

There are many STEM fields that require out of the box thinking on a regular basis. In agriculture, it is often difficult to conduct an exact count of livestock due to the size of the area that the animals inhabit. Similarly, obtaining an accurate count of animals in the wild is a challenge. Population density refers to the application of mathematical modeling to measure a given population within a targeted area or region. As a matter of fact, population density is used often to examine human populations around the globe and is a concept within the realm of social studies as well. In this challenge, student teams will devise a model for counting populations of a given species on Earth and develop a formal presentation of their models for consideration by a panel of experts. As an extension, in science class, students will examine ecosystems and populations of living things (non-human). In social studies, students

Table 6.9 STEM Road Map – Seventh-Grade Innovation & Progress: Space Travel

NGSS Performance Objectives	Common Core		21st Century Skills
	Mathematics	Language Arts	
MS-PS1-1 MS-PS1-5	CCSS.Math.Practices MP1, MP2, MP3, MP4, MP5, MP6, MP7	Reading Standards CCSS.ELA. RI.7.8 RI.7.9	21st Century Themes: Global Awareness Environmental Literacy Health Literacy
MS-ESS1-2 MS-ESS1-3	CCSS.Math.Content.7.NS.A.1 CCSS.Math.Content.7.NS.A.2 CCSS.Math.Content.7.NS.A.3	Writing Standards CCSS.ELA. W.7.1, W.7.1a W.7.1b, W.7.1c W.7.1e, W.7.2, W.7.2a, W.7.2b, W.7.2d, W.7.3.c W.7.6 W.7.7 W.7.8 W.7.9	Learning and Innovation Skills: Creativity and Innovation Critical Thinking and Problem Solving Communication and Collaboration
MS-PS4-2	CCSS.Math.Content.7.EE.A.1	Speaking Standards CCSS.ELA. SL.7.1, SL.7.1a, SL.7.1b, SL.7.1c, SL.7.1d, SL.7.3, SL.7.4, SL.7.5	Information, Media, and Technology Skills: Information Literacy Media Literacy ICT Literacy
	CCSS.Math.Content.7.EE.B.3 CCSS.Math.Content.7.EE.B.4a		Life and Career Skills: Flexibility and Adaptability Initiative and Self-Direction Social and Cross Cultural Skills Productivity and Accountability Leadership and Responsibility
	CCSS.M.Content.7.RP.A.1 CCSS.M.Content.7.RP.A.2		

will explore global populations and relationships between population density and access to goods/services with an economic and geographical lens. In English/language arts, students will read relevant literature focused on the aforementioned issues and apply new knowledge to their model (see Table 6.11). This module is in press as of November 2020.

Table 6.10 STEM Road Map – Seventh-Grade Represented World: Genetic Disorders

NGSS Performance Objectives	Common Core		21st Century Skills
	Mathematics	Language Arts	
MS-LS4-6	CCSS.Math.Practices M1, M2, M3, M4, M5	Reading Standards CCSS.ELA. RI.7.8 RI.7.9	21st Century Themes: Global Awareness Environmental Literacy
MS-LS3-1 MS-LS3-2	CCSS.Math. Content.7.NS.A.3	Writing Standards CCSS.ELA. W.7.1, W.7.1a W.7.1b, W.7.1c W.7.1e W.7.2, W.7.2a, W.7.2b, W.7.2d, W.7.3.c W.7.6 W.7.7 W.7.8 W.7.9	Learning and Innovation Skills: Creativity and Innovation Critical Thinking and Problem Solving Communication and Collaboration
MS-LS4-3 MS-LS4-4	CCSS.Math. Content.7.SP.A.1 CCSS.Math. Content.7.SP.A.2	Speaking Standards CCSS.ELA. SL.7.1, SL.7.1a, SL.7.1b, SL.7.1c, SL.7.1d, SL.7.3, SL.7.4, SL.7.5	Information, Media, and Technology Skills: Information Literacy Media Literacy ICT Literacy
MS-LS1-4			Life and Career Skills: Flexibility and Adaptability Initiative and Self-Direction Social and Cross Cultural Skills Productivity and Accountability Leadership and Responsibility

Optimizing the Human Experience: Genetically Modified Organisms

There are many nations on Earth that struggle each day with access to a sufficient food supply. The challenges of generating adequate food supply and developing pest resistant plants sparked the field of genetically modified organisms (GMOs). However, there are growing concerns

Table 6.11 STEM Road Map – Sustainable Systems: Population Density

NGSS Performance Objectives	Common Core Mathematics	Language Arts	21st Century Skills
MS-LS1-6 MS-LS1-7	CCSS.Math.Practices MP1, MP2, MP3, MP4, MP5, MP6, MP8	Reading Standards CCSS.ELA. RI.7.8 RI.7.9	21st Century Themes: Global Awareness Environmental Literacy Civic Literacy Financial, Economic, Business, and Entrepreneurial Literacy Health Literacy
MS-LS2-1 MS-LS2-2 MS-LS2–3 MS-LS2–4	CCSS.Math.Content.7.SP.A.1 CCSS.Math.Content.7.SP.A.2	Writing Standards CCSS.ELA. W.7.1, W.7.1a W.7.1b, W.7.1c W.7.1e W.7.2, W.7.2a, W.7.2b, W.7.2d, W.7.3.c W.7.6 W.7.7 W.7.8 W.7.9	Learning and Innovation Skills: Creativity and Innovation Critical Thinking and Problem Solving Communication and Collaboration
	CCSS.Math.Content.7.SP.B.4	Speaking Standards CCSS.ELA. SL.7.1, SL.7.1a, SL.7.1b, SL.7.1c, SL.7.1d, SL.7.3, SL.7.4, SL.7.5	Information, Media, and Technology Skills: Information Literacy Media Literacy ICT Literacy
			Life and Career Skills: Flexibility and Adaptability Initiative and Self-Direction Social and Cross Cultural Skills Productivity and Accountability Leadership and Responsibility

about the impact of genetically engineered plants and animals on human health. In this seventh-grade social studies-led module, student teams will investigate the pros and cons of GMOs and will develop a documentary focused on communicating the health, social, and economic aspects of GMO production and consumption. In science, students will learn about genetic factors that influence the growth of organisms as well as basic cell structure and function. Students will explore the costs and benefits of GMO use in mathematics while developing mathematical models to grow further understanding. Finally, students will work in English/language arts on the development of the communication that will be the basis of the documentary and learn how to persuasively relay their ideas in a convincing manner (Table 6.12).

Table 6.12 STEM Road Map – Seventh-Grade Optimizing the Human Experience: Genetically Modified Organisms

NGSS Performance Objectives	Common Core		21st Century Skills
	Mathematics	Language Arts	
MS-LS1-2 MS-LS1-3 MS-LS1-5 MS-LS1-8	CCSS.M.Practices MP1, MP3	Reading Standards CCSS.ELA. RI.7.1 RI.7.4 RI.7.8 RI.7.9	21st Century Themes: Global Awareness Health Literacy Civic Literacy
MS-LS2-5	CCSS.Math. Content.7.RP.A.2c	Writing Standards CCSS.ELA. W.7.1, W.7.1a W.7.1b, W.7.1c W.7.1e, W.7.2, W.7.2a, W.7.2b, W.7.2d W.7.3.c W.7.6 W.7.7 W.7.8 W.7.9	Learning and Innovation Skills: Creativity and Innovation Critical Thinking and Problem Solving Communication and Collaboration
MS-LS4-5	CCSS.Math. Content.7.NS.A.1d CCSS.Math. Content.7.NS.A.3	Speaking and Listening Standards CCSS.ELA. SL.7.1, SL.7.1a, SL.7.1b, SL.7.1c, SL.7.1d, SL.7.2, SL.7.3, SL.7.4, SL.7.5, SL.7.6	Information, Media, and Technology Skills: Information Literacy Media Literacy ICT Literacy
	CCSS.Math. Content.7.EE.B.3		Life and Career Skills: Flexibility and Adaptability Initiative and Self-Direction Social and Cross Cultural Skills Productivity and Accountability Leadership and Responsibility

Sample STEM Careers in the Seventh-Grade STEM Road Map

Biomedical engineers analyze and design solutions to problems in biology and medicine, with the goal of improving the quality and effectiveness of patient care. Biomedical engineers work in manufacturing, universities, hospitals, research facilities of companies, and educational and medical institutions. Biomedical engineers typically need a bachelor's degree in biomedical engineering from an accredited program to enter the occupation. Alternatively, they can get a bachelor's degree in a different field of engineering and then either get a graduate degree in biomedical engineering or get on-the-job training in biomedical engineering.

Microbiologists study the growth, development, and other characteristics of microscopic organisms like bacteria. Microbiologists work in laboratories and offices where they conduct experiments. A bachelor's degree in microbiology or a closely related field is needed for entry-level positions.

Food scientists work to maintain agricultural productivity and food safety. Most food scientists work in research universities, industry, or the federal government in laboratories, offices, and the field. Food scientists need to have earned at least a bachelor's degree, but many have master's degrees and PhD's.

Environmental scientists use their knowledge of natural sciences to protect the environment. They identify problems and find solutions that protect the health of the environment and the people living in it. Environmental scientists often work in laboratories and offices, but also spend time in the environment they're protecting. Environmental scientists need at least a bachelor's degree in natural science.

Cost estimators collect and analyze data to estimate the time, money, resources, and labor required for product manufacturing, construction projects, or services. Some specialize in a particular industry or product type. Although cost estimators generally work in central offices, they often visit factory floors or construction sites. A bachelor's degree is generally needed for entering the field.

Aerospace engineers design aircraft, spacecraft, satellites, and missiles. They also test prototypes to make sure that they function according to design. Aerospace engineers are employed in industries whose workers design or build aircraft, missiles, systems for national defense, or spacecraft. Aerospace engineers are employed primarily in analysis and design, manufacturing, industries that perform research and development, and the federal government. Aerospace engineers must have a bachelor's degree in aerospace engineering or another field of engineering or science related to aerospace systems. Some aerospace engineers work on projects that are related to national defense and thus require security clearances.

Database administrators use software to store and organize data, such as financial information and customer shipping records. They make sure that data are available to users and are secure from unauthorized access. Database administrators work in many types of industries including insurance companies, banks, and hospitals. A bachelor's degree in information or computer-related subjects is commonly required.

Logisticians analyze and coordinate an organization's supply chain (i.e., the system that moves a product from supplier to consumer). They manage the entire life cycle of a product, which includes how a product is acquired, distributed, allocated, and delivered. Logisticians work in nearly every industry. The job can be stressful due to the fast pace of logistical work. Although an associate degree may be sufficient for some logistician jobs, a bachelor's degree is typically required for most positions.

Economists study the production and distribution of resources, goods, and services by researching trends, analyzing data, and evaluating economic issues. Although the majority of economists work independently in an office, some collaborate with other economists and statisticians. Most economists need a master's or doctoral degree; however, some entry-level positions (especially in the federal government) require a bachelor's degree.

The STEM Road Map for Eighth Grade

The eighth-grade year will engage students in exploring STEM Road Map theme-generated topics that also align with grade-level academic content standards (e.g. Common Core and Next Generation Science Standards) which include *Earth on the Move, Medicine, Learning from the Past, Minimizing our Impact,* and *The Role of the Sun in Life on Earth*. Each of these topics is organized around a challenge/problem or project that student teams are assigned to tackle in the course of learning necessary content and skills in the various disciplines (see Table 6.13).

Cause and Effect: The Changing Earth

Our dynamic Earth that we inhabit consists of plates of crust that make up the lithosphere. Over time, these plates, which float on a sea of molten lava underneath, have moved ever so slowly. This continual movement has resulted in some observable changes and events on Earth, including earthquakes, volcanic eruptions, and mountain formation. In the eighth grade, students will evaluate existing theories and data available to propose a model consisting of an alternate explanation for plate movement. This module is led by science, where students will examine various aspects of plate tectonic theory. Mathematical practices and modeling will be emphasized in this module, along with how the movement of the Earth has impacted communities for decades (social studies) including the recent (2014) eruption of Kilauea in Hawaii. In English/language

Table 6.13 Eighth-Grade STEM Road Map Themes, Topics, and Problems/Challenges

STEM Theme	Topic	Problem/Challenge
Cause and Effect	The Changing Earth LEAD Science	Student teams will develop and propose a model based on evidence with an alternative explanation for the changes observed on Earth (plate tectonic theory).
Innovation and Progress	Medicine LEAD English/LA	Student teams will propose an alternative course of treatment for a persistent disease/disorder in humans or animals (health care engineering)
Represented World	Improving Bridge Design LEAD Mathematics	Given the current state of infrastructure decay, student teams will develop a decision model for the local Department of Transportation on how to choose what type of bridge to construct when provided with information about the span length, application, use information, etc.
Sustainable Systems	The Speed of Green LEAD Social Studies	Student teams will design and develop/modify a prototype of an energy efficient automobile racing vehicle.
Optimizing Human Experience	Going Solar LEAD Science	Student teams will develop a prototype of a machine that would harness thermal energy and convert it for a needed societal use. In order to determine the rate of change, students will use the computational thinking practice of pattern recognition to observe the changes that occur with solar energy.

arts, students will engage in conversations as they evaluate their sources and work to develop the presentation of their model (see Table 6.14). This module has been published in 2020 by NSTA Press. See https://my.nsta.org/resource/121424.

Innovation and Progress: Medicine

Each day, new understandings and innovations are discovered in the field of medicine. Technological advances as well as years of research and development have moved our society forward in the diagnosis and approaches for mitigating medical issues. It is important for students to learn about the extensive work that has been conducted in this area and to also learn that many of the treatments of the future have yet

Table 6.14 STEM Road Map – Eighth-Grade Cause and Effect: Earth on the Move

NGSS Performance Objectives	Common Core		21st Century Skills
	Mathematics	Language Arts	
MS-ESS2-1 MS-ESS2-2 MS-ESS2–3	CCSS.Math.Practices MP1, MP2, MP4	Reading Standards CCSS.ELA. RL.8.1 RI.8.9	21st Century Themes: Global Awareness Environmental Literacy
	CCSS.Math. Content.8.EE.A.4 CCSS.Math. Content.8.EE.B.5	Writing Standards CCSS.ELA. RW.8.1, RW.8.1a RW.8.1b, RW.8.1c RW.8.1e RW.8.2, RW.8.2b RW.8.2c, RW.8.2d RW.8.3a, RW.8.3d, RW.8.6 RW.8.7 RW.8.8	Learning and Innovation Skills: Creativity and Innovation Critical Thinking and Problem Solving Communication and Collaboration
	CCSS.Math. Content.8.G.C.9	Speaking and Listening Standards CCSS.ELA. SL.8.1, SL.8.1a, SL.8.1b, SL.8.1c, SL.8.1d, SL.8.2 SL.8.3 SL.8.4 SL.8.5 SL.8.6	Information, Media, and Technology Skills: Information Literacy Media Literacy ICT Literacy
	CCSS.Math. Content.8.SP.A.1 CCSS.Math. Content.8.SP.A.4		Life and Career Skills: Flexibility and Adaptability Initiative and Self-Direction Social and Cross Cultural Skills Productivity and Accountability Leadership and Responsibility

to be revealed. In this module, eighth graders will choose a persistent disease and/or disorder in humans and conduct research to propose an alternative course of treatment. This module is led by English/language arts where students will focus on reading technical reports focused on medicine and the challenges with access to appropriate treatments for humans in various parts of the world. In social studies, students will learn about the inequity in access to healthcare in developing countries. In science, students will examine chemical structures and molecules to learn more about the chemistry behind drug discovery. There are many instances of periodicity in chemistry, and students will use algorithmic thinking to identify steps in chemical interactions that can be applied elsewhere. In mathematics, students will work to solve equations and convert fractions as applied in the field of medicine (see Table 6.15).

Table 6.15 STEM Road Map – Eighth-Grade Innovation and Progress: Medicine

NGSS Performance Objectives	Common Core		21st Century Skills
	Mathematics	Language Arts	
MS-LS1-3 MS-LS1-5	CCSS.Math.Practices MP1, MP2, MP3, MP4, MP5	Reading Standards CCSS.ELA. RL.8.1 RI.8.9	21st Century Themes: Global Awareness Health Literacy
MS-LS4-5	CCSS.Math.Content.8.EE.A.1	Writing Standards CCSS.ELA. RW.8.1, RW.8.1a RW.8.1b, RW.8.1c RW.8.1e RW.8.2, RW.8.2b RW.8.2c, RW.8.2d RW.8.3a, RW.8.3d, RW.8.6 RW.8.7 RW.8.8	Learning and Innovation Skills: Creativity and Innovation Critical Thinking and Problem Solving Communication and Collaboration
	CCSS.Math.Content.8.EE.B.5	Speaking and Listening Standards CCSS.ELA. SL.8.1, SL.8.1a, SL.8.1b, SL.8.1c, SL.8.1d SL.8.2 SL.8.3 SL.8.4 SL.8.5 SL.8.6	Information, Media, and Technology Skills: Information Literacy Media Literacy ICT Literacy
	CCSS.Math.Content.8.EE.C.7b		Life and Career Skills: Flexibility and Adaptability Initiative and Self-Direction Social and Cross Cultural Skills Productivity and Accountability Leadership and Responsibility

Represented World: Improving Bridge Design

This unit will focus on addressing the real problems of today's society through the lens of the past. In science, students will examine observable changes in rocks and fossils to interpret the past. The challenge for this module is led by mathematics and is focused on infrastructure decay, specifically the state of bridges in the United States. With recent bridge collapses (i.e., Minnesota bridge), much debate has ensued regarding the maintenance of bridges and, when building, examining designs that will prove to be more sustainable over time. Student teams will develop a decision model grounded in engineering for the local Department of Transportation on how to select bridge design aligned with appropriate span length, application, use information, and other important data. In social studies, students will learn about how infrastructure such as roads and bridges has helped to move their geographic region forward. In English/language arts, students will work to develop a written proposal that articulates key components of their decision model (see Table 6.16). This module has been published in 2018 by NSTA Press. See https://my.nsta.org/resource/113534.

Sustainable Systems: The Speed of Green

As our world continues to move forward with new innovations and solutions to challenges, a delicate balance must be maintained to ensure the footprint on our Earth and the natural resources and surroundings is minimized. There are thousands of STEM careers that are tied directly or indirectly to preserving our environment. Increasingly, debates in the United States have focused on alternative forms of energy and sources for food and water. In eighth grade, students will consider the potential role of various renewable and non-renewable energy sources in transportation, with an emphasis on the auto industry. In this science and social studies-led module, student teams will each develop a plan for a competitive automobile racing team to fuel a vehicle with minimal environmental impact. In science, students will investigate the chemical reactions involved in combustion and in producing plant-based fuels, the effects of the use of carbon-based and other fuels on the biosphere and recognize the difference between finite and renewable resources. In social studies, students will explore the concept of resource scarcity and investigate the history of energy usage in transportation and current efforts to reduce the impact of auto industry innovations for fuel efficiency and conservation. In mathematics, students will explore various units used to measure energy and will interpret charts and graphs. In English language arts, students will read a variety of texts and online sources of information and create persuasive arguments via essays and presentations and will explore and create biographical literature. The unit will culminate with a

Table 6.16 STEM Road Map – Eighth-Grade Represented World: Learning from the Past

NGSS Performance Objectives	Common Core Mathematics	Common Core Language Arts	21st Century Skills
MS-ESS1-4	CCSS.Math.Practices MP1, MP2, MP3, MP4, MP5, MP6, MP7, MP8	Reading Standards CCSS.ELA. RL.8.1 RI.8.9	21st Century Themes: Global Awareness Environmental Literacy Financial, Economic, Business, and Entrepreneurial Literacy
MS-LS4-1 MS-LS4-2	CCSS.Math. Content.8.EE.A.1	Writing Standards CCSS.ELA. RW.8.1, RW.8.1a RW.8.1b, RW.8.1c RW.8.1e RW.8.2, RW.8.2b RW.8.2c, RW.8.2d RW.8.3a, RW.8.3d, RW.8.6 RW.8.7 RW.8.8	Learning and Innovation Skills: Creativity and Innovation Critical Thinking and Problem Solving Communication and Collaboration
	CCSS.Math. Content.8.EE.B.5	Speaking and Listening Standards CCSS.ELA. SL.8.1, SL.8.1a, SL.8.1b, SL.8.1c, SL.8.1d SL.8.2 SL.8.3 SL.8.4 SL.8.5 SL.8.6	Information, Media, and Technology Skills: Information Literacy Media Literacy ICT Literacy
	CCSS.Math. Content.8.EE.C.7b		
	CCSS.Math. Content.8.F.B.5		Life and Career Skills: Flexibility and Adaptability Initiative and Self-Direction Social and Cross Cultural Skills Productivity and Accountability Leadership and Responsibility

challenge in which student teams each create long-term plans for an automobile racing vehicle that minimizes its environmental impact. Students will design and build a prototype or model of one-design innovation that will support this goal and will create a persuasive video presentation presenting their plans and prototypes or models (see Table 6.17).

Table 6.17 STEM Road Map – Eighth-Grade Sustainable Systems: Minimizing our Impact

NGSS Performance Objectives	Common Core		21st Century Skills
	Mathematics	*Language Arts*	
MS-ESS3-3 MS-ESS3-4	CCSS.Math.Practices MP1, MP2, MP3, MP4, MP5	Reading Standards CCSS.ELA. RL.8.1 RI.8.9	21st Century Themes: Global Awareness Environmental Literacy
MS-PS1-3	CCSS.Math.Content.8.F.B.4 CCSS.Math.Content.8.F.B.5	Writing Standards CCSS.ELA. RW.8.1, RW.8.1a RW.8.1b, RW.8.1c RW.8.1e RW.8.2, RW.8.2b RW.8.2c, RW.8.2d RW.8.3a, RW.8.3d RW.8.6 RW.8.7 RW.8.8	Learning and Innovation Skills: Creativity and Innovation Critical Thinking and Problem Solving Communication and Collaboration
		Speaking and Listening Standards CCSS.ELA. SL.8.1, SL.8.1a, SL.8.1b, SL.8.1c, SL.8.1d SL.8.2 SL.8.3 SL.8.4 SL.8.5 SL.8.6	Information, Media, and Technology Skills: Information Literacy Media Literacy ICT Literacy
			Life and Career Skills: Flexibility and Adaptability Initiative and Self-Direction Social and Cross Cultural Skills Productivity and Accountability Leadership and Responsibility

Optimizing the Human Experience: Going Solar

The final module in eighth grade will focus on the longstanding role of the Sun in life on Earth. This will include learning about how the Sun has been important in cultural ways, seasonal considerations, and as a primary source of sustaining life on Earth. Student teams will be asked to utilize engineering design to develop a prototype of a machine that can harness thermal energy and convert it for a needed use of society. The lead discipline for this module is science, and students will learn specifically about thermal energy and will research potential uses for this resource. In mathematics, for example, students will construct a function to model a linear relationship to determine the rate of change. In order to determine the rate of change, students will use the computational thinking practice of pattern recognition to observe the changes that occur with solar energy. In social studies, students will explore cultural and geographical connections to the sun and how this has impacted various populations around the globe on a daily basis. In English/language arts, students will develop their writing skills through crafting a paper on the importance of exploring solar energy as a potential source of energy for the future.

Sample STEM Careers in the Eighth-Grade STEM Road Map

Medical sonographers use special equipment to assess and diagnose various medical conditions. Most medical sonographers work in hospitals though some might work in doctor's offices. A bachelor's degree, as well as a formal certificate in medical sonography, is required to be a medical sonographer.

Construction managers plan, coordinate, budget, and supervise construction projects from early development to completion. Although many construction managers work from a main office, most work out of a field office at the construction site where they monitor the project and make daily decisions about construction activities. Employers increasingly prefer candidates with both work experience and a bachelor's degree in a construction-related field (i.e. construction management). However, some construction managers may qualify by working many years in a construction trade. Certification, although not required, is becoming increasingly important.

Carpenters construct and repair building frameworks and structures – such as stairways, doorframes, partitions, and rafters – made from wood and other materials. They also may install kitchen cabinets, siding, and drywall. Because carpenters are involved in many types of construction from building highways and bridges to installing kitchen cabinets, they may work both indoors and outdoors. Although most carpenters learn their trade through a formal apprenticeship, some learn on the job, starting as a helper.

Table 6.18 STEM Road Map – Eighth-Grade Optimizing the Human Experience: The Role of the Sun in Life

NGSS Performance Objectives	Common Core Mathematics	Common Core Language Arts	21st Century Skills
MS-PS1-4 MS-PS1-6	CCSS.Math.Practices MP1, MP2, MP3, MP5	Reading Standards CCSS.ELA. RL.8.1 RI.8.9	21st Century Themes: Global Awareness Environmental Literacy Financial, Economic, Business, and Entrepreneurial Literacy
MS-PS3-3	CCSS.Math.Content.8.F.B.4	Writing Standards CCSS.ELA. RW.8.1, RW.8.1a RW.8.1b, RW.8.1c RW.8.1e RW.8.2, RW.8.2b RW.8.2c, RW.8.2d RW.8.3a, RW.8.3d, RW.8.6 RW.8.7 RW.8.8	Learning and Innovation Skills: Creativity and Innovation Critical Thinking and Problem Solving Communication and Collaboration
	CCSS.Math.Content.8.EE.C.8c CCSS.Math.Content.8.EE.C.7b	Speaking and Listening Standards CCSS.ELA. SL.8.1, SL.8.1a, SL.8.1b, SL.8.1c, SL.8.1d, SL.8.2, SL.8.3, SL.8.4, SL.8.5, SL.8.6	Information, Media, and Technology Skills: Information Literacy Media Literacy ICT Literacy
	CCSS.Math.Content.8.EE.B.5		Life and Career Skills: Flexibility and Adaptability Initiative and Self-Direction Social and Cross Cultural Skills Productivity and Accountability Leadership and Responsibility

Software developers are the creative minds behind computer programs. Some develop applications that allow people to do specific tasks on a computer or other device. Others develop the underlying systems that run the devices or control networks. Many software developers work for computer systems design and related services firms or software publishers. Others work in computer and electronic product manufacturing industries. Software developers usually have a bachelor's degree in computer science and strong computer-programming skills.

Lobbyists (political scientists) research and analyze political ideas, policies, political trends, and related issues in order to work with senators and congresspersons to pass policies and laws regarding certain issues. Lobbyists sometimes work overtime to finish reports and meet deadlines. Entry-level education required for this position is a master's degree.

Historians research, analyze, interpret, and present the past by studying a variety of historical documents and sources. They work in government agencies, museums, archives, historical societies, research organizations, and consulting firms. Some must travel to carry out research. Most historian positions require a master's degree; some research positions require a doctoral degree.

Semiconductor processors are workers who oversee the manufacturing process of solar cells. Semiconductors act as conductors of electricity and semiconductor processors oversee their manufacture including the repair and maintenance of machinery. They test completed cells and perform diagnostic tests to make sure the cells work properly. Most production workers are trained on the job and gain expertise with experience; however, some positions may require formal training programs or apprenticeships or college degrees for production managers.

Geoscientists study the physical aspects of the Earth, such as its composition, structure, and processes to learn about its past, present, and future. Most split their time between working in offices and labs and working outdoors. Doing research and investigations outdoors is commonly called fieldwork and can require extensive travel to remote locations and irregular working hours. Most geoscientist jobs require at least a bachelor's degree. In several states, geoscientists may need a license to offer their services to the public.

Chapter Summary

This chapter presented the STEM Road Map for grades 6–8 as an engaging, real-world approach to integration of core content areas for implementation in middle school. With the use of this tool, instruction can be transformed into coordinated modules of instruction which require teams of students to grapple with global and local challenges and problems as they master the content for their grade level, along with skills and habits of mind necessary for success in careers of the future. In the next chapter, the spiraling approach of the STEM Road Map will continue with a presentation of an integrated approach for delivery of traditional high school coursework.

Sample Module

A complete STEM Road Map seventh-grade Transportation – Motorsports module is included in the Appendix of this book. Several of the modules are available now as books that have been published by the National Science Teaching Association (NSTA). The books that are available now include:

- *Amusement of the Future* (sixth)
- *Packaging Design* (sixth)
- *Human Impacts on Our Climate* (sixth)
- *Population Density* (seventh)
- *Genetically Modified Organisms* (seventh)
- *Improving Bridge Design* (eighth)
- *Speed of Green* (eighth)

To learn more about the STEM Road Map Curriculum Series and the available titles, visit https://www.nsta.org/publications/press/stemroad-map.aspx.

7 The STEM Road Map for Grades 9–12

Erin E. Peters-Burton, Padmanabhan Seshaiyer, Stephen R. Burton, Jennifer Drake-Patrick, and Carla C. Johnson

Overview of 9–12 STEM Road Map

This chapter will provide a detailed overview of the integrated STEM Road Map for the high school grade levels 9–12. The STEM Road Map for grades 9–12 continues to be anchored in the overarching five STEM themes which include: (a) Cause and Effect, (b) Innovation and Progress, (c) The Represented World, (d) Sustainable Systems, and (e) Optimizing the Human Experience. Each STEM Road Map theme is designed as a five-week sequence including integrated instruction where the theme and associated problem or project is implemented across core content areas. High school teachers may not be as familiar with integrated content as elementary teachers; therefore, guidance for ways to integrate different disciplines is included in this chapter. For the high school level STEM Road Map, we provide a sample five-week module for 11th grade which includes the complete module and all instructional and assessment materials necessary. It is understood that high school teachers have areas of specialization such as earth science, chemistry, biology, and physics, in addition to their understanding of science in general. This is taken into account in designing the STEM Road Map, which focuses mainly on earth science in 9th grade, biology in 10th grade, chemistry in 11th grade, and physics in 12th grade. However, the curriculum in the STEM Road Map is flexible and can be moved from year to year based on the need of the students, teachers, and school organizations.

The STEM Road Map for grades 9–12 is designed to be delivered in an integrated fashion. High school teachers may take several approaches to integrating content throughout instruction such as enlisting colleagues from other disciplines (e.g. science, social studies, mathematics, language arts) to insert instruction into the teacher's lessons or to have colleagues continue to build the theme through lessons in other classrooms to reinforce learning.

There are three ways to approach integration at the high school level, depending on the type of class scheduling offered at the school. First, if the school schedule is organized by department and not teams, the teacher integrating the course can take on the roles of the other content area teachers. For this approach, some of the curriculum leading up to and including the challenge may need to be streamlined to fit into a reasonable timeframe. In this first approach, the teacher doing the course integration must take on the responsibility of the other content areas. The teacher could also take a second approach and ask the other content area teachers to guest teach in his or her classroom. The curriculum leading up to and including the challenge would also have to be streamlined for this second approach. The third approach can occur if the school schedule allows for co-teaching or teamed teaching. In this approach, all of the content teachers take equal responsibility in their classes for the curriculum leading up to and including the challenge.

The STEM Road Map and associated STEM Road Map modules reflect an integration of Common Core Mathematics, Common Core English/Language Arts, Next Generation Science Standards, and the 21st Century Skills Framework and should be delivered by one lead teacher with other content areas making distinct ties to the project within their own curriculums as suggested in the maps and associated modules. Implementation of the STEM Road Map at the high school level is critical because high school students are equipped with ample background knowledge and skills positioning them to make rich contributions and connections while engaging with Problem-Based Learning (PBL) scenarios. Working across disciplines enhances students' ability to gain deeper conceptual understanding of the content, and the PBLs encourage students to apply and evaluate their learning within the context of the real-world STEM projects. Even in the high school grades (9–12), all content areas (including art and music) play an important role in the inclusive, integrated STEM approach.

STEM Themes in the 9–12 STEM Road Map

The five overarching STEM themes continue to be reinforced and spiraled within the 9–12 STEM Road Map. *Cause and Effect* is the real-world STEM theme that consists of the dynamic relationships between various phenomena in the world. Students in grades 9–12 will explore formation of the earth, biodiversity, conservation of matter in the universe, and electromagnetic radiation within this STEM theme.

Innovation and Progress relate to the various landmark developments driven by human ingenuity that have moved our society and understandings forward across generations. At the high school level, topics in the STEM Road Map within the theme of Innovation and Progress include erosion and weathering management, environmental management, designing new materials, and communications technologies.

The Represented World will take a look at the various models that humans have developed to make sense of the world around them. Students will explore topics including global models and their uses, modeling ecosystems, modeling energy in chemistry, and use of models for prediction.

In the *Sustainable Systems* STEM theme, students will be engaged in challenges including vital systems of the earth, survival and reproduction, chemistry of plants, and human influence on the earth's energy flow.

The STEM Road Map theme of *Optimizing the Human Experience* focuses on innovations that have improved the quality of life. Students in grades 9–12 will investigate evaluating human impact on nature, rebuilding the natural environment, developing and maintaining resources, and natural occurrences and their impact on humans.

Each of these topics will immerse high school students in an authentic, problem and project-based curriculum that spans across traditional content barriers, bringing engineering and technological design, scientific inquiry, mathematical reasoning to life through the process of developing potential prototypes for future innovations. Further, 21st Century Skills such as critical thinking, creativity, communication, collaboration, information, and media literacy will be emphasized daily within the STEM Road Map as students continue to develop their skills in leadership and responsibility for their own learning.

The STEM Road Map 9–12 builds on skills and knowledge learned in elementary and middle school by including compelling topics that require analysis, synthesis, and evaluation of information in order to reach a conclusion. Additionally, the STEM Road Map 9–12 gives students responsibility for their own learning and promotes meaningful learning in authentic, relevant contexts that help students connect their existing knowledge with new knowledge and skills.

The STEM Road Map for Ninth Grade

In middle school, students learned about *Amusement Parks, Human Impacts on Our Climate, Communication, Global Water Quality, Natural Hazards, Transportation, Space Travel, Genetic Disorders, Populations, and Genetically Modified Organisms (GMOs), Earth on the Move, Medicine, Learning from the Past, Minimizing our Impact, and The Role of the Sun in Life on Earth*. The skills and knowledge that students acquired through their work in middle school will continue to be built upon by iteratively connecting topics to the five themes. In ninth grade, students will explore STEM Road Map Theme inspired topics that align with grade-level academic content standards (e.g. Common Core and Next Generation Science Standards). The topics for ninth grade include: *Formation of the Earth, Erosion and Weathering Management, Global Models and their Uses, Vital Systems of the Earth, and Evaluating Human Impact on Nature*. Each topic is organized around a challenge/problem or project that student teams are assigned to tackle in the course of learning necessary content and skills in the various disciplines (see Table 7.1).

Table 7.1 Ninth-Grade STEM Road Map Themes, Topics, and Problems/Challenges

STEM Theme	Topic	Problem/Challenge
Cause and Effect	Formation of the Earth LEAD Science	Student teams will create a multimedia production for use by an environmental consulting firm that relates how Earth's internal processes operate, including interactions involving water, to evidence from ancient Earth materials, meteorites, and other planetary surfaces to explain how Earth's formation and early history have led to current processes on the Earth.
Innovation and Progress	Disruptive Forces of the Earth LEAD Social Studies	Student teams will use the computational thinking practice of pattern recognition to determine the long-term effects of their chosen disruptive force challenge. Teams will engage in research on erosion and weathering and propose a management plan, weighing the costs and benefits of the long-term implications, and communicate their ideas through a policy paper.
The Represented World	Computer-Based Mapping for Addressing Pandemics and Global Challenges LEAD Science	Students will research highly complex topics and synthesize the information into a meaningful and understandable infographic generated from computer-based mapping models. In this project, students will first explore a myriad of major global challenges, such as pandemics (COVID-19 and SARS), identifying one global challenge that they will further research in detail. Students will use the computational thinking practice of abstraction when doing this research because students must understand what ancillary information is and what is core information. Student teams will communicate posed solutions to these problems using computer modeling to examine ways to mitigate spread of the virus. After refining and making meaning from this information, teams are to communicate it succinctly in an infographic.

(Continued)

Table 7.1 (Continued)

STEM Theme	Topic	Problem/Challenge
Sustainable Systems	Vital Systems of the Earth LEAD ELA	Systems are interconnected and often one change can influence the whole. In this challenge, students will be assigned to investigate one change within the Earth's surface system that has had various impacts associated with it. Students will use the computational thinking practice of pattern recognition in this PBL to identify connections across systems. Teams will explore this on a local and global scale and will develop a documentary video that will detail the pros and cons of this change.
Optimizing the Human Experience	Evaluating Human Impact on Nature LEAD Social Studies or Science or English/ Language Arts	Student teams will be challenged to develop a prototype or a model of a technological innovation that could reduce the impact of human activities on natural systems, including a detailed plan for testing the prototype and taking the innovation to market. Students will use the computational thinking practice of algorithmic thinking when they learn to apply the criteria and constraints across different contexts as they brainstorm and work through the prototype.

Cause and Effect: Formation of the Earth

By ninth grade, students most likely have formed some ideas about the early history of the Earth but have not yet connected the theory about the formation of the Earth to the current processes taking place. In this project, students will use their prior knowledge of early Earth formation, but will learn more about the progression of processes such as interactions involving water, erosion, transportation, deposition, convection, and Earth's materials to explain how the formation and early history have led to current processes on the Earth to develop a multimedia presentation for an environmental consulting firm to foresee how events and processes affect the Earth's surface. Students will apply the computation thinking practice of decomposition in this module by observing processes of geomorphology and breaking these processes down into their components (erosion, transportation, deposition). Students will research how geologists and other Earth science professionals gather information about very old and very slow phenomena in order to formulate and support their claims about the connections between formation and current processes. Two motivational components are built into this project: connecting to prior knowledge and designing a multimedia project for an environmental consulting firm. Not only will students acquire new knowledge on the topic, but they will also learn new skills in communicating the information effectively through different media such as audio, video, diagrams, and narrative (see Table 7.2).

Table 7.2 STEM Road Map – Ninth-Grade Cause and Effect Theme: Formation of the Earth

NGSS Performance Objectives	Common Core Mathematics	Common Core Language Arts	21st Century Skills
HS-PS2-1	CCSS.Math.Practices MP1, MP3, MP5, MP6, MP8	Reading Standards CCSS.ELA. RI.9-10.1 RI.9-10.2 RI.9-10.7 RI.9-10.8 RI.9-10.10	21st Century Themes: Global Awareness
HS-ESS1-2 HS-ESS1-6	CCSS.Math.Content. HSN-VM.B.4b	Writing Standards CCSS.ELA. W.9-10.1a, W.9-10.1b, W.9-10.1c, W.9-10.1d, W.9-10.1e W.9-10.2a, W.9-10.2b, W.9-10.2c, W.9-10.2d, W.9-10.2e, W.9-10.2f W.9-10.4 W.9-10.6 W.9-10.8 W.9-10.10	Learning and Innovation Skills: Creativity and Innovation Critical Thinking and Problem Solving Communication and Collaboration
HS-ESS2-1 HS-ESS2-5	CCSS.Math.Content. HSN-VM.B.4c	Speaking and Listening Standards CCSS.ELA. SL.9-10.2 SL.9-10.4 SL.9-10.5 SL.9-10.6	Information, Media, and Technology Skills: Information Literacy Media Literacy ICT Literacy
HS-ETS-3	CCSS.Math.Content. HSN-VM.B.4a	Language Standards CCSS.ELA. L.9-10.2 L.9-10.6	Life and Career Skills: Flexibility and Adaptability Initiative and Self-Direction Social and Cross Cultural Skills Productivity and Accountability Leadership and Responsibility

Innovation and Progress: Disruptive Forces of the Earth

In the Disruptive Forces of the Earth PBL, students will apply what they learned about erosion, transportation, and deposition of Earth's materials from the Cause and Effect PBL. This project is inspired from conservation organizations around the world and intends to raise students' awareness of a little-known problem, the conservation, and management of the movement of soils to prevent landslides. Erosion, although normally occurring as part of the natural order, has become detrimental to the environment due to long-term human impact. Storms that were once harmless now leave the land damaged and vulnerable. Tourism is affected by rapid sand erosion from beaches. For example, the state of Delaware must pump a million pounds of sand back onto the beach each year to maintain the same size beach. California is facing unprecedented erosion and is having to build barriers to save houses on the coast. Rain, which seems harmless enough, has become so acidic in areas that it wears away statues. The increasing rate of landslides has stolen nutrients from agricultural environments, which can in turn cause problems with food distribution. Students will use the computational thinking practice of pattern recognition to determine the long-term effects of their chosen issue. Once students research these and other current problems with erosion and weathering, they will propose a management plan, weighing the costs and benefits of the long-term implications, and communicate their ideas through a policy paper. Writing a policy brief can pose a new challenge for students because it is a way of writing that may not be familiar; however, the task can lay the foundation for students to be advocates for future issues (see Table 7.3).

The Represented World: Computer-Based Mapping for Addressing Pandemics and Global Challenges

In the Represented World PBL, students will discover the vast information about interactions of processes on the Earth and how it impacts human life. Students will research highly complex topics and synthesize the information into a meaningful and understandable infographic generated from computer-based mapping models. In this project, students will first explore a myriad of major global challenges, such as pandemics (COVID-19 and SARS), identifying one global challenge that they will further research in detail. Students will use the computational thinking practice of abstraction when doing this research because students must understand what ancillary information is and what is core information. After finding causes of the problems based on evidence, students will communicate posed solutions to these problems using computer modeling to examine ways to mitigate the spread of the virus. In analyzing the solutions, students will also need to examine societal implications, identifying the differences between needs and wants of different members of society.

Table 7.3 STEM Road Map – Ninth-Grade Innovation and Progress: Erosion and Weathering

NGSS Performance Objectives	Common Core		21st Century Skills
	Mathematics	Language Arts	
HS-PS2-3	CCSS.Math.Practices MP1, MP3	Reading Standards CCSS.ELA. RI.9-10.1 RI.9-10.2 RI.9-10.3 RI.9-10.4 RI.9-10.6 RI.9-10.8 RI.9-10.10	21st Century Themes: Global Awareness Environmental Literacy
HS-ETS-2	CCSS.Math.Content. HSA-CED.A.3	Writing Standards CCSS.ELA. W.9-10.1a., W.9-10.1b, W.9-10.1c, W.9-10.1d, W.9-10.1e W.9-10.2a, W.9-10.2b, W.9-10.2c, W.9-10.2d, W.9-10.2e, W.9-10.2f W.9-10.4 W.9-10.5 W.9-10.7 W.9-10.8 W.9-10.9a, W.9-10..9b W.9-10.10	Learning and Innovation Skills: Creativity and Innovation Critical Thinking and Problem Solving Communication and Collaboration
HS-PS2-3	CCSS.Math.Content. HSN-CN.B.6	Speaking and Listening Standards CCSS.ELA. SL.9-10.1a, SL.9-10.1b, SL.9-10.1c, SL.9-10.1d SL.9-10.2 SL.9-10.4	Information, Media, and Technology Skills: Information Literacy Media Literacy ICT Literacy
		Language Standards CCSS.ELA. L.9-10.1a, L.9-10.1b L.9-10.2a, L.9-10.2b, L.9-10.2c L.9-10.3a SL.9-10.6	Life and Career Skills: Flexibility and Adaptability Initiative and Self-Direction Social and Cross Cultural Skills Productivity and Accountability Leadership and Responsibility

After refining and making meaning from this information, students are to communicate succinctly in an infographic. Information graphics or infographics present graphic visual representations of complex information, data, or knowledge quickly, clearly, and concisely (see Table 7.4).

Table 7.4 STEM Road Map – Ninth-Grade Represented World: Global Models and their Uses

NGSS Performance Objectives	Common Core		21st Century Skills
	Mathematics	*Language Arts*	
HS-ESS1-4 HS-ESS1-5	CCSS.Math.Practices MP1, MP2, MP3, MP4, MP5, MP6, MP8	Reading Standards CCSS.ELA. RI.9-10.1 RI.9-10.2 RI.9-10.3 RI.9-10.4 RI.9-10.6 RI.9-10.8 RI.9-10.10	21st Century Themes: Global Awareness Civic Literacy Environmental Literacy
HS-ETS-1	CCSS.Math.Content. HS-IF.B.5	Writing Standards CCSS.ELA. W.9-10.2a, W.9-10.2b, W.9-10.2c, W.9-10.2d, W.9-10.2e, W.9-10.2f W.9-10.5 W.9-10.6 W.9-10.8 W.9-10.9a, W.9-10.9b	Learning and Innovation Skills: Creativity and Innovation Critical Thinking and Problem Solving Communication and Collaboration
HS-ESS2-6	CCSS.Math.Content. HS-BF.B.4	Speaking and Listening Standards CCSS.ELA. SL.9-10.1a, SL.9-10.1b, SL.9-10.1c, SL.9-10.1d SL.9-10.2 SL.9-10.4 SL.9-10.5	Information, Media, and Technology Skills: Information Literacy Media Literacy ICT Literacy
HS-ESS3-5	CCSS.Math.Content. HSN-Q.A.2	Language Standards CCSS.ELA. L.9-10.3 L.9-10.4a-d L.9-10.5 L.9-10.6	Life and Career Skills: Flexibility and Adaptability Initiative and Self-Direction Social and Cross Cultural Skills Productivity and Accountability Leadership and Responsibility

Research shows that infographics can improve student cognition by enhancing students' abilities to see patterns and trends (Card, 2009; Heer, Bostock, & Ogievetsky, 2010).

Sustainable Systems: Vital Systems of the Earth

The ninth-grade students participating in the prior PBLs have gained valuable knowledge about complex systems including linkages between the formation of the Earth and current processes, erosion and weathering impacts on the environment, and cycling of carbon. The PBL in the Sustainable Systems theme builds on that knowledge to challenge students to predict future implications when one part of a surface system on the Earth changes (atmosphere, geosphere, hydrosphere, biosphere) and the impacts that this has on other components. Students will use the computational thinking practice of pattern recognition in this PBL to identify connections across systems. A broad understanding of systems theory underpins this work, emphasizing interactions between different subsystems and humans. To accomplish this task, students must first find all parts of the system and how they interact, then predict the interaction effects of the system on other components over a progression of time, as well as the feedback systems that are cyclic in the system. Students will demonstrate what they know about spacial and temporal scale by completing the PBL scenario of creating a documentary detailing the pros and cons of the interactions (see Table 7.5).

Optimizing the Human Experience: Evaluating Human Impact on Nature

The PBL for Optimizing the Human Experience theme, Evaluating Human Impact on Nature, again extends the knowledge discovered in the PBLs from the prior themes' PBL experiences. However, this PBL poses a different challenge that involves the engineering design process, which is defined in this chapter by the NASA model (National Aeronautic and Space Administration, 2014). Now that students have proficient knowledge in the systems on Earth, how they interact, how one change might influence other changes, and how to communicate complex information in an understandable way, students will be compelled in this PBL to pose a real solution and implement the solution in a societal context. Students will work in teams to identify a problem regarding the negative impact of human activities on natural systems. Then, they will identify criteria and constraints, brainstorm possible solutions, generate ideas, explore possibilities, select an approach, and build a prototype or model of a technological innovation that can help solve this program. Students will use the computational thinking practice of algorithmic thinking when they learn to apply the criteria and constraints across different contexts

Table 7.5 STEM Road Map – Ninth-Grade Sustainable Systems: Vital Systems of the Earth

NGSS Performance Objectives	Common Core		21st Century Skills
	Mathematics	Language Arts	
HS-PS3-1	CCSS.Math.Practices MP1, MP3, MP8	Reading Standards: CCSS.ELA. RI.9-10.1 RI.9-10.2 RI.9-10.3 RI.9-10.4 RI.9-10.6 RI.9-10.8 RI.9-10.10	21st Century Themes: Global Awareness Civic Literacy Environmental Literacy
HS-LS1-6	CCSS.Math.Content. HSA-REI.A.1	Writing Standards CCSS.ELA. W.9-10.1a, W.9-10.1b, W.9-10.1c, W.9-10.1d, W.9-10.1e W.9-10.2a, W.9-10.2b, W.9-10.2c, W.9-10.2d, W.9-10.2e, W.9-10.2f W.9-10.4 W.9-10.5 W.9-10.7 W.9-10.8 W.9-10.9a, W.9-10.9b W.9-10.10	Learning and Innovation Skills: Creativity and Innovation Critical Thinking and Problem Solving Communication and Collaboration
HS-LS2-5	CCSS.Math.Content. HS-IF.C.7c	Speaking and Listening Standards CCSS.ELA. SL.9-10.1a, SL.9-10.1b, SL..9-10.1c, SL.9-10.1d SL.9-10.2 SL.9-10.4 SL.9-10.6	Information, Media, and Technology Skills: Information Literacy Media Literacy ICT Literacy
HS-ESS2-2 HS-ESS2-3 HS-ESS2-7	CCSS.Math.Content. HSA-CED.A.4	Language Standards CCSS.ELA. L.9-10.1a, L.9-10.1b L.9-10.2a, L.9-10.2b, L.9-10.2c L.9-10.3a	Life and Career Skills: Flexibility and Adaptability Initiative and Self-Direction Social and Cross Cultural Skills Productivity and Accountability Leadership and Responsibility
HS-ETS-1			

as they brainstorm and work through the prototype. They will design how they might go about testing their product to ensure proof-of-concept and redesign based on the results of their testing. Finally, when students choose a variation of their innovation that balances benefits and risks, then they must design a way to market the innovation and convince the general public to use the innovation (see Table 7.6).

Table 7.6 STEM Road Map – Ninth-Grade Optimizing the Human Experience: Evaluating Human Impact on Nature

NGSS Performance Objectives	Common Core		21st Century Skills
	Mathematics	Language Arts	
HS-ESS3-4	CCSS.Math.Practices MP1, MP3, MP5, MP6, MP7, MP8	Reading Standards CCSS.ELA. RI.9-10.1 RI.9-10.2 RI.9-10.3 RI.9-10.4 RI.9-10.6 RI.9-10.8 RI.9-10.10	21st Century Themes: Global Awareness Financial, Economic, Business, and Entrepreneurial Literacy Civic Literacy Environmental Literacy
HS-LS4-6	CCSS.Math.Content. HS-BF.B.3	Writing Standards CCSS.ELA. W.9-10.2a, W.9-10.2b, W.9-10.2c, W.9-10.2d, W.9-10.2e, W.9-10.2f W.9-10.5 W.9-10.6 W.9-10.8 W.9-10.9a, W.9-10.9b	Learning and Innovation Skills: Creativity and Innovation Critical Thinking and Problem Solving Communication and Collaboration
HS-ETS-4	CCSS.Math.Content. HS-BF.A.1	Speaking and Listening Standards CCSS.ELA. SL.9-10.1a, SL.9-10.1b, SL.9-10.1c, SL.9-10.1d SL.9-10.2 SL.9-10.4 SL.9-10.5	Information, Media, and Technology Skills: Information Literacy Media Literacy ICT Literacy
		Language Standards CCSS.ELA. L.9-10.3 L.9-10.4a-d L.9-10.5 L.9-10.6	Life and Career Skills: Flexibility and Adaptability Initiative and Self-Direction Social and Cross Cultural Skills Productivity and Accountability Leadership and Responsibility

Sample STEM Careers in the Ninth-Grade STEM Road Map

Ninth grade is an ideal time for students to explore career possibilities. The website, O*Net Online (www.onetonline.org), based on the US Department of Labor statistics, is a valuable tool to learn about career options. The website explores a wide range of occupations and shows types of tasks the professionals are expected to perform, skills and education needed, the tools and technologies used in the field, work styles that best suit the profession, and the wage and employment trends for the occupation. Occupations can be searched on this site by keyword, career cluster, industry, level of education and experience (Job Zone), amount of expected growth of the industry (Bright Outlook), jobs in the green economy sector, groups of occupations based upon work performed, or by STEM discipline. A key word search of "Earth systems" brings up 544 occupations; the most relevant listed as Earth drillers (except oil and gas), atmospheric, Earth, marine, and space science teachers, geographers (bright outlook indication), geoscientists (green job), and construction laborers (bright outlook and green job).

The STEM Road Map for Tenth Grade

In the tenth grade, students will continue to explore STEM Road Map Theme inspired topics that align with grade-level academic content standards (e.g. Common Core and Next Generation Science Standards). The topics for the tenth grade include: *Healthy Living, Environmental Management, Modeling Ecosystems, Survival and Reproduction,* and *Rebuilding the Natural Environment.* Each topic is organized around a challenge/problem or project that student teams are assigned to tackle in the course of learning necessary content and skills in the various disciplines (see Table 7.7).

Cause and Effect: Healthy Living

As Americans, we are bombarded with information about living a healthier lifestyle daily through advertisements, the news, magazines, and food labels. Sometimes, this information is conflicting and at many times, confusing. In order to become more informed citizens, students are going to look at living a healthy lifestyle in a way that is not common, by looking at healthy living at a cellular level. In this project, students can research at the cellular level why healthy eating and exercise result in optimal conditions for health. Students can explore why certain plant, animal, and industry-produced foods can be either healthy or unhealthy. Students can examine whole organism metabolism from a cellular perspective, reflecting on how exercise is beneficial in a healthy lifestyle. Students will use the computational thinking practice of decomposition when they examine the whole organism and break down the parts of

Table 7.7 Tenth-Grade STEM Road Map Themes, Topics, and Problems/Challenges

STEM Theme	Topic	Problem/Challenge
Cause and Effect	Healthy Living LEAD Science or English/Language Arts	Student teams will address the problem of obesity in the United States though conducting research, interviewing key stakeholders locally, and developing a video documentary and associated print materials to promote healthy living habits. Teams will present their work to local city or county officials in an effort to inform policy.
Innovation and Progress	Resource Management LEAD Science	Student teams will design methods to keep track of the relationships of management of natural resources, sustaining human, plant, and animal populations, and maintaining biodiversity. Students can use a range of tools to help them computationally manage the resources, from an electronic spreadsheet to designing a simulation, embracing the computational thinking aspect of automation.
Represented World	Modeling Ecosystems LEAD Science	Design, build, test, and rebuild a self-sustaining ecosystem accompanied with a video that presents how complex interactions in the ecosystem are as well as the fragility of the ecosystem. Students will use the computational thinking practice of pattern recognition to seek commonalities and uniqueness within their chosen ecosystem.
Sustainable Systems	Survival and Reproduction LEAD Mathematics	Student teams are challenged to create an app (or storyboard for the app) based on probability and statistics that models the phenomena that organisms with an advantageous heritable trait tend to increase in proportion to organisms lacking the trait. Students will use the computation thinking practice of algorithmic thinking to apply the logic of statics to the lineage of their organism.
Optimizing Human Experience	Rebuilding the Natural Environment LEAD Social Studies or Science	Student teams are challenged to create a new renewable energy company with a specific focus on an innovative way to create energy in a cost-effective manner. Teams will research current renewable sources of energy then design and pitch a company that will provide an innovative renewable energy product. Teams will need to create a model for the change in energy consumption in the United States or the world if your company is successful.

that system into components at the cellular level. Students can find out why certain plants are healthy for us to eat while others are poisonous (these plants are often in the same genus) as well as finding out what the food industry creates and how that might affect the animals and us involved on a cellular level. Students will continue their extensive research on this subject by interviewing key stakeholders locally such as school nutritionists or doctors. To summarize all of the findings in a coherent way, students will communicate what they have learned through a documentary and associated print materials that espouse the practices they have found. Students will present their final products to local officials in an effort to inform policy (see Table 7.8). This module has been published in 2020 by NSTA Press. See https://my.nsta.org/resource/121548.

Innovation and Progress: Resource Management

Managing resources is a life skill that all students can learn and refine and can be applied to a range of topics from core academic ones such as using natural resources to daily household topics such as saving for college. In this project, students will design methods to keep track of the relationships of management of natural resources, sustaining human, plant, and animal populations, and maintaining biodiversity. Students can use a range of tools to help them computationally manage the resources, from an electronic spreadsheet to designing a simulation, embracing the computational thinking aspect of automation. An example scenario can entail an opportunity where a small wetland conservation organization has an interest in stopping the development of a four-lane highway bridge over the wetland. In doing so, the organization must first develop an inventory of what is sustained in the wetland and how the systems work to sustain life and the environment. The organization must then also determine what portions of the systems that are maintained will be affected and determine the short and long-term implications of building the highway. Teachers are encouraged to partner with local conservation organizations and ask professionals in the organization to come and hear the presentations of the students. In enlisting local community members to evaluate the students' work authentically, students may become engaged and volunteer for conservation management activities outside of the classroom (see Table 7.9).

Represented World: Modeling Ecosystems

According to the IPCC Fifth Assessment Report of the United Nations Intergovernmental Panel on Climate Change (2014) citing over 6,000 peer-reviewed scientific studies, increasing global temperature means that ecosystems will change. Lessened snow cover, receding glaciers, rising sea levels, and weather changes influence ecosystems, causing some species to be forced out of their habitats, while other species flourish. In this PBL,

Table 7.8 STEM Road Map – Tenth-Grade Cause and Effect: Healthy Living

NGSS Performance Objectives	Common Core		21st Century Skills
	Mathematics	Language Arts	
HS-LS1-1 HS-LS1-7	CCSS.Math.Practices MP1, MP3, MP8	Reading Standards CCSS.ELA. RI.9-10.1 RI.9-10.2 RI.9-10.3 RI.9-10.4 RI.9-10.5 RI.9-10.6 RI.9-10.7 RI.9-10.8 RI.9-10.9 RI.9-10.10	21st Century Themes: Health Literacy Environmental Literacy
HS-ETS-4	CCSS.Math.Content. HS-BF.B.4a CCSS.Math.Content. HS-BF.B.4b CCSS.Math.Content. HS-BF.B.4c	Writing Standards CCSS.ELA. W.9-10.1a, W.9-10.1b, W.9-10.1c, W.9-10.1d, W.9-10.1e W.9-10.2a, W.9-10.2b, W.9-10.2c, W.9-10.2d, W.9-10.2e, W.9-10.2f W.9-10.3a, W.9-10.3b, W.9-10.3c, W.9-10.3d, W.9-10.3e W.9-10.4 W.9-10.5 W.9-10.6 W.9-10.7 W.9-10.8 W.9-10.9a, W.9-10.9b W.9-10.10	Learning and Innovation Skills: Creativity and Innovation Critical Thinking and Problem Solving Communication and Collaboration
HS-LS2-3	CCSS.Math.Content. HS-SSE.A.1 CCSS.Math.Content. HS-SSE.A.1a	Speaking and Listening Standards CCSS.ELA. SL.9-10.2 SL.9-10.3 Language Standards CCSS.ELA. L.9-10.3 L.9-10.4a-d L.9-10.5 L.9-10.6	Information, Media, and Technology Skills: Information Literacy Media Literacy ICT Literacy Life and Career Skills: Flexibility and Adaptability Initiative and Self-Direction Social and Cross Cultural Skills Productivity and Accountability Leadership and Responsibility

Table 7.9 STEM Road Map – Tenth-Grade Innovation and Progress: Environmental Management

NGSS Performance Objectives	Common Core		21st Century Skills
	Mathematics	Language Arts	
HS-ESS3-3	CCSS.Math.Practices MP1, MP2, MP3, MP4, MP5, MP6, MP7, MP8	Reading Standards CCSS.ELA. RI.9-10.1 RI.9-10.2 RI.9-10.3 RI.9-10.4 RI.9-10.6 RI.9-10.8 RI.9-10.10	21st Century Themes: Global Awareness Health Literacy Environmental Literacy
HS-ETS-1	CCSS.Math.Content. HS-LE.A.1b CCSS.Math.Content. HS-LE.A.1c	Writing Standards CCSS.ELA. W.9-10.2a, W.9-10.2b, W.9-10.2c, W.9-10.2d, W.9-10.2e, W.9-10.2f W.9-10.5 W.9-10.6 W.9-10.8 W.9-10.9a, W.9-10.9b	Learning and Innovation Skills: Creativity and Innovation Critical Thinking and Problem Solving Communication and Collaboration
	CCSS.Math.Content. HS-LE.A.2	Speaking and Listening Standards CCSS.ELA. SL.9-10.1a, SL.9-10.1b, SL.9-10.1c, SL.9-10.1d SL.9-10.2 SL.9-10.4 SL.9-10.5	Information, Media, and Technology Skills: Information Literacy Media Literacy ICT Literacy
	CCSS.Math.Content. HS-CED.A.1	Language Standards CCSS.ELA. L.9-10.3 L.9-10.4a-d L.9-10.5 L.9-10.6	Life and Career Skills: Flexibility and Adaptability Initiative and Self-Direction Social and Cross Cultural Skills Productivity and Accountability Leadership and Responsibility

students will create a short documentary about the complex interactions in a chosen ecosystem and explain how the trends of current changes, if unchecked, can lead to a new ecosystem. Students will use the computational thinking practice of pattern recognition to seek commonalities and uniqueness within their chosen ecosystem. Some examples of ecosystem changes include terrestrial ecosystems and biodiversity where warming of 3°C, relative to 1990 levels, it is likely that global terrestrial vegetation would become a net source of carbon and a 4°C increase globally would lead to major extinctions (Schneider et al., 2007); marine ecosystems and biodiversity where a warming of 2°C above 1990 levels would result in mass mortality of coral reefs globally (Schneider et al., 2007); and freshwater ecosystems where a 4°C increase in global mean temperature by 2100 (relative to 1990–2000) would cause the extinction of many species of freshwater fish (Schneider et al., 2007). Producing a documentary allows students to be creative while still convincing an audience of a change in ecosystems based on empirical data. Students will acquire yet another style of communication through writing narrative scripts for the video and learn to associate images with words and use graphs effectively to communicate change (see Table 7.10).

Sustainable Systems: Survival and Reproduction

In the Age of Information, we deal with making sense of large amounts of information that is constantly streamed through media outlets available 24 hours a day. Students have the world of information at their fingertips by performing an electronic data search of anything they desire through their smart phone or computer. Being able to interpret statistics is a key part of making sense of the information we receive in the modern world. For example, people use statistics to interpret weather forecasts, predict disease and emergency situations, making health and medical decisions, engaging in political campaigns, considering insurance options, and making decisions on consumer choices. In this PBL, students will gain skills in probability and statistics by using this basis to design an app or a storyboard of an app that mimics the phenomena that organisms with advantageous heritable traits tend to increase in proportion to those lacking the trait. Students will use the computation thinking practice of algorithmic thinking to apply the logic of statics to the lineage of their organism. In doing so, students must first learn about the types of traits that might be advantageous regarding group behavior, individual behavior, and/or environmental factors. Then, students will apply this knowledge to develop a systematic and logical app (or storyboard for an app) that utilizes this knowledge and how the trait might affect populations over time (see Table 7.11).

Optimizing the Human Experience: Rebuilding the Natural Environment

The inclusion of the category of "Green Economy Sector" in *The O*-Online Data Base* (2014) is a strong indication that future businesses will need to consider not only their financial progress but also their positive contributions to the human experience, including careers that focus on rebuilding the natural environment. In this PBL, students will connect to their prior knowledge about energy production and the effects

Table 7.10 STEM Road Map – Tenth-Grade Represented World: Modeling Ecosystems

NGSS Performance Objectives	Common Core Mathematics	Common Core Language Arts	21st Century Skills
HS-LS2-1 HS-LS2-2 HS-LS2-4 HS-LS2-6	CCSS.Math.Practices MP1, MP3, MP8	Reading Standards CCSS.ELA. RI.9-10.1 RI.9-10.2 RI.9-10.7 RI.9-10.8 RI.9-10.10	21st Century Themes: Global Awareness Civic Literacy Health Literacy Environmental Literacy
HS-LS3-3	CCSS.Math.Content. HS-BF.A.2	Writing Standards CCSS.ELA. W.9-10.1a, W.9-10.1b, W.9-10.1c, W.9-10.1d, W.9-10.1e W.9-10.2a, W.9-10.2b, W.9-10.2c, W.9-10.2d, W.9-10.2e, W.9-10.2f W.9-10.4 W.9-10.6 W.9-10.8 W.9-10.10	Learning and Innovation Skills: Creativity and Innovation Critical Thinking and Problem Solving Communication and Collaboration
HS-ETS-4	CCSS.Math.Content. HS-SSE.B.3	Speaking Standards CCSS.ELA. SL.9-10.2 SL.9-10.4 SL.9-10.5 SL.9-10.6	Information, Media, and Technology Skills: Information Literacy Media Literacy ICT Literacy
		Language Standards CCSS.ELA. L.9-10.2 L.9-10.6	Life and Career Skills: Flexibility and Adaptability Initiative and Self-Direction Social and Cross Cultural Skills Productivity and Accountability Leadership and Responsibility

of this process on the natural environment to create innovations in renewable sources of energy based on research evidence in a cost-effective way. Various skills from different academic disciplines are integrated into this PBL by requiring the students to design a company based on their innovative idea and to develop a pitch for the marketability of the company, focusing on how the innovation will optimize human experiences while being mindful of the natural environment. Students will use

Table 7.11 STEM Road Map – Sustainable Systems: Survival and Reproduction

NGSS Performance Objectives	Common Core Mathematics	Language Arts	21st Century Skills
HS-LS1-2 HS-LS1-3 HS-LS1-4	CCSS.Math.Practices MP1, MP2, MP3, MP4, MP5, MP6, MP7, MP8	Reading Standards CCSS.ELA. RI.9-10.1 RI.9-10.2 RI.9-10.3 RI.9-10.4 RI.9-10.6 RI.9-10.8 RI.9-10.10	21st Century Themes: Global Awareness Environmental Literacy
HS-LS2-8	CCSS.Math.Content. HSS-ID.A.1, CCSS. Math.Content. HSS-ID.A.2 CCSS.Math.Content. HSS-ID.A.3 CCSS.Math.Content. HSS-ID.A.4	Writing Standards CCSS.ELA. W.9-10.2a, W.9-10.2b, W.9-10.2c, W.9-10.2d, W.9-10.2e, W.9-10.2f W.9-10.5 W.9-10.6 W.9-10.8 W.9-10.9a, W.9-10.9b	Learning and Innovation Skills: Creativity and Innovation Critical Thinking and Problem Solving Communication and Collaboration
HS-LS4-1 HS-LS4-3	CCSS.Math.Content. HSS-ID.B.5 CCSS. Math.Content. HSS-ID.B.6 CCSS.Math.Content. HSS-ID.B.6a CCSS. Math.Content. HSS-ID.B.6b	Speaking Standards CCSS.ELA. SL.9-10.1a, SL.9-10.1b, SL.9-10.1c, SL.9-10.1d SL.9-10.2 SL.9-10.4 SL.9-10.5	Information, Media, and Technology Skills: Information Literacy Media Literacy ICT Literacy
HS-ETS-3	CCSS.Math.Content. HSA-SSE.B4	Language Standards CCSS.ELA. L.9-10.3 L.9-10.4a-d L.9-10.5 L.9-10.6	Life and Career Skills: Flexibility and Adaptability Initiative and Self-Direction Social and Cross Cultural Skills Productivity and Accountability Leadership and Responsibility

the computational thinking practice of abstraction to cull out the unnecessary or extraneous information to present only the critical information. Further, students will have to use predictive skills to consider how their innovation will affect energy consumption and the implications of this consumption over a long period of time. In effect, students will be thinking about making the world a better place, being able to make a career from this idea, and finding ways to sustain progress and conservation in the same effort (see Table 7.12).

Table 7.12 STEM Road Map – Tenth-Grade Optimizing the Human Experience: Rebuilding the Natural Environment

NGSS Performance Objectives	Common Core		21st Century Skills
	Mathematics	Language Arts	
HS-PS3-3	CCSS.Math.Practices MP1, MP3, MP5, MP6, MP7, MP8	Reading Standards CCSS.ELA. RI.9-10.1 RI.9-10.2 RI.9-10.3 RI.9-10.4 RI.9-10.6 RI.9-10.8 RI.9-10.10	21st Century Themes: Global Awareness Financial, Economic, Business, and Entrepreneurial Literacy Civic Literacy Environmental Literacy
HS-LS2-7	CCSS.Math.Content. HS-BF.A.1a	Writing Standards CCSS.ELA. W.9-10.1a. W.9-10.1b, W.9-10.1c, W.9-10.1d, W.9-10.1e W.9-10.2a, W.9-10.2b, W.9-10.2c, W.9-10.2d, W.9-10.2e, W.9-10.2f W.9-10.4 W.9-10.5 W.9-10.7 W.9-10.8 W.9-10.9a, W.9-10.9b W.9-10.10	Learning and Innovation Skills: Creativity and Innovation Critical Thinking and Problem Solving Communication and Collaboration
HS-ETS-3		Speaking Standards CCSS.ELA. SL.9-10.1a, SL.9-10.1b, SL.9-10.1c, SL.9-10.1d SL.9-10.2 SL.9-10.4	Information, Media, and Technology Skills: Information Literacy Media Literacy ICT Literacy
		Language Standards CCSS.ELA-Literacy.L.9-10.1a-b, L.9-10.2a-c, L.9-10.3a, SL.9-10.6	Life and Career Skills: Flexibility and Adaptability Initiative and Self-Direction Social and Cross Cultural Skills Productivity and Accountability Leadership and Responsibility

Sample STEM Careers in the Tenth-Grade STEM Road Map

As mentioned previously, the inclusion of the categories of *"Green Economy Sector"* and *"STEM Discipline"* fields on the opening page of *the O*Online* occupations data base (2014) is a strong indication that these types of jobs are going to be central to a tenth-grader's career in the future. Categories of jobs listed under Green Economy Sector include Agriculture and Forestry, Energy and Carbon Capture and Storage, Energy Efficiency, Energy Trading, Environment Protection, Government and Regulatory Administration, Green Construction, Manufacturing, Recycling and Waste Reduction, Renewable and Energy Generation, Research, Design and Consulting Services, and Transportation. STEM Discipline occupations on the website include Chemistry, Computer Science, Engineering, Environmental Science, Geosciences, Mathematics, Life Sciences, and Physics/Astronomy. Researching the tasks, abilities, and education needed for these occupations has potential to motivate students to see that the study of integrated fields and learning through PBL can help them prepare for careers of the future. Teachers can scaffold these understandings for students by demonstrating how the learning tasks students are performing in the PBLs are identical to the list of tasks and abilities recognized by the Department of Labor to be successful in the various STEM fields.

The STEM Road Map for 11th Grade

The 11th-grade year will engage students in exploring STEM Road Map Theme generated topics that also align with grade-level academic content standards (e.g. Common Core and Next Generation Science Standards) which include: *Standing on the Shoulders of Giants, Construction Materials, Radioactivity, Green Building Rooftops,* and *Mineral Resources.* Each of these topics is organized around a challenge/problem or project that student teams are assigned to tackle in the course of learning necessary content and skills in the various disciplines (see Table 7.13).

Cause and Effect: Standing on the Shoulders of Giants

The phrase "Standing on the Shoulders of Giants" is attributed to Isaac Newton when he was giving a speech to the Academies, "If I have seen further it is by standing on the shoulders of giants." However, the metaphor was first recorded in the 12th century and attributed to Bernard of Chartres (Merton, 1965). Regardless, the meaning remains the same and refers to building your work on the work of others and is an acknowledgement that even the most unique work has a foundation in others' ideas. Traditional textbooks rarely refer to how scientists build from other work and often communicate the contrary that scientists think of

Table 7.13 Eleventh-Grade STEM Road Map Themes, Topics, and Problems/Challenges

STEM Theme	Topic	Problem/Challenge
Cause and Effect	Standing on the Shoulders of Giants LEAD Social Sudies	Student teams are challenged to create a museum display prototype that explains how chemists, beginning with Galileo Galilei, have discovered the world around them including patterns of chemical properties, creation of the periodic table, rates of reactions, and large-scale production of chemicals in modern day. Conclude the book with proposals about how chemistry may create better living conditions in the future.
Innovation and Progress	Construction Materials LEAD Science	Student teams are challenged to use knowledge of molecular-level structure to examine the collapse of the World Trade Center twin towers and develop a proposal and prototype for new or improved building materials that could be incorporated into the design of future high-rise buildings in US cities.
Represented World	Radioactivity LEAD Mathematics	Student teams will be challenged to construct a scale model of the atom (virtually or physically) that will illustrate fission, fusion, and radioactive decay. Teams will also prepare a persuasive essay indicating potential future uses or dangers of the energy sources.
Sustainable Systems	Rooftop Gardening LEAD Social Studies/ Science	Student teams will work together to plan and implement a rooftop mini garden in their community. This project will include securing sponsors for the mini garden, developing a business model to sustain the work and constructing a marketing plan to make the products available to families in the community. Students will use the computational thinking practice of abstraction when they do research to separate out the essential information from the noise. A key component of this work will be developing the green plan.
Optimizing Human Experience	Mineral Resources LEAD Science English/ Language Arts	Student teams will develop an op-ed article for a local publication or website that will evaluate competing design solutions for developing, managing, and utilizing mineral resources based on cost-benefit ratios with both qualitative and quantitative criteria. Student teams will use the computational thinking practice of abstraction to focus on the key information to communicate.

ideas just from a stroke of brilliance (e.g. Newton being hit on the head with an apple and conjuring the law of gravitation in an afternoon). The intention of this PBL is to help students to see how ideas continue to be elaborated by continuing research over time by having students create a prototype of an interactive museum installation that explains the progression of ideas about the nature of matter. The product, an interactive museum display, gives students enough latitude to have technical detail, while still needing to be scaled down from their own work, thus giving students an opportunity to synthesize their own research. Students will use the computational thinking practice of decomposition to break down the different components of a museum display and the role of each component. The museum installation topics begin with Galileo Galilei, then proceeds through various experiments with chemical properties of matter, creation of the periodic table, rates of reactions, gas laws, and large-scale production of chemicals in modern day. The objective for students is to link the ideas that are traditionally presented as singular genius events, thus demonstrating that everyone is capable of being a scientist and that scientists work as collaborators. The book should conclude with proposals about how chemistry may create better living conditions in the future, which requires students to think about how the past informs the future (see Table 7.14).

Innovation and Progress: Construction Materials

In our busy modern world, it is easy to overlook the building blocks of the magnificent engineering feats such as bridges, roadways, and the various materials used in constructing buildings of all shapes and sizes. The purpose of this PBL is to guide students to learn about how construction materials are made, the specifications that are necessary in different types of construction, and why these materials work the way they do at a molecular level through an investigation of the collapse of the World Trade Center twin towers. The field of structural materials science is a robust one, and recent movements in the field are rapidly developing in biomaterials (Boom time for Biomaterials, 2009). In addition to building awareness of engineering achievements and learning about new advances in materials science, students will also learn about how failures inform future work, particularly in engineering, by using the computational thinking practice of pattern recognition as they look at how different failed engineering designs led to innovations. The proposal format of the product of this PBL will assist in building student skills in technical writing and should include detailed and coherent information, allowing for some creativity while still upholding rigorous accuracy in describing the natural (science and mathematics) and designed (engineering and technology) world (see Table 7.15). This module has been published in 2017 by NSTA Press. See https://my.nsta.org/resource/110052.

The STEM Road Map for Grades 9–12 157

Table 7.14 STEM Road Map – 11th-Grade Cause and Effect: Standing on the Shoulders of Giants

NGSS Performance Objectives	Common Core		21st Century Skills
	Mathematics	Language Arts	
HS-PS1-2 HS-PS1-5 HS-PS1-6	CCSS.Math.Practices MP1, MP2, MP3, MP4, MP5, MP6, MP7, MP8	Reading Standards CCSS.ELA. RI.11-12.1 RI.11-12.2 RI.11-12.3 RI.11-12.4 RI.11-12.5 RI.11-12.7 RI.11-12.8 RI.11-12.10	21st Century Themes: Global Awareness Civic Literacy
HS-ESS1-3	CCSS.Math.Content. HSN-Q.A.1	Writing Standards CCSS.ELA. W.11-12.1a-e W.11-12.2a-f W.11-12.3a-e W.11-12.4 W.11-12.5 W.11-12.6 W.11-12.7 W.11-12.8 W.11-12.9a-b W.11-12.10	Learning and Innovation Skills: Creativity and Innovation Critical Thinking and Problem Solving Communication and Collaboration
HS-ETS-1		Speaking Standards CCSS.ELA. SL11-12.1a-e SL11-12.2 SL11-12.5	Information, Media, and Technology Skills: Information Literacy Media Literacy ICT Literacy
		Language Standards CCSS.ELA. L.11-12.1a-b L.11-12.2a-b L.11-12.3a L.11-12.4a-e L.11-12.5a-b L.11-12.6	Life and Career Skills: Flexibility and Adaptability Initiative and Self-Direction Social and Cross Cultural Skills Productivity and Accountability Leadership and Responsibility

Represented World: Radioactivity

Radioactivity has been a subject of interest in developing clean energy sources since the early 1930s. Fission is widely used in thermonuclear power generation, although there are serious complications with

Table 7.15 STEM Road Map – 11th-Grade Innovation and Progress: Construction Materials

NGSS Performance Objectives	Common Core Mathematics	Common Core Language Arts	21st Century Skills
HS-PS2-6	CCSS.Math.Practices MP1, MP2, MP3, MP4, MP5, MP6, MP7, MP8	Reading Standards CCSS.ELA. RI.11-12.1 RI.11-12.2 RI.11-12.3 RI.11-12.4 RI.11-12.5 RI.11-12.7 RI.11-12.8	21st Century Themes: Global Awareness Financial, Economic, Business, and Entrepreneurial Literacy Environmental Literacy
HS-ETS-3	CCSS.Math.Content. HS-LE.B.5	Writing Standards CCSS.ELA. W.11-12.1a, W.11-12.b, W.11-12.1c, W.11-12.1d, W.11-12.1e W.11-12.2a, W.11-12.2b, W.11-12.2c, W.11-12.2d, W.11-12.2e, W.11-12.2f	Learning and Innovation Skills: Creativity and Innovation Critical Thinking and Problem Solving Communication and Collaboration
	CCSS.Math.Content. HSA-CED.A.2	Speaking Standards CCSS.ELA. SL.11-12.1a, SL.11-12.1b, SL.11-12.1c, SL.11-12.1d SL.11-12.2 SL.11-12.3 SL.11-12.4 SL.11-12.5 SL.11-12.6	Information, Media, and Technology Skills: Information Literacy Media Literacy ICT Literacy
		Language Standards CCSS.ELA. L.11-12.1a L.11-12.4 L.11-12.5 L.11-12.6	Life and Career Skills: Flexibility and Adaptability Initiative and Self-Direction Social and Cross Cultural Skills Productivity and Accountability Leadership and Responsibility

containment and waste that still need to be worked through. Fusion has been the focus of famous projects such as the Manhattan Project, and although research has been conducted on fusion for the past 80 years, generating more energy out of the reaction than is being put into the reaction has not yet been overcome, rendering it a somewhat useless energy source. However, scientists, mathematicians, and engineers continue to explore these phenomena in hopes of a breakthrough (see National Ignition Facility in Livermore, California, and International Thermonuclear Experimental Reactor in the south of France). In this PBL student, teams are to look deeply into the processes of fission, fusion, and radioactive decay to develop a physical or virtual scale model of the changes in composition of the nucleus and the amounts of energy released. Students will employ the computational thinking practice of pattern recognition to see similarities and differences in fission, fusion, and radioactive decay as they make sense of the information. Scale modeling in this PBL could take on many flexible forms such as visual modeling, computer simulations, and mathematical modeling, all of which focus on the relative interactions in the descriptions of the phenomena. Teams will also prepare a persuasive essay indicating potential future uses or dangers of the energy sources (see Table 7.16). This module has been published in 2019 by NSTA Press. See https://my.nsta.org/resource/116903.

Sustainable Systems: Rooftop Gardening

Placing plants on rooftops of urban buildings (or urban greening) has long been regarded as a way to aesthetically add more green space to areas dominated by concrete as well as contributing waste diversion, managing stormwater run-off, curbing urban heat island effects, and improving air quality. Additional benefits include increasing energy efficiency, fire retardation (Köehler, 2004), reduction of electromagnetic radiation (Herman, 2003), and noise reduction (Peck and Callaghan, 1999). Clearly, green building rooftops have various beneficial outcomes. The purpose of this PBL is to build an awareness of the phenomena of transfer of energy. Student teams will work together to plan and build a rooftop mini garden at their school or in their community. This project will include securing sponsors for the mini garden, developing a business model to sustain the work, constructing a marketing plan to make the products available to families in the community. Students will use the computational thinking practice of abstraction when they do research to separate out the essential information from the noise. A key component of this work will be developing the green plan. Not only will students learn about the environmental pros and cons of developing a rooftop mini garden, but they will learn core 21st century skills in development of the business model and market plan to connect to the community (see Table 7.17).

Optimizing the Human Experience: Mineral Resources

Citizens in a democratic society have a responsibility to contribute to the good of the community and to be knowledgeable about controversial

Table 7.16 STEM Road Map – 11th-Grade Represented World: Radioactivity

NGSS Performance Objectives	Common Core		21st Century Skills
	Mathematics	Language Arts	
HS-PS1-1 HS-PS1-4 HS-PS1-7 HS-PS1-8	CCSS.Math.Practices MP1, MP3, MP4, MP7	Reading Standards CCSS.ELA. RI.11-12.1 RI.11-12.2 RI.11-12.3 RI.11-12.4 RI.11-12.5 RI.11-12.7 RI.11-12.8	21st Century Themes: Global Awareness Financial, Economic, Business, and Entrepreneurial Literacy Civic Literacy Environmental Literacy
HS-ETS-2	CCSS.Math.Content. HS-IF.B.4 CCSS.Math.Content HS-IF.B.6	Writing Standards CCSS.ELA. W.11-12.1a, W.11-12.1b, W.11-12.1c, W.11-12.1d, W.11-12.1e W.11-12.2a, W.11-12.2b, W.11-12.2c, W.11-12.2d, W.11-12.2e, W.11-12.2f	Learning and Innovation Skills: Creativity and Innovation Critical Thinking and Problem Solving Communication and Collaboration
	CCSS.Math.Content. HSA-APR.D.6	Speaking Standards CCSS.ELA. SL.11-12.1a, SL.11-12.1b, SL.11-12.1c, SL.11-12.1d SL.11-12.2 SL.11-12.3 SL.11-12.4 SL.11-12.5 SL.11-12.6	Information, Media, and Technology Skills: Information Literacy Media Literacy ICT Literacy
		Language Standards CCSS.ELA. L.11-12.1a, L.11-12.1b L.11-12.4 L.11-12.5 L.11-12.6	Life and Career Skills: Flexibility and Adaptability Initiative and Self-Direction Social and Cross Cultural Skills Productivity and Accountability Leadership and Responsibility

subjects. It is of particular importance in a democracy that its citizens be able to make decisions based on evidence and to be able to distinguish between a reliable and an unreliable resource. The purpose of this PBL is to give students an opportunity to write an opinion article based on evidence that is designed to be published in a newspaper and to convince

Table 7.17 STEM Road Map – 11th Grade Sustainable Systems: Green Building Rooftops

NGSS Performance Objectives	Common Core Mathematics	Common Core Language Arts	21st Century Skills
HS-PS3-1 HS-PS3-2	CCSS.Math.Practices MP1, MP3, MP7	Reading Standards CCSS.ELA. RI.11-12.1 RI.11-12.2 RI.11-12.3 RI.11-12.4 RI.11-12.6 RI.11-12.8 RI.11-12.10	21st Century Themes: Global Awareness Financial, Economic, Business, and Entrepreneurial Literacy Civic Literacy Health Literacy Environmental Literacy
HS-LS1-5	CCSS.Math.Content. HS-BF.A.1c	Writing Standards CCSS.ELA. W.11-12.2a, W.11-12.2b, W.11-12.2c, W.11-12.2d, W.11-12.2e, W.11-12.2f W.11-12.5 W.11-12.6 W.11-12.8 W.11-12.9a, W.11-12.9b	Learning and Innovation Skills: Creativity and Innovation Critical Thinking and Problem Solving Communication and Collaboration
HS-ETS-4	CCSS.Math.Content. HS-IF.C.7 CCSS.Math.Content. HS-IF.C.8	Speaking Standards CCSS.ELA. SL.11-12.1a, SL.11-12.1b, SL.11-12.1c, SL.11-12.1d SL.11-12.2 SL.11-12.4 SL.11-12.5	Information, Media, and Technology Skills: Information Literacy Media Literacy ICT Literacy
		Language Standards CCSS.ELA. L.11-12.3a L.11-12.4a-d L.11-12.5a-b L.11-12.6	Life and Career Skills: Flexibility and Adaptability Initiative and Self-Direction Social and Cross Cultural Skills Productivity and Accountability Leadership and Responsibility

readers of the effectiveness of a particular design solution for developing, managing, and utilizing mineral resources. In this, PBL students will use the computational thinking practice of abstraction to focus on the key information to communicate. In this activity, students will find reliable qualitative and quantitative resources to present a cost-benefit analysis

Table 7.18 STEM Road Map – 11th-Grade Optimizing the Human Experience: Mineral Resources

NGSS Performance Objectives	Common Core Mathematics	Common Core Language Arts	21st Century Skills
HS-PS3-3	CCSS.Math.Practices MP1, MP2, MP3, MP4, MP5, MP6, MP7, MP8	Reading Standards CCSS.ELA. RI.11-12.1 RI.11-12.2 RI.11-12.3 RI.11-12.4 RI.11-12.6 RI.11-12.8 RI.11-12.10	21st Century Themes: Global Awareness Financial, Economic, Business, and Entrepreneurial Literacy Civic Literacy Environmental Literacy
HS-ESS3-2	CCSS.Math.Content. HSA-REI.D.10 CCSS.Math.Content. HSA-REI.D.11	Writing Standards CCSS.ELA. W.11-12.1a, W.11-12.1b, W.11-12.1c, W.11-12.1d, W.11-12.1e W.11-12.2a, W.11-12.2b, W.11-12.2c, W.11-12.2d, W.11-12.2e, W.11-12.2f W.11-12.4 W.11-12.5 W.11-12.7 W.11-12.8 W.11-12.9a, W.11-12.9b W.11-12.10	Learning and Innovation Skills: Creativity and Innovation Critical Thinking and Problem Solving Communication and Collaboration
HS-ETS-1		Speaking Standards CCSS.ELA. SL.11-12.1a, SL.11-12.1b SL.11-12.2 SL.11-12.4	Information, Media, and Technology Skills: Information Literacy Media Literacy ICT Literacy
		Language Standards CCSS.ELA. L.11-12.1a-b L.11-12.2a-c L.11-12.3a L.11-12.6	Life and Career Skills: Flexibility and Adaptability Initiative and Self-Direction Social and Cross Cultural Skills Productivity and Accountability Leadership and Responsibility

for their chosen mineral resource. *The USGS Mineral Resources Program* (MRP) is an excellent resource of scientific information for objective resource assessments and research results on mineral potential, production, consumption, and environmental effects (see Table 7.18).

Sample STEM Careers in the 11th-Grade STEM Road Map

The variety of integrated content and contexts in the PBLs taught during 11th grade offer a foundation to explore a variety of careers. In doing so, teachers may want to choose a specific career and go through the various indicators of tasks and abilities and education during this year because 11th graders will need to begin thinking about narrowing down and preparing for college or a career. For example, a keyword search for "radioactivity" in *the O*Online* website mentioned in the ninth and tenth-grade STEM Careers sections yields the career of a nuclear engineer, marked with a green job notation. Examples of the ten tasks listed on the website that a nuclear engineer would perform are:

- Perform experiments that will provide information about acceptable methods of nuclear material usage, nuclear fuel reclamation, or waste disposal.
- Conduct tests of nuclear fuel behavior and cycles or performance of nuclear machinery and equipment to optimize performance of existing plants.
- Keep abreast of developments and changes in the nuclear field by reading technical journals or by independent study and research.

Tools used in this job include desktop computers, facial shields, nuclear reactor control rod systems, nuclear tools, and respirators. Knowledge required to be a nuclear engineer as indicated on the website include understanding of engineering, chemistry, mathematics, physics, design, computers, public safety, security, administration, and management. Skills of a nuclear engineer include active listening, critical thinking, operations analysis, reading comprehension, speaking, science, systems analysis, writing, complex problem solving, and monitoring. A nuclear engineer would also need the qualities of problem sensitivity, oral comprehension, oral expression, written comprehension, inductive reasoning, category flexibility, and prioritizing. In the PBLs taught during 11th grade, teachers may want to actively incorporate the knowledge, skills, and abilities of a particular career relevant to the problem and have students indicate when they are enacting those qualities. In doing so, teachers may help students identify with a career that they might not have previously considered.

The STEM Road Map for 12th Grade

The 12th-grade year will engage students in exploring STEM Road Map Theme generated topics that also align with grade-level academic content standards (e.g. *Common Core and Next Generation Science Standards*) which include *The Business of Amusement Parks, Creating the Next Smart Phone, Car Crashes, Creating Global Bonds,* and *Dealing with Natural Catastrophes.* Each of these topics is organized around a challenge/problem or project that student teams are assigned to tackle in the course of learning necessary content and skills in the various disciplines (see Table 7.19).

Cause and Effect: Amusement Park Management

People who design and build amusement park rides are constantly innovating. In this PBL, student teams are challenged to create a prototype for an amusement park ride powered by a combination of electricity and magnetism. Teams can acquire information about the current innovations of transportation where magnets and electricity can optimize velocity and fuel economy such as the superconducting Maglev train in Central Japan. Teams will create a marketing and financial plan for the ride as well as a detailed risk assessment to ensure safety. Researching and designing such innovative amusement park rides involve the topics of electromagnetic radiation, electricity and magnetism, motors, generators, and transformers. Students will use the computational thinking practice of abstraction to find the essence of the ideas to apply to their business plan. Additionally, student teams will need to develop a theme for the ride based on its characteristics including a marketing and financial plan. Finally, student teams need to consider the safety innovations that must accompany any thrill ride. As the PBL is created, students will need to design the ride, which requires not only knowledge of facts but also an overall understanding of how the facts fit together (see Table 7.20).

Innovation and Progress: Innovating the Next Smartphone

Progressive technology development is characterized by responding to the needs of the users. Successful technologies are designed to upgrade product performance and improve product solutions with more effective techniques of analysis of users' needs. This PBL captures these principles in the overall objective to create a model or prototype of an upgraded smart phone that is based on a needs analysis. Smartphones are ubiquitous and perhaps indispensable in students' lives, but they may not have much of an idea of how they work. In order to accomplish this, students must first understand the basics of wave behavior, transmission, and storage and apply these principles to the current structure of a smartphone, including how a phone is like a transmitter and a receiver of radio waves and the role of cell towers in that transmission. Then, student teams must develop a survey to find out what changes others may want to make to their current smartphone. Students will use the computational thinking practice of pattern recognition to interpret the results of the survey. Once students have an understanding of how a smartphone works, coupled with the knowledge of what other people

Table 7.19 Twelfth-Grade STEM Road Map Themes, Topics, and Problems/Challenges

STEM Theme	Topic	Problem/Challenge
Cause and Effect	Amusement Park Management LEAD Social Studies	Student teams are challenged to create a prototype for an amusement park ride powered by a combination of electricity and magnetism. Students will use the computational thinking practice of abstraction to find the essence of the ideas to apply to their business plan. Teams will create a marketing and financial plan for the ride as well as a detailed risk assessment to ensure safety.
Innovation and Progress	Innovating the Next Smart Phone LEAD Science	Student teams will create a model or a prototype of improvements for a smart phone based on a needs analysis survey of your friends, fellow students, and family. Students will use the computational thinking practice of pattern recognition to interpret the results of the survey and will explain all technical information in a prospectus including wave behavior, transmission, and storage.
Represented World	Car Crashes LEAD Mathematics	Student teams will develop models and mathematical representations of different car crash scenarios to illustrate how analysis of momentum and forces can inform law enforcement how the crashes occurred.
Sustainable Systems	Creating Global Bonds LEAD Social Studies/Science	Student teams are challenged to build and implement an international blog focused on energy consumption and links to climate change. The teams will identify potential school partners in three other countries to join the blog and share ideas. Each team will prepare a presentation and white paper that summarize their findings from international discussions and will provide an argument for one mitigation strategy that could be implemented locally and globally.
Optimizing Human Experience	Navigating Geographical Challenges LEAD Science	Student teams are challenged to create marketing materials including electronic and paper-based that promote the development of new luxury homes built on a fault line. In these materials, student teams will demonstrate pros and cons of these natural hazards and demonstrate the innovative safety features and energy consciousness of the new development.

Table 7.20 STEM Road Map – 12th-Grade Cause and Effect: The Business of Amusement Parks

NGSS Performance Objectives	Common Core Mathematics	Common Core Language Arts	21st Century Skills
HS-PS1-3	CCSS.Math.Practices MP1, MP2, MP3, MP5, MP6, MP8	Reading Standards CCSS.ELA. RI.11-12.1 RI.11-12.2 RI.11-12.3 RI.11-12.4 RI.11-12.6 RI.11-12.8 RI.11-12.10	21st Century Themes: Global Awareness Environmental Literacy
HS-PS2-5	CCSS.Math.Content. HS-LE.A.1, HS-LE.A.1a	Writing Standards CCSS.ELA. W.11-12.2a, W.11-12.2b, W.11-12.2c, W.11-12.2d, W.11-12.2e, W.11-12.2f W.11-12.5 W.11-12.6 W.11-12.8 W.11-12.9a, W.11-12.9b	Learning and Innovation Skills: Creativity and Innovation Critical Thinking and Problem Solving Communication and Collaboration
HS-ESS1-1	CCSS.Math.Content. HS-TF.A.1	Speaking Standards CCSS.ELA. SL.11-12.1a, SL.11-12.1b, SL.11-12.1c, SL.11-12.1d SL.11-12.2 SL.11-12.4 SL.11-12.5	Information, Media, and Technology Skills: Information Literacy Media Literacy ICT Literacy
HS-PS4-3 HS-PS4-4	CCSS.Math.Content. HS-IF.A.2	Language Standards CCSS.ELA. L.11-12.3 L.11-12.4a-d L.11-12.5 L.11-12.6	Life and Career Skills: Flexibility and Adaptability Initiative and Self-Direction Social and Cross Cultural Skills Productivity and Accountability Leadership and Responsibility
HS-ETS-2			

would like in a smartphone, they must create a model or prototype of an improved phone. The presentation of the model or prototype can serve as an assessment of how students show what they know (see Table 7.21).

Table 7.21 STEM Road Map – 12th Grade Innovation and Progress: Creating the Next Smart Phone

NGSS Performance Objectives	Common Core		21st Century Skills
	Mathematics	Language Arts	
HS-PS4-1 HS-PS4-2 HS-PS4-5	CCSS.Math.Practices MP1, MP2, MP3, MP5, MP6, MP8	Reading Standards CCSS.ELA. RI.11-12.1 RI.11-12.2 RI.11-12.3 RI.11-12.4 RI.11-12.5 RI.11-12.7 RI.11-12.8	21st Century Themes: Global Awareness Financial, Economic, Business, and Entrepreneurial Literacy Environmental Literacy
HS-ETS-4	CCSS.Math.Content. HS-LE.A.2 CCSS.Math.Content. HS-LE.A.3	Writing Standards CCSS.ELA. W.11-12.1a, W.11-12.1b, W.11-12.1c, W.11-12.1d, W.11-12.1e W.11-12.2a, W.11-12.2b, W.11-12.2c, W.11-12.2d, W.11-12.2e, W.11-12.2f	Learning and Innovation Skills: Creativity and Innovation Critical Thinking and Problem Solving Communication and Collaboration
		Speaking Standards CCSS.ELA. SL.11-12.1a, SL.11-12.1b, SL.11-12.1c, SL.11-12.1d SL.11-12.2 SL.11-12.3 SL.11-12.4 SL.11-12.5 SL.11-12.6	Information, Media, and Technology Skills: Information Literacy Media Literacy ICT literacy
		Language Standards CCSS.ELA. L.11-12.1a-b L.11-12.4 L.11-12.5 L.11-12.6	Life and Career Skills: Flexibility and Adaptability Initiative and Self-Direction Social and Cross Cultural Skills Productivity and Accountability Leadership and Responsibility

Represented World: Car Crashes

As students in 11th grade are learning to be new drivers, a PBL focusing on analyzing the forces involved in different types of car crashes may be timely and informative. There are many resources available such as videos, simulations, and car manufacturer reports that can help students understand the forces on a driver and passengers as well as impacts on cars. There is even *The Stapp Car Crash Journal* published annually by the Society of Automotive Engineers that provides mathematical modeling and results of many scenarios. This PBL asks student teams to take relevant information from the large range of materials on this well-studied phenomenon and synthesize it into a model and mathematical representation, documenting several different scenarios. Students can investigate car crash variables according to orientations (head-on collision, side swiping), sizes of vehicles (car vs. truck), or variables in momentum (fast vs. slow), in addition to other relevant variables of their choosing. Students will use the computational thinking practice of decomposition to break down the relevant variables to study. The intended audience for communication of the synthesis of information is law enforcement, so students can direct their efforts to inform police at the scene of an accident, provide evidence at a legal trial, or persuade policy makers in traffic laws (see Table 7.22). This module has been published in 2018 by NSTA Press. See https://my.nsta.org/resource/114554.

Sustainable Systems: Creating Global Bonds

In our global economy and technologically oriented world, we are connected in ways that couldn't be imagined in the 1950s. These global connections among people and resources have tremendous benefits, but also carry a great deal of responsibility and negotiation to suit the needs of all members. Therefore, it is imperative that students have educational experiences that require them to interact with people who have different views in a positive way. This PBL challenges student teams to build and implement an international blog focused on energy consumption and links to climate change. The teams will identify potential school partners in three other countries to join the blog and share ideas. In communicating with the school partners, students need to use abstraction to find the core ideas so that they succinctly describe their ideas. Each team will prepare a presentation and white paper that summarize their findings from international discussions and will provide an argument for one mitigation strategy that could be implemented locally and globally. Because the issues of energy flow in the atmosphere, ocean and land that can contribute to climate change are vast and complex; students can form working groups on different aspects of the problems in order for the work to be manageable. The PBL helps students to develop communication and technology skills by creating a blog, discovering and considering all sides of an issue based on evidence, communicating with

Table 7.22 STEM Road Map – 12th-Grade Represented World: Car Crashes

NGSS Performance Objectives	Common Core Mathematics	Common Core Language Arts	21st Century Skills
HS-PS2-2 HS-PS2-4	CCSS.Math.Practices MP1, MP2, MP3, MP4, MP5, MP6, MP7, MP8	Reading Standards CCSS.ELA. RI.11-12.1 RI.11-12.2 RI.11-12.3 RI.11-12.4 RI.11-12.5 RI.11-12.7 RI.11-12.8	21st Century Themes: Global Awareness Civic Awareness Environmental Literacy
HS-ETS-3	CCSS.Math.Content. HS-IF.B.4	Writing Standards CCSS.ELA. W.11-12.1a, W.11-12.1b, W.11-12.1c, W.11-12.1d, W.11-12.1e W.11-12.2a, W.11-12.2b, W.11-12.2c, W.11-12.2d, W.11-12.2e, W.11-12.2f	Learning and Innovation Skills: Creativity and Innovation Critical Thinking and Problem Solving Communication and Collaboration
HS-PS3-5	CCSS.Math.Content. HSN-VM.B.5 CCSS.Math.Content. HSN-VM.B.5a CCSS.Math.Content. HSN-VM.B.5b	Speaking Standards CCSS.ELA. SL.11-12.1a, SL.11-12.1b, SL.11-12.1c, SL.11-12.1d SL.11-12.2 SL.11-12.3 SL.11-12.4 SL.11-12.5 SL.11-12.6	Information, Media, and Technology Skills: Information Literacy Media Literacy ICT Literacy
		Language Standards CCSS.ELA. L.11-12.1a-b L.11-12.4 L.11-12.5 L.11-12.6	Life and Career Skills: Flexibility and Adaptability Initiative and Self-Direction Social and Cross Cultural Skills Productivity and Accountability Leadership and Responsibility

students from other countries in other contexts, and negotiating with other perspectives to author the white paper and presentation constructing an argument for one mitigation strategy (see Table 7.23).

Table 7.23 STEM Road Map – 12th-Grade Sustainable Systems: Creating Global Bonds

NGSS Performance Objectives	Common Core Mathematics	Language Arts	21st Century Skills
HS-PS3-1 HS-PS3-4	CCSS.Math.Practices MP1, MP3, MP8	Reading Standards CCSS.ELA. RI.11-12.1 RI.11-12.2 RI.11-12.3 RI.11-12.4 RI.11-12.5 RI.11-12.7 RI.11-12.8 RI.11-12.10	21st Century Themes: Global Awareness Civic Literacy Health Literacy Environmental Literacy
HS-PS3-4	CCSS.Math.Content. HSA-REI.B.3	Writing Standards CCSS.ELA. W.11-12.1a, W.11-12.1b, W.11-12.1c, W.11-12.1d, W.11-12.1e W.11-12.2a, W.11-12.2b, W.11-12.2c, W.11-12.2d, W.11-12.2e, W.11-12.2f W.11-12.3a, W.11-12.3b, W.11-12.3c, W.11-12.3d, W.11-12.3e W.11-12.4 W.11-12.5 W.11-12.6 W.11-12.7 W.11-12.8 W.11-12.9a-b W.11-12.10	Learning and Innovation Skills: Creativity and Innovation Critical Thinking and Problem Solving Communication and Collaboration
HS-ESS2-4	CCSS.Math.Content. HSA-REI.A.2	Speaking Standards CCSS.ELA. SL.11-12.1a, SL.11-12.1b, SL.11-12.1c, SL.11-12.1d, SL.11-12.1e SL11-12.2 SL11-12.5	Information, Media, and Technology Skills: Information Literacy Media Literacy ICT Literacy
HS-ESS3-6		Language Standards CCSS.ELA. L.11-12.1a-b L.11-12.2a-b L.11-12.3a L.11-12.4a-e L.11-12.5a-b L.11-12.6	Life and Career Skills: Flexibility and Adaptability Initiative and Self-Direction Social and Cross Cultural Skills Productivity and Accountability Leadership and Responsibility

Optimizing the Human Experience: Navigating Geographical Challenges

Throughout the STEM Road Map, there are several PBLs that focus on habitat conservation and natural resources. Therefore, students should have quite a bit of background knowledge from which they can draw and have a sense of the importance of these topics for future generations. In this PBL, students will create a marketing package, both electronic and paper-based, to promote the development of new luxury homes on a fault line explaining the safety features and energy conscious innovations. For example, student teams can focus their efforts on researching how new materials and new building procedures help people be prepared for inevitable earthquakes in San Francisco. Alternatively, students could focus on the patterns of forest fires in California. Students have the opportunity to learn about natural disasters or effects of climate change that they may not have otherwise known about. In their research, students will use decomposition to break down the factors involved in the geographical challenges and to reverse engineer the solution to the problem. The product for this PBL, a marketing plan, is intentionally open-ended to allow for student creativity that may result in a policy such as an emergency evacuation plan, a technology such as an app that tracks information for residents, a structural innovation for buildings or transportation, or other ways to enhance lives of people who must face natural hazards daily. An emphasis on energy needs creates a higher level of rigor for students to accomplish during this PBL. Of course, all of the innovations should be based on evidence (see Table 7.24).

Sample STEM Careers in the 12th-Grade STEM Road Map

The variety of integrated content and contexts in the PBLs taught during 12th grade continue to offer a foundation to explore a variety of careers. In doing so, teachers may want to choose a group of related careers and go through the various indicators of prospect and growth during this year because 12th graders will need to begin thinking about career sustainability for the long term to meet their life goals. On the www.onetonline.org website, there is a category of occupations called "Bright Outlook" which are expected to grow rapidly in the next several years or are new and emerging fields. A search for the "Rapid Growth" occupations, categorized by an employment increase of 22% or more over the next 10 years, yields 112 occupations, and the list of occupations can be downloaded into an electronic spreadsheet with one button click. On the website, teachers can find categories for each occupation, and for the 12th-grade level the interests and work values categories will be detailed as a demonstration of how the information can be used to support the PBLs. The interests under the Bright Outlook occupation of actuary include conventional, investigative, and enterprising. Conventional

Table 7.24 STEM Road Map – 12th-Grade Optimizing the Human Experience: Dealing with Natural Catastrophes

NGSS Performance Objectives	Common Core		21st Century Skills
	Mathematics	Language Arts	
HS-PS3-3	CCSS.Math.Practices MP1, MP2, MP3, MP4, MP5, MP6, MP7, MP8	Reading Standards CCSS.ELA. RI.11-12.1 RI.11-12.2 RI.11-12.3 RI.11-12.4 RI.11-12.5 RI.11-12.7 RI.11-12.8	21st Century Themes: Global Awareness Environmental Literacy
HS-ESS3-1	CCSS.Math.Content. HS-BF.A.1b	Writing Standards CCSS.ELA. W.11-12.1a, W.11-12.1b, W.11-12.1c, W.11-12.1d, W.11-12.1e W.11-12.2a, W.11-12.2b, W.11-12.2c, W.11-12.2d, W.11-12.2e, W.11-12.2f	Learning and Innovation Skills: Creativity and Innovation Critical Thinking and Problem Solving Communication and Collaboration
HS-ETS-2	CCSS.Math.Content. HSA-REI.C.5 CCSS.Math.Content. HSA-REI.C.6 CCSS.Math.Content. HSA-REI.C.7 CCSS.Math.Content. HSA-REI.C.8	Speaking Standards CCSS.ELA. SL.11-12.1a, SL.11-12.1b, SL.11-12.1c, SL.11-12.1d, SL.11-12.2 SL.11-12.3 SL.11-12.4 SL.11-12.5 SL.11-12.6	Information, Media, and Technology Skills: Information Literacy Media Literacy ICT Literacy
		Language Standards CCSS.ELA. L.11-12.1a-b L.11-12.4 L.11-12.5 L.11-12.6	Life and Career Skills: Flexibility and Adaptability Initiative and Self-Direction Social and Cross Cultural Skills Productivity and Accountability Leadership and Responsibility

occupations describe those careers that tend to work with data and details than with broad ideas. Investigative occupations involve searching for evidence and solving problems. Enterprising occupations involve initiating projects. Another characteristic of the job of actuary listed on the website is work values, which include the offer of job security and good working conditions, a feeling of accomplishment, and ability to make your own decisions in this career. The extensive lists and descriptions of characteristics of each occupation supplied by the Department of Labor on this website can help students decide if they would like the types of work that a particular career requires and whether there is growth, maintenance, or decline for positions in the field so that 12th graders can make informed decisions about their future in the workforce.

Summary

This chapter presented the STEM Road Map for grades 9–12 as an engaging, real-world approach to integration of core content areas for implementation in high school. With the use of the ideas presented in the STEM Road Map, instruction can be transformed into coordinated modules of instruction. These modules require teams of students to grapple with global and local challenges and problems as they master the content for their grade level. As students mature through grade levels, the instruction becomes increasingly rigorous, which requires students to develop skills and habits of mind necessary for success in future careers. The spiraling approach of the STEM Road Map is intended to equip students with the skills to be life-long learners who can think flexibly, be informed consumers of information, and be aware of possibilities for their future.

References

Card, S. (2009). Information visualization. In A. Sears & J. A. Jacko (Eds.), *Human-computer interaction: Design issues, solutions, and applications* (pp. 510–543). Boca Raton, FL: CRC Press.

Heer, J., Bostock, M., & Ogievetskey, V. (2010). A tour through the visualization zoo. *Communications of the ACM, 53*(6), 59–67.

Herman, R. (2003). *Green roofs in Germany: Yesterday, today and tomorrow.* Paper presented at 1st North American Green Roof Conference: Greening Rooftops for Sustainable Communities, Chicago IL.

Intergovernmental Panel on Climate Change. (2014). *Fifth assessment report (AR5).* Retrieved from http://www.ipcc.ch/report/ar5/index.shtml

Köehler, M. (2004). Ecological green roofs in Germany. *Journal of the Korea Society for Environmental Restoration and Revegetation Technology, 7*(4), 8–16.

Merton, R. K. (1965). *On the shoulders of giants: A Shandean postscript.* New York: Free Press.

National Aeronautic and Space Administration. (2014). *Engineering design process.* Retrieved from http://www.nasa.gov/audience/foreducators/plant-growth/reference/Eng_Design_5-12.html#.U2utz_ldWSo

Boom time for biomaterials. (2009). *Nature, 8*, 439. Retrieved from http://www.nature.com/nmat/journal/v8/n6/pdf/nmat2451.pdf

Peck, S. W., & Callaghan, C. (1999). *Greenbacks from green roofs: Forging a new industry in Canada*. Montreal: CMHC/SCHL.

Schneider, S. H., et al. (2007). Assessing key vulnerabilities and the risk from climate change. In M. L. Parry et al. (Eds.), *Climate change 2007: Impacts, adaptation and vulnerability* (pp. 779–810). Contribution of Working Group II to the Fourth Assessment Report of the Intergovernmental Panel on Climate Change. New York: Cambridge University Press.

Part III
Building Capacity for STEM

8 Data-Driven STEM Assessment

Toni A. Sondergeld, Kristin L.K. Koskey, Gregory E. Stone, and Erin E. Peters-Burton

Science, technology, engineering, and mathematics (STEM) education is by design multidisciplinary. To provide the most authentic STEM learning environment, instruction should be delivered in an integrated fashion with concepts from across these disciplines infused throughout lessons focused on 21st century skills and themes to address real-world challenges (Johnson, 2013). Assessments of STEM learning, as a result, must align with this instructional approach to elicit valid indicators of student STEM competencies. To assess STEM learning effectively, teachers must adopt and develop a comprehensive assessment system where students are given multiple opportunities to demonstrate their knowledge through varied types of assessments (National Research Council, 2014). Further, assessment data must then be used to inform STEM instructional decision-making which allows for greater student learning to occur (Black & William, 2001; Sondergeld, Bell, & Leusner, 2010).

Assessments are tools teachers can use to determine student knowledge or skill mastery at varying points during instruction. There are three main types of assessments that can be used in a comprehensive assessment system: diagnostic, formative, and summative. These assessment types differ based on time of delivery and purpose of data use. If an assessment is used for diagnostic purposes, evaluation of student knowledge is done before instruction (pre-assessment) to assess students' prior knowledge and skills and evaluate their strengths and weaknesses. The data can then be used to inform lesson planning and differentiated instruction. Formative assessments are administered during instruction to determine what students have learned over a short period of time, often the topic or lesson of the day. When assessment results are used formatively, information on student learning, or data, is used to determine gaps in student learning and remediate or plan future lessons accordingly as well as to inform students of their progress. Diagnostic and formative assessments are both used to inform instruction and provide teachers with direction on what needs to be done next instructionally to move student learning forward, thus grades on diagnostic and formative assessments should not be given as the learning process is still underway. Summative assessments, on the other hand, are given at the end of a larger learning segment (i.e. unit, chapter and grading period) and result in some form of grade to indicate what students have actually learned from instruction.

We strongly support this notion of teachers developing and using comprehensive assessment plans and the results to influence STEM teaching as prescribed by national organizations such as the National Council of Teaching of Mathematics (NCTM) and the National Science Teaching Association (NSTA). However, we also recognize that most teacher preparation programs focus more on instructional methods and less (if any) on specific assessment development and use strategies. In addition, national efforts to address the integration of engineering education do not focus on assessment. Therefore, teachers are challenged with the task of identifying existing or creating high-quality assessments to align with STEM instruction when they may or may not be fully prepared to tackle this job (Mertler & Campbell, 2005; Sondergeld, 2014). Although many districts do provide their teachers with instructional resources that come with pre-made assessments, these assessments all too often fall short in terms of quality, as they were not created by or in conjunction with assessment experts. As such, the purpose of this chapter is to present a practical guide to developing new STEM classroom assessments, and/or modifying current STEM classroom assessments to be better aligned with integrated STEM curriculum and instruction focusing on learning of complex real-world concepts and practices in order to represent student learning in a valid manner. Additionally, in this chapter, we provide guidelines and examples on how to use STEM classroom assessment results.

Standards, Curriculum, Instruction, and Assessment Alignment

Quality STEM classroom assessments need to be purposefully aligned with state standards, classroom curriculum, and instruction. One of the many responsibilities of teachers is to unpack the state standards and use them as a roadmap for developing classroom curriculum – or what is taught in the classroom. Once what is to be taught in the classroom is determined, instructional methods (or how we teach) can be decided upon. Classroom assessments then must align with the three previously mentioned components of a high-quality STEM learning environment in order to validly measure student learning of STEM content. While this relationship is discussed here in a somewhat linear fashion, all three classroom components influence each other while being simultaneously impacted by state standards and should not necessarily be completed in this order (see Figure 8.1). For example, backward design (Wiggins &

```
                    ┌─────────────────────┐
                    │   State Standards   │
                    │                     │
                    │  (Guide for STEM    │
                    │  classroom learning │
                    │     environment)    │
                    └─────────────────────┘
           ┌───────────────┼───────────────┐
           ▼               ▼               ▼
   ┌──────────────┐ ┌──────────────┐ ┌──────────────┐
   │  Curriculum  │◄►│ Instruction  │◄►│  Assessments │
   │              │ │              │ │              │
   │(What STEM    │ │(How we teach │ │(Measure of   │
   │ content is   │ │ STEM content)│ │ student STEM │
   │   taught)    │ │              │ │   learning)  │
   └──────────────┘ └──────────────┘ └──────────────┘
```

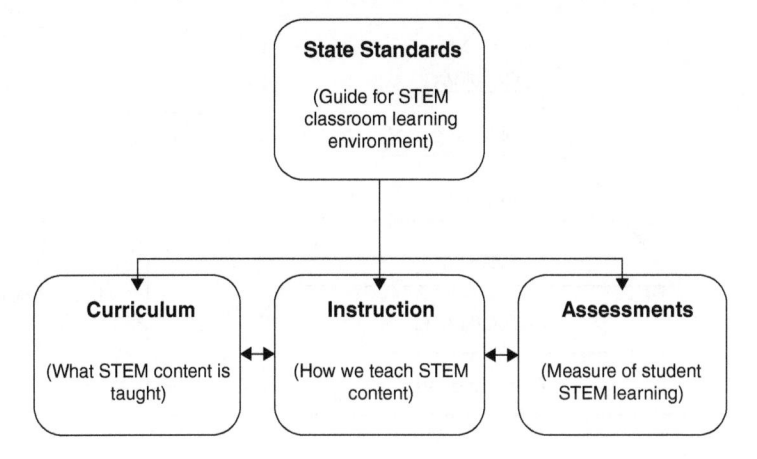

Figure 8.1 Ideal interaction between state standards and classroom curriculum, instruction, and assessments. State standards guide development of the classroom learning environment, and classroom learning environment components all influence each other.

McTighe, 1998) recommends that assessments be designed from what is to be taught, and once the "end point" is clear, activities to facilitate learning be developed.

Learning Objectives (LOs)

While states provide teachers with content standards and many districts give curricular and/or pacing guides to assist with what content to teach at each grade level, this information must be modified – or unpacked – into specific learning objectives which then allow us to directly measure varying levels of student STEM learning. Clearly defined and measurable (or observable) STEM learning objectives (LOs) are critical in linking quality classroom instruction to assessment. LOs tell us *how* we expect students to demonstrate their STEM learning and *what* STEM content we want students to learn.

We begin with an example of a clearly defined and measurable possible STEM LO: *Predicts outcome of single displacement chemical reaction*. "Predicts" tells us *how* we expect students to interact with the content and how we can observe if they do this correctly (our assessment of them). *Outcome of single displacement chemical reaction* specifies *what* content we want students to provide. A well-written, measurable

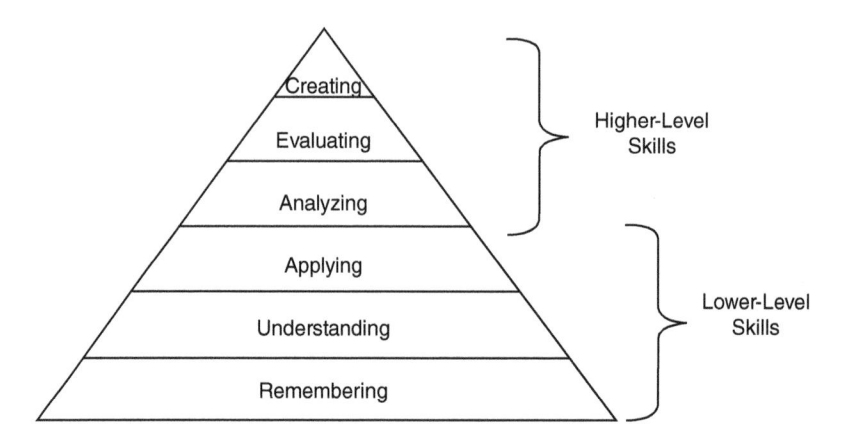

Figure 8.2 Hierarchical structure of Bloom's revised cognitive taxonomy. Lower-level skills need to be mastered before higher-level skills.

LO should be written in a straightforward manner and answer the question: *How* do I want my students to do *what*?

The *how* component of an LO also allows us to identify student *skill level* needed for mastery. In a thorough STEM classroom assessment plan, it is imperative that students are given an opportunity to be assessed at multiple levels of learning. To ensure we capture student learning of lower-level skills (e.g. ability to recall and explain content in own words) and higher-level problem-solving skills (e.g. application of content in new ways), implementing a cognitive taxonomy when developing LOs is essential.

Multiple cognitive taxonomies exist. We choose to focus on Anderson and Krathwohl's (2001) revised version of Bloom's Cognitive Taxonomy (Bloom, Engelhart, Furst, Hill, & Krathwohl, 1956) in our chapter since it is widely used in science education. Regardless of the taxonomic author, taxonomies are classification systems that can be used as a guiding framework when developing LOs. Taxonomies are hierarchical in that students need to master lower-level skills before higher skills. Further, assessment of lower-level skills typically looks different than assessment of higher-level skills because they require different levels of cognitive skill to master. Figure 8.2 illustrates the theoretical hierarchical structure of Bloom's Revised Cognitive Taxonomy.

Bloom's revised taxonomic levels, their respective definitions, measurable keywords (verbs) at each level, and a STEM example are provided in Table 8.1. To implement a comprehensive STEM classroom assessment plan, regardless of the content being covered, some degree of lower-level skills (i.e. Remembering and Understanding) and higher level skills (i.e. Applying, Analyzing, Evaluating, and Creating) should be taught and

Table 8.1 Bloom's Revised Taxonomy Defined with Key Words and STEM Examples

Taxonomic Level	Definition	Sample Key Words	STEM Examples
Creating	Putting parts together into a unique whole.	Compose, Create, Design, Formulate, Generate	Design a chamber that will act as a closed system for a chemical reaction to demonstrate the law of conservation of mass.
Evaluating	Judging the value of a product using specified criteria.	Conclude, Compare, Support, Criticize, Justify	Justify if decomposition reactions are necessary for a healthy ecosystem.
Analyzing	Breaking down material into component parts.	Diagram, Outline, Deduce, Illustrate, Discriminate	Deduce the type of chemical reactions that occur during cellular respiration.
Applying	Using previous knowledge in new and different settings.	Use, Solve, Produce, Compute, Organize	Solve for the limiting reactant that will produce the amount of concrete needed for a parking structure using the chemical reaction equations in the cement hydration process.
Understanding	Grasping the meaning of material.	Explain, Give Examples, Summarize, Paraphrase	Give an example of a common chemical reaction used in manufacturing.
Remembering	Remembering previously learned material.	Define, List, Recall, Identify	List the six types of chemical reactions.

assessed. The degree to which lower and higher-level skills is addressed in a lesson or unit will largely be determined by student grade level, cognitive abilities, and curricular content.

Unpacking Standards to Develop Measurable LOs

Unfortunately, current state and/or national content standards do not typically provide teachers with well-defined and observable LOs. Thus, teachers must unpack their state-adopted content standards in order to instruct and assess STEM learning. When unpacking state-adopted content standards to create functional LOs, there are four basic guidelines to follow:

1. **Content is not an Objective** – an action/skill stating what a student will do along with the content must be identified.

   Poorly Written Example: *Students read lab report.*
   - There is no skill here. Just because students can read the lab report does not mean that they understand what was in the report.

   Better Written Example: *Interpret lab report results.*
   - This demonstrates student learning if they can *interpret* the results.

2. **Focus on Student Behavior** – not on teacher's actions.

   Poorly Written Example: *Teach students six categories of chemical reactions.*
   - This does not indicate that the student has learned anything just because the teacher teaches.

   Better Written Example: *Distinguishes among six types of chemical reactions when given formulas.*
   - Focus is on the student. If students can *distinguish*, they can show what they have learned.

3. **Objectives are Unidimensional** – focus on only ONE concept at a time in an LO.

   Poorly Written Example: *Define and give examples of physical and chemical changes.*
   - This mixes multiple concepts into one LO making it so we cannot clearly assess student learning of this LO well.

   Better Written Example: *LO1: Define physical change. LO2: Give examples of physical changes. LO3: Define chemical change. LO4: Give examples of chemical changes.*
   - Each LO now focuses on only one concept at a time, and we can easily assess which component(s) a student has or has not mastered.

4. **Specify Cognitive Level** – This helps clarify the level that assessment items should target.

   Poorly Written Example: *LO1: List the six types of chemical reactions. LO2: Solve chemical reaction equations. LO3: Deduce the type of chemical reaction when given a formula.*

- LOs are reasonably written; however, the taxonomic level is not provided making it difficult to determine if the appropriate level of learning is occurring for students.

Better Written Example: *LO1: List the six types of chemical reactions (Remembering).*
LO2: Solve chemical reaction equations (Applying). LO3: Deduce the type of chemical reaction when given a formula (Analyzing).
- LOs now have taxonomic level identified showing that there may be a need for additional lower-level LOs to be developed and assessed in addition to the higher-level LOs.

To further illustrate the four basic principles of writing functional LOs, we draw upon the *Next Generation Science Standards (NGSS)*. Our example of unpacking *NGSS* standards comes from the Grades 3–5 Energy section.

4-PS3-4. Apply scientific ideas to design, test, and refine a device that converts energy from one form to another.

This *NGSS* standard combines multiple concepts into one standard and needs to be unpacked into specific measurable LOs in order to more clearly assess student learning of individual concepts. The following LOs offer one example of how this NGSS standard could be unpacked.

LO1: Design a devise that converts energy from one form to another (Application).
LO2: Diagram energy transfer points in devise (Analyzing).
LO3: Explain types of energy transfer occurring in devise (Understanding).
LO4: Use data collected to modify energy conversion devise (Application).
LO4: Justify whether devise transfers energy most efficiently compared to prior developed devise models (Evaluating).

Assessment Tools

Once functional learning objectives are defined, the development of a STEM classroom assessment plan can begin. A reasonable assessment plan in any STEM classroom must be multi-faceted primarily because the learning objectives that govern the classroom are themselves multi-dimensional (National Research Council, 2014). Additionally, a STEM classroom assessment plan should have both pre- and post-assessments to assess student prior knowledge and determine student knowledge growth. Assessments should be selected based on the need to appropriately measure the learning objectives in the most effective and efficient manner. The learning objectives outlined in the STEM classroom tend to reflect the full range of remembering through creating level expectations, and as a result, it is most sensible to employ a full range of objectives (e.g. multiple-choice) and self-constructed (e.g. essay response and performance assessment) item types to measure those expectations. It is incumbent on the teacher to determine (1) which item type is most effective and efficient for the learning objective being measured and (2) how to best develop the item being deployed. In this section, we review the basic decision to be made on deployment of the item type and the fundamental rules associated with writing each type of item.

Items may be divided succinctly into two major categories: objective and self-constructed. These groups depend largely on the intervention of the instructor during the grading process. Objective items may be graded with a key and require little or no grader *interpretation*. Answers are correct or incorrect when considering objective items, no grading rubrics are needed, and no human interpretation, or speculation, is required. Objective items include such item types as multiple-choice, true/false, matching, and, when no partial credit is given, fill-in-the-blank. Self-constructed items must be graded using a rubric and thus require teacher interpretation. The self-constructed category is very broad and includes traditional item types such as essays and short-answers as well as portfolio assessments, practical and performance assessments, projects, papers, and other forms of rubric-graded evaluations. The key difference in the two assessment types is teacher/grader *interpretation*. Interpretation allows for significantly deeper evaluation and understanding. It also encourages bias, distortion, and what measurement experts call error.

Error is anything that impacts accurate assessment of student ability. It is the combination of error and student ability that make up an assessment score. There are endless potential sources of error. of Students may cause error by not getting enough sleep before an assessment or feeling ill during an assessment. Teachers may also introduce error to an assessment score by administering poorly developed items, providing confusing directions, or using inconsistent grading practices.

It is important to realize that assessments in general are simply measurement devices, designed to help teachers understand what students have mastered and what they have not yet succeeded at learning. They are not meant as learning tools per se, although once completed, they may be used as such. As a result, we must treat them as measurement tools. Our goal in assessment is to maximize information (the good information about what students can do with the material) and minimize error (the bad information we cannot control, but can't entirely get rid of, in part because we can never make perfect tests). To do this, we must carefully match the needs of our learning objectives to the assessment types.

For over a century, objective items, particularly well-written multiple-choice items, have demonstrated the ability to capture a great deal of information about student skill while excluding extraneous error. Multiple-choice items get harder to write as we move higher up the cognitive taxonomy. They are easiest when written at the lower levels (remembering, understanding, applying) and very difficult and impossible in some cases at the higher levels of evaluation (analyzing, evaluating, creating). In fact, they are so difficult and time-consuming to write at the higher taxonomic levels that while they may be effective, they are not efficient. It is recommended, therefore, that teachers strongly consider the use of multiple-choice and/or other objective items when assessing LOs at the cognitive taxonomic levels of remembering, understanding, and applying, but consider alternative forms for higher taxonomic levels. Furthermore, applying is a crossover level, where both objective and self-constructed item types are appropriate.

Within this approach, which stresses the most effective and efficient model, self-constructed items, including performance assessments, are best utilized for assessing higher-level skills. The higher-level taxonomic objectives tend to be multifaceted, requiring complex thought processes, steps, and often take the student significant time to complete. They also take significant time to grade. Because of these dimensions, they are far better evaluated using a multidimensional rubric than a simple correct/incorrect key. Further, they allow the teacher to assess student thought process throughout the exercise, encouraging the evaluation of development rather than of outcome alone.

The minimization of error and maximization of student ability information require both the selection of an appropriate item type and the deployment of a reasonably well-written item. Teachers should be cognizant of the simple but clear rules for the development of each type of item. Following these rules will greatly improve the quality of information that emerges from the developed assessments and reduce much of error introduced by poorly-written items. The 20 guidelines compiled below for developing successful multiple-choice items come from our experiences analyzing high- and low-stakes tests in multiple fields of study as well as the literature (see Haladyna, Downing, & Rodriguez, 2002). Assessments written following these guidelines typically produce more reliable results that are a better indication of student skill mastery.

20 Keys to Developing Successful Multiple-Choice Items

1 **Create Plausible and Real Response Options.** When constructing the options, ensure that they are plausible and real. Ensure that the incorrect answers are reasonable but clearly incorrect and do not make up *nonsense* terms. The ultimate goal of the item is to differentiate between those who know the material and those who do not. By adding tricky or unrealistic options, it makes it difficult to differentiate in this regard as students quickly eliminate the unrealistic options.

2 **Alphabetize/Logically Order Answer Options.** By alphabetizing the options (or ordering the numbers in the options) it will ensure that

the answers are random (i.e. there will not be too many of any one particular letter).
3. **Do not Repeat Words in the Answer Options.** If the same word or words appear at the beginning of all the options, move those repeated words (e.g. "the," "a," and "an") into the main stem of the item.
4. **Answer Options should be Independent and Mutually Exclusive.** Selections in Option A, for example, should not appear in Options B or C or D. When this occurs, the items become known as complex multiple-choice items, and the functioning of the item is severely compromised. Each option should be completely unique. If asking about a list or a sequence is desired, then a form of a question other than a multiple-choice item should be selected.
5. **Avoid the Use of Negatives** (e.g. "Which of the following is not" or "All of the following are true except"). While they are easy to write, they are generally confusing and perform poorly. Research indicates that items worded negatively are very confusing and generally cause more error to be measured than ability.
6. **Do Not Teach in the Question.** Examinations are not meant as learning exercises. Let the learning occur before and after the assessment. If the material presented in the question is not directly needed for the question, exclude it.
7. **Refer to a Learning Objective.** Each item must refer to a learning objective. Ensure that all items closely match the objectives. What isn't in the objectives cannot be assessed.
8. **All Items Should Present a Question (a Problem) in the Stem.** Do not make students read the stem (the question) *and* all the options before they figure out what is being asked.
9. **Avoid Using "All of the Above" and "None of the Above."** Items with these options tend to perform poorly, causing greater errors and less information to be measured because they are often the correct answers.
10. **Avoid Biased Language and Cultural References.** Not all students come from a single community or background. Multiple-choice items *should* be biased, but only against those who do not know the material. They should not be biased based on a student's ethnic, gender, or socio-economic background.
11. **Watch the Grammar and Parallel Content of the Options.** Ensure that all options are grammatically correct (particularly if they complete the sentence in the stem of the question) and of parallel content. If the options complete the sentence, make certain each option does so in a grammatically correct fashion. Making certain the options are parallel reduces guessing. For example, if the correct answer is a noun, ensure all options are nouns.

12 **Keep the Lengths of the Options Similar.** If the correct answer is very short or very long compared to the incorrect options, students are often cued to select that answer even if they do not know the content, and this defeats the purpose of the item.
13 **Avoid Using Ambiguous Terms.** Avoid using ambiguous terms like *usually*, *often*, or *rarely*. Be specific. When ambiguous words are used, students are left wondering "how often is *often*?" Instead, specify a percentage or frequency of occurrence.
14 **Avoid Abbreviations.** Avoid abbreviations unless they are standard, should be remembered, and are printed in textbooks. For example, if students are learning about measurement and have learned that cm = centimeters, it would be acceptable to use such an abbreviation in an item.
15 **Do Not Clue the Answer.** Avoid using associations, phrases, or wording that is too similar between the question and the options.
16 **Choose the Incorrect Answers Wisely.** Create incorrect answers based on common error or misconceptions when possible. This helps when diagnosing where the specific problems are in student learning or teacher teaching.
17 **Ensure There is Only One Correct Answer.** Make sure your incorrect options are not partially correct and that your correct answer is by far the *best* answer.
18 **Testing Definitions.** When testing definitions, place the word being assessed in the question (stem) and the multiple definition possibilities as the options.
19 **Be Simple, Direct, and Concise.** Avoid presenting irrelevant information. Ask yourself, "Does the student need this information to answer the question?"
20 **Use a Straightforward Vocabulary.** Do not use $100 words when $25 words will do just fine. The language of each item should be written at the reading level of the lowest student.

Ten Key Questions to Ask When Developing Self-Constructed Assessments

It is important to remember the broad scope that the term self-constructed (or constructed response) encompasses. The purpose of self-constructed items is to allow students to apply their understanding of concepts through free expression without artificial restriction or prompting. In STEM education essays, including math problems where students are asked to "show their work," short answer problems, papers, projects, presentations, so-called authentic assessments, and portfolios are all part of the self-constructed toolkit. While each of these assessment types is clearly different, they all share common elements, including the need for clear and precise directions, and the requirements for rubric usage when grading, discussed later in this chapter. The key principles addressed in this section apply to all self-constructed assessments.

1 *Does the assessment have a clear purpose that specifies the decision that will be made resulting from the assessment?* For example, will the results be used formatively (to provide students with feedback to improve their learning) or summatively (to provide a grade for students)? Will the assessment focus on process, product, or both?
2 *Have the observable aspects of student performance of product that will be judged been identified?* Supply the performance criteria (i.e. the rubric) with the specific, observable standards by which the student performances or products will be assessed. It is preferable to limit the criteria to a reasonable and manageable number.
3 *Can you provide an appropriate setting, where applicable, to complete the task, and ensure that all students can complete the assessment?* Because self-constructed projects are themselves multi-dimensional, the scoring rubric should result in one or more scores that describe the performance.
4 *Does the assessment evaluate an important aspect of the learning objectives, requiring the student to demonstrate more than just facts, lists, definitions, etc.?*
5 *Does the assessment match the learning objectives in terms of performance, emphasis, and weight given to the assignment (e.g. number of points in the grading scheme)?*
6 *Does the assessment require the students to apply their knowledge and skills to solve new and novel problems?*
7 *When viewed in relation to the other assessments in the class, does this assessment measure new information covering the range of content and behavior specified in the learning objectives?*
8 *Is the assessment focused? Does it define a task with specific directions rather than leaving the assignment so broad that almost anything would be acceptable?*
9 *Is the task defined by the assessment within a level of complexity that is appropriate for the intellectual ability and maturity of the particular students?* Make sure the assessment is worded in a way that leads all students to interpret the assignment in the way you intended.
10 *Do the directions make clear all necessary items for completion, i.e. length, purpose, and the basis for evaluation?*

Annotated Example

You are an architect hired to create a new shopping mall.[1] Using the building materials supplied in class, create a model of a shopping mall and demonstrate how you used at least two different geometric principles in constructing the model.[2] (For instance, how would you use geometry to build a perfect square?)³ Show all equations and how they were used to create your model in a short paper (approximately three to five pages.)⁴ You and your fellow architect classmates will present your models and your use of geometry to the class next week.⁵ You will be graded on the correctness of your use of the geometric principles in constructing your model and your in-class explanation.⁶

Notes:

1   Provides the student with a *real-world*, interesting problem, grounded in activity.
2   Indicates that all students will start from a level playing field (e.g. materials supplied in class) and tells the students what they will physically do during the project. In addition, it describes and connects the principles learned with the action.
3   Provides a specific example of the physical/theoretical connection that may be explainable.
4   Provides the student with an understanding of how they will communicate part of their fundamental understanding (i.e. via a paper) and what should be included in that paper. Also provides parameters for the length of the paper.
5   Provides the student with a further understanding of how they will communicate the remainder of their understanding (i.e. via a presentation) and how that presentation should be made.
6   Offers the student insight into how the project will ultimately be graded.

Developing and Using Rubrics

A rubric is a scoring tool for a self-constructed assessment that lists the criteria for a piece of work. Rubrics indicate to all stakeholders "what counts." There are multiple purposes for using rubrics in grading self-constructed assessments. Well-constructed rubrics define *quality* for students and teachers – there is no guessing about what needs to be done to earn full credit on a self-constructed assessment. Quality rubrics allow students to accept more responsibility for their own learning and help students improve their work by using the rubric as a guide when creating and/or revising assessments. Additionally, rubrics help teachers explain *why* students received their grade on somewhat subjectively graded assessments. Most importantly, rubrics are essential for ensuring fair and meaningful results to self-constructed assessments.

Creating Different Types of Rubrics

Regardless of the type of rubric developed, there are two main characteristics rubrics must possess. First, in developing a rubric you need to determine the evaluative criteria. This means deciding which factors or skills will be assessed. Ask yourself, *what are the pieces of the puzzle that need to be graded in this STEM assessment?* Once the evaluative criteria are established, qualitative descriptions of differences need to be formed. For each criteria being graded (piece of the puzzle), a meaningful distinction between possible scores must be provided so students understand what is expected to earn full credit and teachers are able to be consistent with grading. There are two main types of rubrics: analytic and holistic. Checklists may also fall into the category of grading tool for self-constructed assessments but will not be discussed in this chapter.

Analytic Rubrics

When using an analytic rubric, each criterion (or piece of the puzzle) is graded separately. Multiple scales may be used with different point values depending on the importance of the criterion. Analytic rubrics give diagnostic information providing for formative and descriptive feedback for students to use when revising assignments or completing future assessments. However, they often take a considerable amount of time to create and may be tedious to apply. The following illustrates a student task and corresponding analytic rubric (Table 8.2).

Student Task: Describe the concepts of potential and kinetic energy. Give an example of each in your description. Write at least two complete sentences and use your best spelling (6 pts possible).
 Score: _____ /6 pts

Table 8.2 Analytic Rubric Example for Potential/Kinetic Energy Task

1 Student correctly explained terms.	
2 pts	Answer was clear and fully correct with both potential and kinetic energy described properly.
1 pt	One or more parts of the answer were nearly correct, but the student missed a key concept.
0 pts	Student failed to provide the correct answer.
2 Examples were clear and correct.	
2 pts	Both examples were appropriate.
1 pt	One example was appropriate.
0 pts	Neither example was appropriate.
3 Length	
1 pt	Two complete sentences used.
0 pts	Less than two complete sentences used.
4 Grammar/Spelling	
1 pt	Minimal errors that do not impede understanding.
0 pts	So many errors that meaning is unclear.

Holistic Rubrics

This type of rubric evaluates all criteria (pieces of the puzzle) at the same time and applies one scale across the entire rubric. Therefore, a student's score is based on the lowest competency demonstrated across all criteria. Holistic rubrics are often considered more efficient to apply when grading a large number of assessments but lack formative and descriptive feedback. If a holistic rubric is on a scale of 0–4 and a student receives a score of 2, they will not know why they received this score unless specific measures are taken to indicate strengths and weaknesses of the student's work. The following illustrates the same student task as before, but shows a potential holistic rubric for grading the assessment instead (Table 8.3).

Student Task: Describe the concepts of potential and kinetic energy. Give an example of each in your description. Write at least two complete sentences and use your best spelling.
Strengths:
Areas for improvement:

When constructing a rubric, there are some questions you should ask yourself:

- What are the learning objectives? Does the rubric align with these?
- What are the pieces of the puzzle the student is expected to provide (specific attributes to assess)?
- Should an analytic or holistic rubric be used to evaluate the assessment?

Table 8.3 Holistic Rubric Example for Potential/Kinetic Energy Task

4/A	Both potential and kinetic energy are described properly; appropriate examples of both types of energy are provided; writing is clear and well organized into two or more complete sentences.
3/B	Both potential and kinetic energy are described properly; examples of each type of energy are provided but may not be appropriate; writing is clear and organized into two or more complete sentences.
2/C	Potential and kinetic energy are described, but one description may not be completely accurate; examples may not be appropriate; less than two complete sentences are provided; writing needs editing.
1/D	Energy forms are not completely accurate; examples are not provided; less than two complete sentences are provided; writing needs significant editing.
0/F	Essay is not about kinetic and potential energy and/or so many errors in grammar and spelling make meaning impossible to interpret.

- Which criterion is the most to least important? Should these be weighted differently?
- Are all achievement categories clearly distinct, or do they overlap?
- Will the final score produce a meaningful grade representative of the student's ability level?

Sources of Error When Applying Rubrics

Recall that all assessment scores are comprised of two components: student ability and error. Failing to use rubrics in a standard and systematic fashion also adds to error in assessment results. Every assessment will undoubtedly have some degree of error associated with the score regardless of the type, and self-constructed assessments are typically worse in this regard since student scores are subjective as they are based on grader interpretation. Although we can never eliminate error, we can minimize the error by making ourselves aware of common mistakes when scoring self-constructed assessments that lead to additional error in assessment results.

Procedural flaws, or the way we use rubrics, can result in lack of consistency when scoring student work. These flaws can occur when a rubric is being used differently by multiple raters and is called *inconsistent standards*. If a student would earn an "A" if scored by one teacher and a "B" if scored by a different teacher, inconsistent standards are being applied. *Rater drift* is another type of procedural flaw leading to additional error. This drift happens when an individual rater fails to pay attention to the criteria established or changes how they grade the criteria over time.

Personal bias errors are also common when scoring self-constructed assessments. *Changes in topic and prompt* may lead to personal bias errors as the rater may like one topic more than another and resultantly give the topic higher grades. The *carryover effect* may take place when a teacher scores multiple self-constructed responses (e.g. short answer or essay items) by the same student and judgment of response to question 1 impacts judgment of response to question 2. For example, if a student answers question 1 poorly, the rater may have a bad feeling about the student's performance when grading question 2. Finally, the *halo effect* occurs when a teacher grades based on criteria that are not specified in the rubric. For instance, if a student uses good grammar or provides a very lengthy response, their score may be higher even if their response does not fully answer the item.

Scoring suggestions to keep in mind that will help reduce error are as follows:

- Converting rubrics to grades needs to be a logical process rather than a mathematical process. If using a 1–4 point scale, should a 3 on this scale be considered a 75%, which is typically a "C" in most classrooms? Or should a 3 really be viewed as a "B"?

- Score assessments anonymously to reduce personal biases.
- Score essays one topic/item at a time to increase consistency in scoring across students.
- Score subject-matter separately than grammar, spelling, and mechanics.
- Have another set of knowledgeable eyes review your rubric and give you feedback before implementing to ensure rubric is clear and aligned with learning objectives.
- Revise, revise, and revise. The first time a rubric is crafted and used it will typically not work as well as you hope it will – this is quite normal. Make sure you revise your rubric based on the flaws that you experience and try it again! Resources to assist in constructing rubrics are provided in the Appendix.

Using Assessment Results to Inform Decision-Making

The use of assessment results to inform teaching is commonly referred to as Data-Driven Decision-Making (DDDM), which is the "systematic collection, analysis, examination, and interpretation of data to inform practice and policy in educational settings" (Mandinach, 2012, p. 71). Researchers and practitioners have found using student assessment data to inform practice increases student performance (Alwin, 2002; Doyle, 2003; Peterson, 2007; Wayman, 2005). It is for this latter reason that the US Department of Education, particularly the Institute of Education Sciences (IES), emphasizes the use of DDDM at the classroom, school, district, state, and national levels for continuous improvement. After all, there is no point to invest time in administering and completing assessments if we don't use the results in some fashion beyond to determine students' grades. As the US Secretary of Education, Arne Duncan (2009), shared, "Data gives us the roadmap to reform. It tells us where we are, where we need to go, and who is most at risk. What we need to do to teach and how to teach it."

The IES outlined five helpful recommendations on *Using Student Achievement Data to Support Instructional Decision Making* (Hamilton et al., 2009) that apply across content areas including STEM education. The first two recommendations are specific to teachers' use of data at the classroom level: (1) Use assessment data as part of a continuous cycle of instructional improvement and (2) teach students how to analyze their own assessment data to set individualized learning goals.

Building a systematic process that is ongoing and cyclical so that each stage in your process informs the next stage is key to implementing the first recommendation. Many school districts now have a DDDM model outlined for teachers to follow. If your district does not provide such a model, develop a process that works for you. Six qualities should be reflected in this process as outlined in Table 8.4. These qualities are based on what Mandinach (2009) communicates are the six skills teachers

Table 8.4 What to Include in Your DDDM Process

Quality	Helpful Tips	Example
Collect data	Collect multiple forms of data	Performance Expectation MS-PS1-1 from NGSS – Develop models to describe the atomic composition of simple molecules and extended structures: use rubric for ammonia and diamonds (varying complexity) to find level of proficiency of Scientific and Engineering Practices – Developing and Using Models; measure student learning of Disciplinary Core Ideas through objective items such as multiple-choice items: different substances are made of different types of atoms, and Crosscutting Concepts with self-constructed items such as essay prompts: create a visual system to describe types of bonds; there may be overlap of constructs across assessments.
Organize data	Use a technology tool or instructional management system to assist in organizing and analyzing data	Input scores in electronic grade books, spreadsheets, or tablet applications organized by Science and Engineering Practice, Disciplinary Core Idea, and Crosscutting Concept as well as by assessment implementation and by student.
Analyze data		
Summarize data	Identify patterns across the assessment results for individuals and the class as a whole	Look across Disciplinary Core Ideas, Crosscutting Concepts, and Practices for trends in individual student learning, and across rating scales and items for validity.
Synthesize data	Triangulate the results from multiple assessments	Determine proficiency by looking across assessment and across Disciplinary Core Ideas, Crosscutting Concepts, and Practices.
Prioritize next actions	Consider the most pressing next steps to take in instruction and set learning objectives	Determine the level of proficiency for the Performance Expectation, given the Assessment Boundary. If students can develop models of varying complexity accurately, teacher can move on to more sophisticated material. If students can only develop simple models such as methane, teacher should provide scaffolding for more learning opportunities before assessing the Performance Expectation again.

Note. DDDM Process Quality indicators come from Mandinach (2009).

need in order to exercise pedagogical data literacy. These skills involve moving beyond marking how many questions students got correct or incorrect or determining a student's performance level based on rubric ratings. Central to the process is to not only collect multiple forms of assessments but also triangulate the results to make informed decisions about the next steps in instruction.

Hamilton et al. (2009) outline a four-step process that can be used as a guide for implementing the second recommendation to scaffold students to participate in the DDDM process. Analyzing, evaluating, and creating are higher-level thinking skills targeted in the STEM standards, and thus, it is reasonable to expect students to exercise these same skills to move their own learning forward. First is to share the performance criteria with the students, which communicates the student learning outcomes and highlights the most important criteria. In order to monitor their own learning, students need a clear understanding of what they are supposed to be able to do. Second is to provide constructive and efficient feedback. Feedback should not only highlight the areas of strength and weaknesses but also provide specific suggestions on *how* to improve (Brookhart & Nitko, 2015). In order for students to use feedback on objective-type assessments to make decisions about their own learning, rationales should be provided as to *why* each distractor (incorrect option) was incorrect and *why* the correct answer was correct. Oftentimes, feedback includes *what* the correct answer is but not the *why*.

The third and fourth steps relate to scaffolding students to engage in the DDDM process through providing students with tools to learn from their feedback and analyze their own assessment results (Hamilton et al., 2009). To achieve these steps, teachers should assist students in organizing their assessment results to provide a visual for them to track their growth and easily identify their strengths and areas needing improvement. For example, have students document how their answers compared to the correct answer or criterion, why the answer was incorrect or why they did not achieve the highest rating, and what goals they need to set to improve. Students are essentially engaging in their own DDDM parallel to the teacher but asking themselves, "What do my assessment results indicate about my progress towards the learning objectives?," "What are my strengths?," "What areas do I need to improve?," "What goals should I set?," "What are my next steps to meet those goals?"

The types of data you will work with in this process will vary depending on whether the assessment is objective or self-constructed. Objective-type assessments produce "correct" and "incorrect" responses to analyze. If you are comfortable in using Excel or a similar program, an item analysis can be conducted where you score each response as either correct (1) or incorrect (0) to then examine what percentage of students got each item correct and how highly related the students' performance on an item was to their overall score, along with a number of other indicators of item quality.

Another simple way to analyze objective-type assessment data is to simply create a matrix ordering the highest to lowest performing students in the left column and the items across in subsequent columns as illustrated in Figure 8.3. Organizing the data in this fashion can provide quick insight on individual student performance, the performance of the class as a whole, and which items might have unexpected response patterns that might indicate an assessment error (e.g. incorrect key, item not linked with a standard, confusing item wording) or misconception held by the students. Further, linking each item in the matrix with the student learning objective or standard aligned helps to provide a visual to detect patterns in performance related to specific standards to assist in making more targeted decisions on areas of strength and needing improvement. Creating a similar matrix for each item for each response option can further illuminate the performance of the response options such as whether the incorrect options selected at all or need to be replaced with better distractors and what incorrect response options being selected, pointing to what misconception to directly address in instruction.

Rubrics produce more detailed data as to what degree a student mastered each criterion, resulting in useful information to identify patterns in student learning. Organizing rubric ratings in a similar fashion to that illustrated in Figure 8.4 provides an efficient visual of students' areas of strengths and weaknesses to again guide more targeted decisions for the next steps in instruction. The key question to ask is whether the evidence from the assessment corroborates with the evidence from the students' other assessment results. If the results from across assessments corroborate, then you can be more confident in using that data to inform students' learning and make an action plan for the next steps in instruction.

So, what do you once you have collected, organized, summarized, and synthesized the data? The answer to this question depends on the purpose of the assessment. If the purpose is diagnostic, the data are used to inform areas of weaknesses to target and determine what instructional strategies to implement to best target those weaknesses. If the purpose of the assessment is formative, the data are used to inform if students are on track and if not, what next steps are needed to move learning forward. Considerations for each student and the class as a whole should be made. This point is optimal for students to write individual goals as they monitor their own progress or to reflect on their progress using a coding system such as Red = Can't do it, Yellow = Need Improvement or Can do it with Help, and Green = Can do it Without Help. Finally, if the purpose of the assessment is summative, the data are used to inform to what degree the students mastered the learning objectives and whether significant growth occurred from the pre to post-assessment. Summative assessment data can also be used in a formative way, however.

Data-Driven STEM Assessment   197

Student	Assessment Item (Standard Aligned)										Total Score
	1 (PS4.A)	2 (PS4.A)	3¹ (PS4.A)	4 (PS4.B)	5 (PS4.B)	6 (PS4.B)	7² (PS4.C)	8² (PS4.C)	9² (PS4.C)	10² (PS4.C)	
1³	1	1	1	1	1	1	1	1	0	1	9
2	1	1	1	1	1	1	1	0	1	0	8
3	1	1	1	1	1	1	0	1	0	0	7
4	1	1	1	1	1	0	1	0	1	0	7
5	1	1	1	1	1	1	0	1	0	0	7
6	1	1	1	0	1	0	0	0	0	0	4
7	1	1	0	1	0	1	0	0	0	0	4
8	1	0	1	0	1	0	0	0	0	0	3
9	0	1	1	0	0	0	0	0	0	1	2
Total	8	8	8	6	7	5	3	3	2	1	

Figure 8.3 Matrix of results for an objective-type assessment. Students are ordered from highest to lowest performing based on total score.

1 Some students who score low overall get Item 3 correct, perhaps indicating an issue with the item (Is the correct answer cued? Is the key incorrect?).
2 Students, including the high performers, score low on the items aligned to standard PS4.C (Information Technologies and Instrumentation), indicating re-teaching might be needed.
3 Student 1's pattern in scores reveals an anomaly on item 9 (Is the key correct? Does the item wording need improvement? Does this student need additional instruction on this concept?)

Student	Criterion (Standard Aligned)					
	Plan Investigation	Carry Out Investigation	Explain Solutions [1]	Use of Tools to Solve a Specific Problem	Use of a Variety of Methods in Scientific Investigations	Appropriate Use of Tools	Order and Expression of Length of Objects [2]
	(1-PS4-1, 3)	(1-PS4-1, 3)	(1-PS4-4)	(1-PS4-4)	(1-PS4-1)	(MP.5)	(1.MD.A.1, 1.MD.A.2)
1	3	3	3	3	3	3	3
2	3	3	2	3	3	3	3
3	3	3	2	3	3	2	3
4	3	3	2	3	2	2	3
5	3	3	2	3	2	2	3
6	3	3	2	2	2	2	3
7	3	3	2	2	2	2	3
8	3	2	1	2	2	3	3
9	2	2	1	2	2	2	3
Total	26	25	17	24	21	21	27

Figure 8.4 Matrix of results for a self-constructed type assessment. Ratings range from 1 to 3, whereby 1 = Not Evident, 2 = Needs Improvement, and 3 = Proficient.

1 Students have difficulty satisfying the criterion of explain solutions. Even the high performing students need improvement on this criterion.
2 Evidence from this assessment provides one indicator that students mastered ordering and expressing the length of objects.

For instance, the data from an end of the unit summative project can be used to inform next steps in instruction in that even though that unit is concluded, general skills might continue to need to be targeted throughout the academic year. In the example from Figure 8.4, we learned that students need additional scaffolding in explaining solutions, which is a skill that can be targeted across units.

Regardless of the purpose of the assessment, after each assessment, the data can be used to inform students' learning, your instruction, and the quality of the assessment used. When reflecting on the students' learning, ask yourself:

- "Are the students progressing?"
- "What are the students' strengths and weaknesses?"
- "Where are the students in their learning compared to where they are expected to be?"
- "Did the students' performance significantly improve over time?"

When reflecting on your instruction, ask yourself,

- "What remediation is needed?"
- "What modifications are needed in my instruction?"
- "Did I achieve my goals?"

Reflective on the quality of the assessment tool is perhaps the least discussed in the literature but is imperative to improving our assessments in terms of validity and reliability.

Questions to ask when reflecting on the quality of the tool include:

- "Does the data indicate an item needs revisions?"
- "Was the assessment the appropriate difficulty level?"
- "Did the assessment yield the intended information or does the prompt or task need revision?"
- "Was the rubric easy to apply when rating students' products or are revisions needed on the rating scale or descriptors?"
- "Did the students understand the directions?"
- "Was sufficient time provided to complete the assessment?"

Conclusions/Summary

STEM classroom curriculum and instructional strategies are becoming necessarily interdisciplinary and more rigorous to align with real-world challenges. To successfully assess STEM learning, it must be done through the use of a comprehensive assessment plan aligned with clearly defined learning objectives at various cognitive levels, incorporating multiple types of assessments, and using data to inform instructional decision-making. The alignment and integration of a STEM assessment plan will make the process of instruction significantly more fruitful and fulfilling. Using the practical strategies provided in this chapter to develop new STEM assessments or revise current STEM assessments will lead to more effective assessment practices that produce results more representative of actual student ability while minimizing error.

Resources

Resources for Creating and Implementing Rubrics

1. Factors to consider in weighting the criteria: https://www.teachervision.com/teaching-methods-and-management/rubrics/4525.html?detoured=1
2. Electronic tools for creating rubrics:

    a. Rubistar: http://rubistar.4teachers.org
    b. Google Forms:

- http://www.educatorstechnology.com/2013/10/this-is-how-to-create-rubrics-using.html
- https://docs.google.com/templates?type=forms&q=rubric&sort=user&view=public
- http://www.googlegooru.com/tips-for-teachers-using-google-forms-as-grading-rubrics/

Resource for Conducting an Item Analysis

1. Fulcher (2014) provides a free tool for conducting an item analysis at http://languagetesting.info/statistics/excel.html

References

Alwin, L. (2002). The will and the way of data use. *School Administrator, 59*(11), 11.
Anderson, L. W., & Krathwohl, D. R. (Eds.). (2001). *A taxonomy for learning, teaching and assessing: A revision of Bloom's Taxonomy of educational objectives*. New York: Longman.
Black, P., & William, D. (2001). Inside the black box: Raising standards through classroom assessment. *Phi Delta Kappan, 80*(2), 139–148.
Bloom, B. S., Engelhart, M. D., Furst, E. J., Hill, W. H., & Krathwohl, D. R. (1956). *Taxonomy of educational objectives: The classification of educational goals. Handbook I: Cognitive domain*. New York: David McKay Company.
Brookhart, S. M., & Nitko, A. J. (2015). Providing formative feedback. In S. M. Brookhart & A. J. Nitko (Eds.), *Educational assessment of students* (7th ed., pp. 153–165). Boston, MA: Pearson Education.

Doyle, D. P. (2003). Data-driven decision-making: Is it the mantra of the month or does it have staying power? *T.H.E. Journal, 30*, 19–21.

Duncan, A. (2009, June). *Secretary Arne Duncan addresses the fourth annual IES research conference.* Speech made at the Fourth Annual IES Research Conference, Washington, DC. Retrieved from http://www2.ed.gov/news/speeches/2009/06/06082009.html

Fulcher, G. (2014). *Excel spreadsheets for classical test analysis.* Retrieved from http://languagetesting.info/statistics/excel.html

Haladyna, T. M., Downing, S. M., & Rodriguez, M. C. (2002). A review of multiple-choice item-writing guidelines for classroom assessment. *Applied Measurement in Education, 15*(3), 309–334.

Hamilton, L., Halverson, R., Jackson, S., Mandinach, E., Supovitz, J., & Wayman, J. (2009, September). *Using student achievement data to support instructional decision making* (NCEE 2009–4067). Washington, DC: National Center for Education Evaluation and Regional Assistance, Institute of Education Sciences, U.S. Department of Education. Retrieved from http://ies.ed.gov/ncee/wwc/pdf/practice_guides/dddm_pg_092909.pdf

Johnson, C. C. (2013). Conceptualizing integrated STEM education. *School Science and Mathematics, 113*(8), 367–368.

Mandinach, E. B. (2012). A perfect time for data use: Using data-driven decision making to inform practice. *Educational Psychologist, 47*(2), 71–85. doi: 10.1080/00461520.2012.667064

Mertler, C. A., & Campbell, C. (2005). *Measuring teachers' knowledge and application of classroom assessment concepts: Development of the Assessment Literacy Inventory.* Paper presented at the annual meeting of the American Educational Research Association, Montreal, Quebec, Canada.

National Research Council. (2014). *Developing assessments for the Next Generation Science Standards.* Washington, DC: National Academy Press.

Peterson, J. L. (2007). Learning facts: The brave new world of data-informed instruction. *Education Next, 1*, 36–42.

Sondergeld, T. A. (2014). Closing the gap between STEM teacher classroom assessment expectations and skills. *School Science and Mathematics, 114*(4), 151–153.

Sondergeld, T. A., Bell, C. A., & Leusner, D. M. (2010). Understanding how teachers engage in formative assessment. *Teaching and Learning, 24*(2), 72–86.

Wayman, J. C. (2005). Involving teachers in data-driven decision-making: Using computer data systems to support teacher inquiry and reflection. *Journal of Education for Students Placed at Risk, 10*, 295–308. doi: 10.1207/s15327671espr1003_5

Wiggins, G., & McTighe, J. (1998). *Understanding by design.* Alexandria, VA: ASCD.

9 Sociotransformative STEM Education

Alberto J. Rodriguez

Sociocultural and Institutional Factors Affecting Diverse Students' Engagement and Achievement

It has been well established that there are many institutional and sociocultural factors that obstruct culturally diverse students' access and success in our schools. In a 2004 monograph, *Turning Despondency into Hope: Charting New Paths to Improve Students' Achievement and Participation in Science Education* (Rodriguez, 2004), I describe in detail many of these factors. Sadly, the same issues still negatively impact students and the professional lives of teachers today in spite of all the advances we have made in educational research. In the last 60 years we have accumulated a great deal of research and generated many insights for what needs to be done to enhance teacher preparation and increase students' engagement and achievement (Bianchini, Akerson, Barton, Lee, & Rodriguez, 2013). We also know that when there is political will, strong and supportive leadership, and a collaborative environment focused on student success, even schools with limited resources begin to show significant progress (Rodriguez & Zozakiewicz, 2010; Rodriguez, Zozakiewicz, & Yerrick, 2008). Due to space limitations, I will only highlight some of the institutional and sociocultural factors that influence student achievement and provide some suggestions for addressing them.

Institutional Factors

Standardized Testing

There is no question that the punitive accountability that started with the *No Child Left Behind Education Act* (The White House, 2001) and then followed by the *Every Student Succeeds Act* (*ESSA*, DOE, 2015) continues to be a factor aggravating the educational opportunities of all students and driving the professional lives of teachers. This situation is particularly odd because the current Federal Government has not proposed an educational act of its own, and all it has done since 2017 is to transfer the ESSA accountability rules to the states. While none of these educational acts were based on sound educational research, and while a great deal research continues to show the detrimental impact of punitive and mandated testing, science and social studies continue to be pushed aside in elementary schools to make more time for drill and practice before mandated language arts and mathematics testing. Many of us have been denouncing this situation for years, and it was comforting to see this issue taken up by the very committee that developed the conceptual framework for the Next Generation Science Standards:

> Over the past decade, accountability pressures—generated by the focus on student achievement as measured by high-stakes assessments—have heightened the curricular emphasis on mathematics and English/language arts and lowered attention to (and investment in) science, art, and social studies—especially at the elementary school level. In another California study—this one involving elementary school teachers in nine San Francisco Bay area counties—participants indicated that science is the subject area in which they felt the most need of professional development [21]. They also reported that they taught science *less than one hour per week on average* across the elementary school grades—with science instruction being more prevalent in the upper elementary grades than in the K-2 grade band.
> (National Research Council, 2012, p. 282, emphasis mine)

Even the National Science Teachers Association (NSTA, 2018) finally released an official statement regarding the need to provide science instruction equal attention as other elementary school curriculum subjects. Although this acknowledgment is rather subtle and late (seven years after the aforementioned NRC's statement), this is a welcome action toward drawing more attention to the maddening contradictions of having new science standards that call for more integration of science and engineering practices, problem-solving, and creativity, and on the other hand, using high-stake multiple choice science tests to measure growth. Even more puzzling, it is the expectation for students to be well prepared for those mandate science tests while *not* being exposed to science mainly at the K-3 levels because science (and social studies) are pushed aside due to drill and practice of mathematics and language arts tests.

This damaging trend is having a reprehensible impact on the future of science education in this country as thousands of students are denied access to the joy of learning science and to better their understanding of the natural world around them (Rodriguez, 2015a). While the emphasis on standardized testing over learning for understanding is not going away any time soon, some schools have shifted their cultures to focus on students' needs first and not on tests. Through this approach, these schools have sought to focus on providing support to students and their parents and on teacher professional collaboration. The report, *Why Some Schools with Latino Children Beat the Odds and Others Don't* (Waits et al., 2006), is one of several studies that document transformative findings. Similarly, recent studies illustrate how a dynamic and culturally relevant leadership at the school level (McLaughlin, 2020) and culturally and socially relevant teacher development can help overcome social inequalities and deficit thinking (Darling-Hammond, Hyler, and Gardner, 2017; Rodriguez, 2015b). Again, success can be found if we seek to apply what we already know from educational research and if we allow teachers to apply what they learned in their professional programs with a focus on students and not on teaching to the test.

Class Sizes and Access to Equipment and Materials

There have been many reports calling for a reduction of class sizes, especially at the elementary school level, but these essential calls for reform go unheeded. For example, almost three decades ago, the National Commission on Teaching and America's Future (1996) proposed radical re-structuring of our schools to increase the number of teachers, reduce the average class size, reduce the number of other staff, and increase planning time for teachers. It is unfortunate that significant recommendations for enhancing students' learning and enriching teachers' professional lives are only taken up by other countries, but not here. If we closely observe what is happening in other countries like Finland, we can gain new insights. For years, Finnish school children were outperforming the United States (and most other countries) on the Programme for International Student Assessment's (PISA) reading, mathematics, and science international tests (Programme for International Student Assessment, 2018). Finnish educators attributed their success on research-based pedagogy and curriculum centered on the child's emotional and cognitive growth. Finland has no mandated tests (except for one at the end of high school), no rankings, or competition among schools or students, but they do have smaller class sizes, well-prepared, supported, respected, and well-paid teachers; and government officials in charge of education who are actually educators themselves (Hancok, 2011). Interestingly, even though Finland's scores have been dropping since 2015, Finland still outperforms the United States. While there is much debate about the reasons for Finland's decline in performance, one major factor could be the country's challenged economy and the increasing challenges (and opportunities) of meeting the needs of new immigrants, refugees, and other displaced peoples due to political turmoil.

Now, returning to the US context, with the advent of the *Next Generation Science Standards* (NGSS) (Achieve, 2013) and the emphasis on the integration of science with engineering practices, one cannot avoid the questions: How are schools (especially in economically disadvantaged neighborhoods) going to secure the equipment and materials to carry out hands-on engineering activities as expected by the NGSS? How can these integrated STEM activities be carried out in commonly large classes? Where are teachers going to receive the necessary professional development? (Rodriguez, 2015a).

Weiss, Banilower, McMahon, and Smith (2001) conducted a national survey with approximately 6,000 teachers, and they found that most teachers saw the lack of appropriate resources and equipment as serious issues influencing the teaching of science and mathematics. In fact:

> inadequate funds for purchasing equipment and supplies was labeled as a serious problem by 25–35% of the respondents [teaching at the elementary, middle and high school levels], inadequate facilities by 20–28%, and lack of materials for individualized instruction by 16–27%.
> (p. 101)

Most teachers also stated that they did not have time to plan and/or discuss issues related to the teaching of science and mathematics. This study was replicated in 2013 with 7,752 teachers, and additional questions were included to assess teachers' conceptions of preparedness to teach science and engineering practices (Banilower et al., 2013). The researchers found that only 4% of participating teachers "felt very well prepared to teach engineering" and only 39% "felt very well prepared to teach science." The survey was administered again four years later with the participation of 7,600 teachers (Banilower et al., 2018), and the results were strikingly similar. Only 31% of the participating teachers "felt very well prepared to teach science," and 3% "felt very well prepared to teach engineering."

According to Banilower et al.'s (2018) report, "Elementary science teachers stand out for the relative paucity of professional development in science or science teaching, with fewer than about 60 percent having participated in the last three years" (p. 47).

In an extensive critique of the NGSS, I argued that these new set of science standards will most likely fail to have the kind of impact expected unless explicit steps are taken for securing appropriate funding for professional development, equipment, and materials, reducing class sizes and for addressing issues of equity and diversity more consistently (Rodriguez, 2015a).

Sociocultural Factors

Parent Involvement

Even though parent involvement in their children's education has been associated with increased student achievement (Hill & Tyson, 2009), there are very few studies that focus on increasing parent involvement in science education. We are conducting a review of significant reports and metastudies in order to identify "what works" and draw common strategies that could be applied and further investigated in the science classroom context. So far, it has been interesting to realize that taken-for-granted assumptions about what constitutes effective parent involvement must be dispelled. For example, traditional forms of parent involvement, such as participating in school activities, or assisting with homework, do not have as large an impact on student achievement as just simply parents having high aspirations for their children. That is, when parents make explicit to their children that they wish them to do well and to stay in school, this has a more significant impact on student achievement (Rodriguez, Collins-Parks, & Garza, 2013). Similarly, it has been consistently shown that peers and siblings play a significant and positive role on academic achievement. Horn (1998) conducted a comprehensive analysis using data from the *National Education Longitudinal Study*, which originally involved the participation of 25,000 eighth-graders. At-risk students whose peers expressed a strong interest in learning activities had 70% higher odds of pursuing higher education in four-year colleges and almost 2.5 times the odds of enrolling in any post-secondary institutions. In addition, students at risk who reported that their friends planned to attend college had six times higher odds of doing the same.

There are several other strategies for increasing parent involvement, and subsequently, student achievement that could be easily transferred to STEM education contexts (Jeynes, 2007). What we need then is the political will to start enacting insights drawn from educational research in general and further investigate what works in STEM education contexts specifically.

Second-Language Learners

The cultural and linguistic diversity of the United States continues to increase dramatically. The US Census Bureau (2012) projected that the Anglo student population will become a minority as early as 2022. Some school districts are realizing that the cultural and linguistic make up of their schools can also change very rapidly as families seek better employment opportunities in non-urban areas as well. More than ever, teachers face challenging demands – the pressure of standardized testing, new sets of standards with a focus on engineering and scientific practices, and continuing low material and professional support for meeting the needs of an increasing language learners' population. On the bright side, a growing body of research highlights various pedagogical strategies that have a significant impact on English Language Learners' (ELLs) achievement and engagement. While class sizes, standardized testing, and lack of professional development and material support are major institutional factors that must be addressed as discussed above, some of these pedagogical strategies provide windows of success for teachers and students. Buxton and Lee (2014) reviewed various studies that showed that at the core of increased achievement and participation of ELLs was higher teacher expectations, responsive support, and using the students' cultural backgrounds as resources in the classroom. In other words, instead of using a deficit approach to work with ELLs, Buxton and Lee (2014) found that "learning to recognize and value diverse views of the natural world can simultaneously promote academic achievement and strengthen ELLs' cultural and linguistic identities" (p. 208).

Another important myth that must be dismissed when working with ELLs is the notion that they must develop specific language skills in English before being exposed to more challenging science instruction. Buxton and Lee (2014) found several studies that showed that a focus on hands-on and minds-on instruction with an emphasis on academic and language literacy development in English benefits ELLs. My own research in culturally diverse classrooms and that of many others confirms this (Rodriguez, 2015b; Rodriguez & Zozakiewicz, 2010; Rodriguez, Zozakiewicz, & Yerrick, 2008; Tolbert, Knox, & Salinas, 2019). We also found that the same hands-on, minds-on inquiry-based activities increased engagement and scientific discourse in the classroom for all students. Students even learned pedagogical discourse; that is, we were surprised to learn during focus group interviews that some students tended to accurately name the pedagogical strategies we used when describing what they found most useful and engaging (e.g. concept mapping; predict, observe, and explain; problem-solving scenarios).

Teacher Resistance to Pedagogical and Ideological Change

Two major factors influencing any progress we might eventually make on STEM education in the United States remain seldom acknowledged: teacher resistance to pedagogical change and resistance to ideological change (Rodriguez, 2005). Resistance to ideological change has to do with an individual's inability to change his/her beliefs and values systems in response to specific social contexts. For example, some pre- and in-service teachers believe in a kind of "rugged individualism" that has worked well for them and their families as members of the predominant culture. Through this ideological lens, they believe that if students from diverse backgrounds spoke English only, "worked hard enough," or had "caring parents," they would do well in school (Rodriguez, 2015b, 1998). This is just one example of many, but it is enough to appreciate that a teacher could have the best preparation in learning theory, content, and pedagogy, but if he or she has not been well prepared to be a more culturally inclusive, respectful, and responsive teacher, this individual would likely not be able to establish a productive professional relationship with students and their parents. The other side of this coin is resistance to pedagogical change. This is defined as the resistance to changing one's perceptions of what constitutes being an effective teacher in today's schools (Rodriguez, 2005). Thus, a pre-service teacher who has mainly been exposed to traditional and transmissive pedagogy for 12–16 years and then is exposed to student-centered, hands-on, culturally relevant pedagogy for 15 weeks in a science methods course – while observing regular teachers implement transmissive approaches during student teacher – is most likely to end up mimicking what appears to be "the safest practices." I have observed this pattern throughout my career as a teacher-educator and researcher in all the universities where I have taught. While resistance to pedagogical change may be by choice (e.g. seeking not to antagonize whatever relevant practice might exist in a particular school), or lack of pedagogical knowledge and practice (e.g. fear of losing control of students during hands-on activities), one cannot blame novice teachers (Rodriguez, 2015b, 2002). We must ask instead teacher educators, policy makers, and school district administrators why are we not using what we have learned from over six decades of research? Why are we still preparing teachers in contradictory contexts? That is, what pre-service teachers learn in methods courses is not what they observe during student teaching, and it is likely not what they will be able to implement once they graduate (given the institutional and social factors described thus far). If we are truly interested in having an impact on teachers' practice, on school administrators, and on policy makers, as well as in making the general public more aware of the importance on culturally and socially relevant teaching and learning, we must conduct more systemic research. That is, studies that critically and purposely connect insights gathered from cross-cultural education research with those gathered from social constructivist research. Findings from these types of projects may help us develop a collective sense of direction for establishing meaningful change at multiple levels. Toward this end, I describe next an alternative framework, sociotransformative constructivism (sTc), that I have found useful in guiding my work with teachers and their students in culturally diverse contexts.

Moving toward Sociotransformative STEM Education

sTc is a theoretical framework that merges social constructivism (as a theory of learning) with critical cross-cultural education (as theory of social justice) (Rodriguez & Morrison, 2019; Rodriguez, 2015b, 1998). While findings from these two fields of inquiry continue to be presented as separate and disconnected as discussed earlier, sTc argues that the individual's cultural, social, historical, and academic locations cannot be separated from the what (curriculum), how (pedagogy), why (policies), for whom (students), and by who (teachers). It is this artificial disconnect amongst the individual, context, teacher, curriculum, and policies that generates one of the principal factors perpetuating the achievement gap and the lack of impact of research on teachers' practice and on students' learning.

According to sTc, learning to teach for diversity and for understanding can be accomplished by enacting four interconnected components: The dialogic conversation, authentic activity, metacognition, and reflexivity. Due to space constraints, these terms will only be explained briefly. The dialogic conversation uses Bakhtin's (1986) notions of speech genre. This involves engaging in a deeper kind of dialogic exchange through which the goal is to understand not only what is being said but also the reasons (emotional tone, ideological, and conceptual positions) the speaker chooses to use in a specific context. Thus, developing trust amongst teachers and students is paramount to establishing a productive learning community in which students' identities and cultural experiences are valued. The next component of sTc is authentic activity. Just as the name implies, this aspect involves hands-on, minds-on activities that are also socio-culturally relevant and tied to the everyday life of the learner. The third element is metacognition. This term is defined as the learner's awareness and control of how he or she learns (Idol and Jones, 1991). Thus, metacognition can be used as a powerful tool to encourage learners to become more reflective about each other's preferred learning patterns, and how these interact in preventing or assisting in learning new concepts. The final element, reflexivity, involves becoming critically aware of how one's own cultural background, socioeconomic status, belief systems, values, education, and skills influence what we consider important to teach/learn. Through reflexivity, one becomes more aware of how issues of power determine who has access to education and to better opportunities in life and the role each one of us plays in enriching a pluralist society. This aspect of sTc is particularly useful in today's schools because learners are also urged to reflect on the social and cultural relevance of what they are being asked to consume and/or produce as knowledge as well as the role students could play in advancing knowledge (Brown, 2017).

sTc has been used successfully in various projects involving grades 4–12 pre- and in-service teachers and their students (Rodriguez, 2015b, 2010, 2008), and a teacher practice observation protocol is being developed that should help scale up teacher professional development projects using the sTc framework. While sTc is not being suggested as "a theory of everything," this alternative approach illustrates how social constructivism and cross-cultural education could be merged purposely and critically to teach for understanding and for diversity. Given the current interest for a more integrated curriculum and the emphasis on STEM education as required by the NGSS, sTc might provide an effective vehicle for making STEM education more culturally and socially relevant.

210　*Alberto J. Rodriguez*

By using this approach, students are exposed to real-world applications of technology and scientific tools to carry out meaningful problem-based activities. These activities could have a local, national, or global focus, but they will always be firmly grounded on students' interests and sociocultural contexts.

The next section provides a brief description of how a popular activity can be altered to enact sociotransformative STEM education. The various elements of sTc explained above are described in parentheses in the following narrative to facilitate understanding.

Making Ice Cream to Teach Sociotransformative STEM Education

Ice cream making is a popular activity carried out in classrooms at any level; however, this hands-on activity is usually done by using a recipe-centered, minds-off approach, with little connection to understanding science content and STEM practices. Below, I describe how this activity can be done to integrate sociotransformative STEM education.

Before instruction, the teacher organizes the classroom into three learning centers: Science and Mathematics, Engineering and Technology, and Analysis and Write up. The number of centers is dependent of the size of the class, and students will rotate to each center. It would be ideal for students to complete the science and math center first, but the reality of most schools will prevent this from happening due to limited access to lab equipment and supplies. Therefore, we have found that learning centers are a powerful pedagogical strategy for maximizing resources and keeping students engaged.

Next, students are organized in teams of three according to mixed ability, cultural diversity, and same sex (note that gender is different than sex. Sex is a biological construct [male, female, or both]; whereas gender is a social construct and represented in a multiple ways). Same sex grouping is an excellent way to encourage girls to get more involved in science and manipulate equipment. Also, by placing girls who are more assertive with other girls who are not creates opportunities for intergroup modeling (*Reflexivity*: paying attention to issues of equity and diversity in your specific context).

Begin the activity with a Predict, Observe, and Explain (POE) to activate students' curiosity and prior knowledge. Ask the students to discuss in their teams what would happen when a half cup of milk, one half teaspoon of vanilla, and one tablespoon of sugar are added into a pint size zip-lock bag (do not actually add the ingredients yet – allow students to visualize). Similarly, ask students what would happen when you add two cups of ice and six tablespoons of salt into a one-gallon zip-lock bag. After students make some predictions, ask what will happen to the milk, sugar, and vanilla mixture when the pint-size bag is added to the gallon size bag (containing the ice and salt mixture). Students must write their predictions on their POE sheet including arguments to support their predictions. Encourage students to be *MetaThinkers* and remind them to look at a previously made poster on the wall that has the following questions: *How did you come up with that idea? Tell me more about what you were thinking. Show me what you mean.* (*Metacognition*: by reminding students about the MetaThinkers poster, the teachers are encouraging students to reflect on how they and their partners learn and how they come up with ideas and arguments to support their thinking. *Dialogic Conversation*: By organizing students in mixed ability, ethnicity, and same sex, the teacher is promoting opportunities for students and the teachers to share their cultural experiences and learn more than just STEM content. For example, a student from the east coast studying in southern California may share that he has seen salt trucks come out during heavy snowfalls and that salt melts the ice. Other students from rural areas may share that they have seen old-fashion ice cream makers at county fairs and so on. Allowing students to bring their prior knowledge and experiences in a supportive environment of trust enables productive dialogic conversations).

After listening to the students' predictions and arguments, allow students to conduct the activity making sure at least one person will be in charge of recording observations. If the school has access to probeware, such as the Vernier CBL units and temperatures probes (http://www.vernier.com), this technology is an excellent way to demonstrate in real time, and graphically, the dramatic changes in temperature as salt is added to the ice and as the smaller bag with the vanilla, sugar, and milk mixture is placed in the large bag. The teacher should work with one group of girls using either regular thermometers and plotting the change in temperature or with the Vernier probes. This data will be used for discussion later. For this example, let's assume that we are using Vernier probes and that one temperature probe was placed inside the milk mixture and the other was placed in the ice/salt mixture. By connecting the CBL unit to a laptop and projecting the changes in temperature for the whole class to see, students are often shocked to discover how dramatically the temperatures drop in both bags, but a lot more in the salt and ice mixture. This activity also creates a discrepant event – a phenomenon that is opposite to what is commonly known or accepted. In other words, the teacher should ask why is it that the temperature of the ice and salt mixture is below the freezing point, yet it has turned into a liquid. How is it possible that it is so cold that the milk mixture turned into a solid (ice cream), yet the salt and ice mixture is a liquid? These questions and the graph showing the changes in temperature on the screen generate a lot of debate. [*Authentic Activity*: Students are carrying out an authentic inquiry activity similar to the work that scientists do. While they were following a given procedure at first (for making ice cream), now they have generated their own questions for investigating further and seeking to gather evidence to support their hypothesis].

The teacher should print a copy of the graph created with Vernier probes (or with the regular thermometers) for each group and allow students to discuss their observations (while they enjoy their ice cream). Students in the Science and Mathematics Center will be required to come up with a hypothesis to explain what they saw and prove it by conducting their own experiment. For example, we have found that some children think that the fat in the milk has something to do with this phenomenon. Others believe that the salt makes the ice "colder," so they design experiments that remove the salt variable. It is important to note here that by this point, students are being encouraged to use scientific discourse. New terms such as mixtures, variables, hypothesis, chemical change, freezing point, and physical change can now be explained "in use." In other words, using Dewey's approach that we learn best by doing, students can better appreciate what these important terms mean through direct experiences. Also, at this juncture, the teacher should point to a large poster of a Word Wall that includes key terms in English and Spanish. The Word Wall is pedagogical strategy that ELLs find very useful. In our project classrooms, students become used to writing key terms and definitions in their science journals without being told to do so. (*Authentic Activity, Dialogic Conversation, Reflexivity*). Most of the teachers with whom we work explain that they always introduce all the key terms and concepts first before doing an activity. They feel that students must have a "foundation" first before they can understand what is expected of them. We argued that this is a transmissive approach that assumes students come with no prior knowledge or experiences into the classroom. Also, we ask our participating teachers to consider how this traditional approach tends to make science really boring and detached from students' lives. By the time teachers finish lecturing and asking students to write down definitions, students are so uninterested that it becomes difficult to capture their interest again. We have found that the approach described herein keeps students excited and engaged in scientific discourse.

In the Science and Mathematics Center, students are also being encouraged to integrate mathematics computation and concepts by asking them to be aware of the units of measurement. Since the United States is one of three countries in the world that still uses the Imperial System of Measurement, and all children who might come from different countries will know only metric, the activity includes a requirement to use measurements only in the system the student knows least. In addition, by asking students to closely examine and interpret the graph created, students use higher-order thinking and apply these insights for developing their hypothesis and arguments. Again, students are encouraged to be MetaThinkers throughout the whole process so that they can better understand how they and their peers construct knowledge (*Metacognition, Authentic Activity*).

Students are allowed to test their hypothesis by carrying out their own experiments. The teacher can have groups of students rotate to the center that has the Vernier probes or thermometers, depending on what is available. In our projects, elementary to high school students quickly become quite proficient in the use of the Vernier probeware and require very little assistance (Rodriguez, 2015b).

In the Engineering and Technology Center, students are required to invent an ice cream-making device with the following constraints: (1) The device must be environmentally friendly – no gas or electricity power; (2) no hard hand cranking like the old-fashioned ice cream machines and cannot allow their hands to get cold like in the previous activity; (3) the device must produce enough ice cream for three people (1/2 cup or 120 ml for each); and (4) the device must be cost-effective; i.e. yield a profit so that the proceeds could be donated to the Heifer International Project for Ending Hunger and Poverty (http://www.heifer.org). The top three devices which best met the given criteria will be selected for a school-wide fundraiser. Given the space constraints, the details of this activity cannot be described here. However, one possible example is students modifying a bicycle so that the back wheel can be used to rotate a large coffee can. Inside this coffee can, a smaller coffee can (containing the milk, vanilla, and sugar mixture) is placed, and the ice and salt is poured around it. The challenge students might face is deciding whether the large can should be attached horizontally or vertically to the wheel and what kind of gears must be designed to meet the job. This is an excellent opportunity to work with a local business specializing in gear manufacturing (e.g. www.oerlikon.com/fairfield/en). The handout for this activity (see link) includes a table to assist students in integrating their mathematics skills to figure out the new recipe and estimate costs. Students are also given a graphic organizer of the engineering design process so that they can keep in mind how engineering is iterative, collaborative, and guided by certain constraints that must be met to create a successful product.

Students are of course required to first draw their design and consider all possible options. They are urged to involve their parents and siblings in their projects as well as members of the business community to assist in the fund raising event. [All elements of sTc are included in this center, but it is important to highlight metacognition and reflexivity here. Students are again asked to be MetaThinkers and carefully listen and probe each other's thinking to better understand how they individually and collectively construct new knowledge. In addition, through reflexivity, students are made aware of their privilege position; that is, essentially "playing" with a source of food we often take for granted. By engaging in a dialogic conversation and helping students understand that many people in the United States (46 million or 1 in every 6) live in poverty, the class could discuss ways to help address this issue locally and/or globally. One approach is contributing to the Heifer International Project through the proposed fund raising activity. The goal is to help students recognize that they have agency and power to effect positive social change individually and/or collectively. It is important to note that the proposed engineering and fund raising projects are real. Too many "engineering" activities involve pretend projects (e.g. build a bridge) in artificial contexts (e.g. for the poor people in X-country). These approaches trivialize other people's real struggles and fail to acknowledge their own efforts to effect change for themselves. In addition, "pretend projects" do nothing to help students recognize and develop their potential as agents of change (Brown, 2017; Rodriguez, 2017; Tolbert, Schindel, & Rodriguez, 2018)].

In terms of making this activity more culturally inclusive and relevant, I have already mentioned how the groups were organized, how the STEM content was set up to be more hands-on, minds on, and inquiry-based, and how students' choices and voices were included as they select their own experiments and engineering designs. In addition, this activity can be made more culturally inclusive by the teacher adding a brief interactive discussion on the history of ice cream while introducing the learning centers component of the activity. This is also an excellent place to further contextualize the activity and use students' prior knowledge and cultural experiences. For example, the teacher should ask what experiences students have with ice and ice cream making. The teacher should show a picture of Nancy Johnson, a woman from New York who first invented the ice cream machine in 1843, and whose design is still used in modern versions today. There are several websites with (often conflicting) information about the history of ice cream that should be shared with students. They should also be encouraged to investigate these websites and decide which are the most reliable sources and why. Finally, the Analysis and Write Up Center is simply a space to where students can rotate to continue investigating answers to their questions and figuring out where to gather the resources they need to test their engineering design.

In terms of the STEM content knowledge covered during the activity, the reader would note that main concepts are not "lectured" to students, but *experienced* in use. After the students have tested their experiments, the teacher could explain that the freezing point depression of water is due to the salt. This means that the physical property of water to always freeze at 32°F/0°C no longer applies because it is not just water anymore (it is a mixture of salt and water). If students are high school students, the teacher could explain the thermodynamics of water molecules and how they interact with the sodium and chlorine ions to lower the freezing point of water. Teachers could also choose to discuss the states of matter and/or the nature of science with this activity. Regarding mathematics concepts and skills, students graph and interpret data, conduct unit conversions, and make estimates and various other computations. The engineering process is enacted with their design and construction project, and technology is also integrated with the Vernier probes to gather and interpret data. In addition, technology is created to make their devices work (like the gears or connector needed to attach the coffee can to the bicycle wheel). In short, this activity allows teachers to stress various STEM concepts according to their desired learning objectives.

Conclusion

In this chapter, some of the key institutional and sociocultural factors that continue to obstruct equal opportunities for the access and success of culturally diverse students were highlighted. In the last 60 years, we have gathered a great deal of insights from educational research that remains unheeded by policy makers and administrators. While politics and political slogans seem to drive national educational policies, researchers are partially to blame as we continue to mainly write and publish our work for own community of practice and not for those on whom we base our work – teachers and students.

Regardless of what framework(s), we might end up choosing to guide educational research, teaching, curriculum, and/or policy, one fact is certain. We cannot afford to continue responding to pervasive inequalities in our increasingly culturally diverse schools with well-intended policies, political slogans, or with research that has no impact on teaching practice or on student learning (Tolbert et al., 2018). It is imperative that we find our way out of this dangerous quagmire – we need a sense of collective direction, and we need a compass. sTc is one possible framework amongst others that may provide us with a common sense of purpose to systematically integrate cross-cultural education with social constructivism. In this way, we could simultaneously tackle the achievement gap and rekindle students' excitement about STEM education connected to their everyday lives.

References

Achieve, Inc. (2013). *The next generation science standards.* Retrieved March 2014 http://www.nextgenscience.org
Bakhtin, M. M. (1987). *Speech genres and other late essays.* Austin, TX: University of Texas Press.
Banilower, E. R., Smith, P. S., Malzahn, K. A., Plumley, C. L., Gordon, E. M., & Hayes, M. L. (2018). *Report of the 2018 NSSME+.* Chapel Hill, NC: Horizon Research, Inc.
Banilower, E. R., Smith, P. S., Weiss, I. R., Malzahn, K. A., Campbell, K. M., & Weis, A. M. (2013). *Report of the 2012 national survey of science and mathematics education.* Chapel Hill, NC: Horizon Research, Inc.

Bianchini, J. A., Akerson, V. L., Barton, A., Lee, O., & Rodriguez, A. J., (2013). *Moving the equity agenda forward: Equity research, practice, and policy in science education*. New York, NY: Springer.

Brown, J. (2017). A metasynthesis of the complementarity of culturally responsive and inquiry-based science education in K-12 settings: Implications for advancing equitable science teaching and learning. *Journal of Research in Science Teaching, 57*, 1143–1173.

Buxton, C., & Lee, O. (2014). English learners in science education. In Lederman, N.G. & Abell, S.K. (Eds.), *Handbook of research on science education* (2nd ed., pp. 204–222). New York: Taylor Francis.

Darling-Hammond, L., Hyler, M. E., & Gardner, M. (2017). *Effective teacher professional development*. Palo Alto, CA: Learning Policy Institute. This report can be found online at https://learningpolicyinstitute.org/product/teacher-prof-dev

Hancok, L. (2011). Why are Finland's schools successful? The country's achievements in education have other nations, especially the United States, doing their homework. *Smithsonian Magazine* downloaded May 2020. https://www.smithsonianmag.com/innovation/why-are-finlands-schools-successful-49859555/

Hill, N. E., & Tyson, D. F. (2009). Parental involvement in middle school: A meta-analytic assessment of the strategies that promote achievement. *Developmental Psychology, 45*(3), 740–763.

Horn, L. (1998). *Confronting the odds: Students at risk and the pipeline to higher education*. Washington, DC: National Center for Education Statistics.

Idol, L., & Jones, F. (1991). *Educational values and cognitive instruction*. New York: Erlbaum Associates.

Jeynes, W. H. (2007). The relationship between parental involvement and urban secondary school academic achievement: A meta-analysis. *Urban Education, 42*(1), 82–110.

McLaughlin, D. V. (2020). Exemplary leadership in diverse cultural contexts. In R. Papa (Ed.), *Handbook on promoting social justice in education* (pp. 2–16). Netherlands: Springer. https://doi.org/10.1007/978-3-319-74078-2_77-1

National Commission on Teaching and America's Future. (1996). *What matters most: Teaching for America's future*. New York: Author (www.nctaf.org).

National Research Council. (2012). *A framework for K-12 science education*. Washington, DC: National Academy Press.

National Science Teaching Association (2018). Position Statement: Elementary School Science. Retrieved [11/20/2020] from https://www.nsta.org/nstas-official-positions/elementary-school-science

Programme for International Student Assessment. (2018). *Organisation for Economic Co-operation and Development (OECD)*. https://www.oecd.org/pisa/

Rodriguez, A. J. (1998). Strategies for counterresistance: Toward sociotransformative constructivism and learning to teach science for diversity and for understanding. *Journal of Research in Science Teaching, 35*(6), 589–622.

Rodriguez, A. J. (2002). Using sociotransformative constructivism to teach for understanding in diverse classrooms: A beginning teacher's journey. *American Educational Research Journal, 39*(4), 1017–1045.

Rodriguez, A. J. (2004). *Turning despondency into hope: Charting new paths to improve students' achievement and participation in science education.* Tallahassee, FL: Southeast Eisenhower Regional Consortium for Mathematics and Science Education @ SERVE.www.serve.org/Eisenhower.
Rodriguez, A. J. (2005). Using sociotransformative constructivism to respond to teachers' resistance to ideological and pedagogical change. In A. J. Rodriguez & R. Kitchen (Eds.), *Preparing mathematics and science teachers to teach for diversity: Promising strategies for transformative pedagogy* (pp. 33–48). Mahwah, NJ: Lawrence Erlbaum Associates.
Rodriguez, A. J. (2008). *The multiple faces of agency: Innovative strategies for effecting change in urban school contexts.* Rotterdam, Netherlands: SENSE Publishing.
Rodriguez, A. J. (2010). *Science education as a pathway to teaching language literacy.* Rotterdam, Netherlands: SENSE Publishing.
Rodriguez, A. J. (2015a) What about a dimension of equity, engagement and diversity practices? A critique of the next generation science standards. *Journal of Research in Science Teaching, 52*(7), 1031–1051.
Rodriguez, A. J. (2015b). Managing sociocultural and institutional challenges through sociotransformative constructivism: A longitudinal case study of a high school science teacher. *Journal of Research in Science Teaching, 52*(4), 448–460.
Rodriguez, A. J. (2017). How do we prepare for and respond to students' evoked emotions when addressing real social inequalities through engineering activities? *Theory into Practice (Special Issue on Disciplinary Literacies),56*(04), 263–270.
Rodriguez, A. J., & Morrison, D. (2019). Expanding and enacting transformative meanings of equity, diversity and social justice in science education. *Cultural Studies in Science Education, 14*, 265–281. https://doi.org/10.1007/s11422-019-09938-7
Rodriguez, A. J., & Zozakiewicz, C. (2010). Facilitating the integration of multiple literacies through science education and learning technologies. In A. J. Rodriguez (Ed.), *Science education as a pathway to teaching language literacy* (pp. 23–45). Rotterdam, Netherlands: SENSE Publishing.
Rodriguez, A. J., Collins-Parks, T., & Garza, J. (2013). Interpreting research on parent involvement and connecting it to the science classroom. *Theory into Practice, 52*(1), 51–58.
Rodriguez, A. J., Zozakiewicz, C., & Yerrick, R. (2008). Students acting as change agents in culturally diverse schools. In A. J. Rodriguez (Ed.), *The multiple faces of agency: innovative strategies for effecting change in urban school contexts* (pp. 47–72). Rotterdam, Netherlands: SENSE Publishing.
Tolbert, S., Knox, C., & Salinas, I. (2019). Framing, adapting, and applying: Learning to contextualize science activity in multilingual science classrooms. *Research in Science Education, 49*, 1069–1085. https://doi.org/10.1007/s11165-019-9854-8
Tolbert, S., Schindel, A., & Rodriguez, A. J. (2018). Relevance and relational responsibility in justice-oriented science education research. *Science Education.* DOI:10.1002/sce.21446
U.S. Census Bureau. (2012). *Statistical abstract of the United States, 2012.* Washington, DC: Government Printing Office.

Waits, M. J., Campbell, H. E., Gau, R., Jacobs, E., Rex, T., & Hess, R. (2006). *Why some schools with Latino/a children beat the odds and why others don't*. Morrison Institute for Public Policy School of Public Affairs, College of Public Programs. Phoenix: Arizona State University.

Weiss, I. R., Banilower, E. R., McMahon, K. C., & Smith, P. S. (2001). *Report of the 2000 national survey of science and mathematics education*. Chapel Hill, NC: Horizon Research, Inc.

10 Effective STEM Professional Development

Carla C. Johnson and Toni A. Sondergeld

Need for Professional Development to Change Practice

As the knowledge base on educational reform and improving teacher quality has grown over the past decade (e.g. Darling-Hammond, 2010; Desimone, 2009; Loucks-Horsley, Love, Stiles, Mundry, & Hewson, 2007; Putnam & Borko, 1997), it has become more evident that traditional professional development (PD) formats do not result in sustained improvement of teacher practice and/or student learning. Fortunately, we know a great deal about what types of PD experiences translate into changes in teacher practice that are linked to growth in student learning of STEM content and skills. Desimone (2009) conducted an extensive review of published research in this area and developed a *Core Conceptual Framework for Professional Development* that included five key components that were consistently connected to programs that produced results in either teacher or student outcomes. *The Core Conceptual Framework for Professional Development* requires collective participation, active learning, coherence with policy, extended duration, and a focus on learning new skills in the context of building content knowledge. Each of the five components will be described in detail in the following paragraphs.

Collective Participation

The likelihood of teacher participation in PD resulting in change in teacher practice is increased when more than one teacher from any given school is included in the opportunity (e.g. Desimone, Porter, Garet, Yoon, & Birman, 2002; Johnson, Kahle, & Fargo, 2007). Further, collective participation also improves the sustainability of change in teacher practice (e.g. Johnson, Fargo, & Kahle, 2010). Teacher PD is very constructivist in nature, as teachers attend workshops with other teachers and engage in discourse about their practice as they consider new strategies and grow their understanding of content and of their own students. Unfortunately, less effort is placed on keeping participants connected following the PD, and participants devote their time to implementing new practice and have little availability to reach out to those outside of their school/district.

Recently, more PD programs have purposefully required teams of teachers to participate, and the results have indicated that informal and formal professional learning communities are established. With teams of teachers from the same district or building participating, teachers have an in-house support system for implementing what are often challenging changes to their pedagogy. Collective participation ensures more buy-in to the reform on the school level and provides much needed support to improve the odds of achieving intended outcomes for teachers and students of the STEM PD program. Collective participation is also a key component of PD focused on achieving integrated STEM instruction. Teams of teachers should be provided time to plan together as well as learning together and reflecting on implementation of integrated STEM curriculum (such as the STEM Road Map). Therefore, collective participation during PD and also during school planning time is a critical component for adoption of the STEM Road Map curriculum.

Active Learning

Active learning experiences within PD have been strongly linked to positive teacher outcomes (Banilower & Shimkus, 2004; Darling-Hammond, 1997; Johnson, 2011; Johnson & Fargo, 2010). Moving from a teacher-centered classroom toward implementing PBL and integrated STEM requires opportunities for teachers to experience the curriculum they will deliver and acquire the new content and skills in the context of the learner. Therefore, traditional PD format of "sit and get" focus is not adequate, and in many cases, these types of PD result in little to no change in practice. Active learning should comprise at least 80% of the duration of the PD program. The PD facilitators should model the use of skills as they deliver new content to participants. Teachers should grapple with trying to solve the same problems their students will be presented with and should also be engaged in reflecting on how the new activities and/or curriculum might look in their own classes and what types of accommodations will be necessary to meet the needs of all learners. Next, participants should have opportunities to practice delivery of new instructional models and content with their peers in the PD setting.

Coherence

PD programs have the best chance of impact on teacher and student outcomes when the goals of the PD program are aligned with policies at the school, district, and state levels as well as existing teacher beliefs regarding STEM. This is an area of challenge for some programs, as a focus on STEM or the teaching of science is sometimes not a priority of a school and/or district due to increased high-stakes testing pressures in the areas of mathematics and language arts (e.g. Fullan, 1993; Johnson, 2013). However, for schools intending to adopt the STEM Road Map as their curricular guide and tool for delivery of instruction, alignment should be a non-issue. One challenge in the area of coherence may be existing teacher beliefs regarding the integration of STEM across the curriculum, which is required for STEM Road Map implementation. Schools/districts moving forward with the STEM Road Map should spend some time in discussing the benefits of integration with teachers and provide support for teachers to learn and implement new PBL pedagogy as well as time to plan with other teachers.

Duration

We have learned a great deal regarding the duration of PD programs over the past decade and now understand that for change in practice to take place, over 80 hours of PD must occur (e.g. Banilower, Heck, & Weiss, 2007; Cohen & Hill, 2001; Fullan, 1993; Guskey, 1994; Johnson & Fargo, 2010; Supovitz & Turner, 2001). Further, these contact hours should be spread across at least one academic year of implementation to provide support for teachers as they are using the new pedagogical content knowledge with their own students and reflecting on the outcomes. Formats that have been used in many settings include 5–10 days of PD in the summer followed by monthly sessions on Saturday or a released day from school. This allows the PD facilitator to provide just-in-time support for teachers who may be struggling with implementation or may need to have critical feedback from their peers on how things are working in their classrooms. The duration of PD for teachers who are using the STEM Road Map curriculum will also be essential to be delivered in this format to provide opportunities as described above but also to allow for teams of teachers to plan for delivery of the various PBLs across the school year.

Content Knowledge

At the elementary and middle school levels, a focus on content knowledge within PD has been fairly a routine as most teachers in these grades do not have a bachelor's degree in the specific content area. Research has shown that the most effective PD programs include new strategies taught within the context of the content that will be delivered (e.g. Gonzales et al., 2003). The STEM Road Map will require teachers to be familiar with some content (big ideas) from other disciplines in order to engage in discourse with their students regarding their work on associated projects/problems. Therefore, PD focused on enabling teachers to implement the STEM Road Map curriculum modules should have a clear and purposeful focus on STEM content knowledge included in each grade-level's curriculum.

Further Support for the Components of Effective STEM PD

In addition to Desimone's (2009) study, the Core Conceptual Framework for PD was examined in statewide implementations of STEM PD programs funded through Race to the Top. The programs ranged from K-2 focus on mathematics, science, and/or engineering and literacy to elementary, middle, and high school STEM-focused PD. While specific content covered in the PD varied, all programs were required to be developed with a focus on collective participation, active learning, coherence with policy, extended duration, and building content knowledge in order to be funded by the state initiative. The overall findings revealed a positive impact overall for the state on enabling teacher quality in STEM areas to be significantly improved as a result of participation in the program (i.e. Johnson, Sondergeld, & Walton, 2017). More specifically, analysis of content knowledge assessments, surveys, and direct teacher instructional observations data revealed that implementing the Core Conceptual Framework for PD in each of the STEM PD programs led to significant increases in teacher content knowledge, beliefs, and attitudes toward reformed-based STEM instruction and implementation of reform-based methods in the classroom across programs. Thus, regardless of the grade level or content area focused on, the Core Conceptual Framework for PD demonstrated its effectiveness for promoting PD that improved teacher quality.

Data-Driven PD

When teachers are learning to implement new instructional practices learned through PD, there is a need for regular and collaborative formative evaluation (Guskey, 1997; Joyce & Showers, 2002). Formative evaluation means teachers, administrators, and/or PD providers collaboratively examine standardized and informal data sources to inform the direction and assess the effectiveness of PD implementation. Data such as standardized test results, classroom pre-post assessments, student and teacher surveys, and teacher observations should all be used to inform decisions about PD.

Using data to drive decision-making in PD should not be linear in fashion. Rather, it needs to be a cycle of inquiry used to provide information to PD developers and educators. This reflective cycle should be continuous and focus on questions such as "What are the most effective strategies for improving student learning?" and "What are the instructional needs of my classroom?" (Hayes & Robnolt, 2010). With these questions in mind goals about classroom instruction, student achievement can be collaboratively developed based on the data. To promote this process of data-driven PD, time needs to be set aside for teachers and PD developers to meet and discuss data, goals, and the direction of PD. This also means that PD developers must be flexible enough to modify PD content based on student and teacher needs that become evident through this data-driven process. Further, this process must be structured in such a way that teachers are taught to collect, interpret, and use data since this is not typically a skill teachers learn in their traditional educational training.

Using a data-driven PD process allows teachers to take greater ownership over their learning and implementation of the PD. When teachers are involved in data-based discussions about the effectiveness of the PD in their classrooms, they typically develop greater buy-in to the initiative. Resultantly, higher levels of teacher buy-in have been shown to produce increased levels of implementation fidelity and greater chance of long-term initiative sustainability (Datnow & Stringfield, 2000).

Creating Individual STEM Professional Development Plans

In today's era of accountability, many states and districts require teachers to create yearly Individual Professional Development Plans (IPDP) (e.g. Massachusetts, New Jersey, Ohio, Vermont). Oftentimes these IPDPs are a component of the state's teacher evaluation system, and templates for completing an IPDP are frequently provided by states or school districts. Regardless of the state, IPDPs serve as a tool to help teachers meet their professional learning needs with the ultimate goal of improving student learning. Additionally, IPDPs need to be aligned with state standards for teacher learning and continuous improvement and are typically evaluated for their effectiveness in providing teachers with the skills needed to be effective in the classroom.

To develop a STEM IPDP, a five-step process should be undertaken: (1) determine PD needs, (2) set goals, (3) identify resources, (4) develop timeline, and (5) reflection. First, educators must self-assess to determine their PD needs. Student data along with professional experiences should be used to drive this stage. For instance, if a teacher notices her students are struggling in a particular area of the curriculum or realizes that due to content standard changes in the state she will be teaching something new that she is not entirely confident about, she might choose either of these areas to look for PD citing these reasons as justification.

Once area(s) of needed PD are identified, specific goals related to the STEM PD content should be set. These goals must also align with state PD standards and need to be measurable. One might think of a teacher's STEM PD goal as similar to a student-learning target. With this in mind, a specific STEM PD goal might be something like "Incorporate more integrated STEM project-based learning into my classroom instruction." This is measurable in that the teacher can actually track if they do this or not by comparing what they did over the last few years to after they receive integrated STEM PBL PD.

After specific PD goals are established, teachers need to identify where they will be able to obtain the PD aligned with their goals and determine a timeline for completion. Often, school districts do not have the resources or expertise on staff to enable delivery of individualized PD plans. Teachers should explore their local universities and regional education centers as a source of potential professional growth opportunities. Also, with the increasing emphasis and focus on STEM, many informal education agencies and business/industry partners have sponsored workshops and learning experiences for teachers and students. As we shared in this chapter, it is important to build a plan for your PD that includes the key components of effective PD. Therefore, when you develop your plan, you should build a collection of experiences that are related that extend across the academic year that include both short-term and long-term goals to be achieved throughout the STEM PD.

Finally, teachers should be reflective of their STEM PD experiences since their IPDP is most likely a component of their state's teacher evaluation system. STEM PD reflection should be a continuous process whereby teachers are examining their own confidence and beliefs about their new teaching content and practices they are learning. It is also critical that teachers consider the impact of their new STEM PD on their student learning as measured by classroom assessments, standardized tests, and/or student attitudes toward doing STEM class work. Establishing a reflective feedback loop between STEM PD and teacher/student outcomes allows school leaders to ensure that educators are receiving the tools needed to be successful in the promoting student learning.

Planning for District-Level STEM PD

District-level PD should be planned strategically with as much attention to individualization for teachers as possible. It is clear that district-level PD has historically focused on very general skills and strategies that all teachers are required to attend. Unfortunately, this approach does not tend to the need for teachers to continue to advance their individual pedagogical content expertise and most who attend these types of sessions do not feel connected and likely do not receive the intended benefit of the PD.

Schools that are planning to move toward implementing a grade level, program, or whole school STEM focus should approach PD for the staff in a very collaborative manner. First and foremost, teachers will need collaborative planning time beyond any time that is provided within the PD to ensure success of implementation. This is something that school leaders should consider and make necessary accommodations for up front. Second, the PD should be structured to provide an interactive overview of integrated STEM. Allow for teachers to participate in a STEM learning environment and to discuss how they individually and as a discipline fit within an integrated STEM context. Third, intensive PD on problem and/or project-based learning and engineering design thinking should be provided to ensure that all teachers are empowered to utilize this STEM pedagogy. Finally, a majority of the PD time should be focused on providing teacher teams facilitated time to develop PBL topics and modules aligned with the context of the school or to take STEM Road Map modules and modify as necessary. Encourage teachers to implement at least two modules (four to five weeks) in the first year of the STEM effort. Finally, district-level PD should include a focus on measuring individual teacher growth and the transformation of the learning environment. Encourage teachers to engage in reflection on their practice and to use this as a way to iteratively develop their own PD plans for the future.

References

Banilower, E. R., Heck, D. J., & Weiss, I. R. (2007). Can professional development make the vision of the standards a reality? The impact of the National Science Foundation's Local Systemic Change through Teacher Enhancement Initiative. *Journal of Research in Science Teaching, 44*(3), 375–395.

Banilower, E. R. & Shimkus, E. (2004). *Professional development observation study.* Chapel Hill, NC: Horizon Research.

Cohen, D. K., & Hill, H. C. (2001). *Learning policy: When state education reform works.* New Haven, CT: Yale University Press.

Darling-Hammond, L. (1997). *Doing what matters most: Investing in quality teaching.* New York, NY: National Commission on Teaching and America's Future.

Darling-Hammond, L. (2010). *The flat world and education: How America's commitment to equity will determine our future.* New York, NY: Teachers College Press.

Datnow, A., & Stringfield, S. (2000). Working together for reliable school reform. *Journal of Education for Students Placed at Risk, 5*(1), 183–204.

Desimone, L. M. (2009). Improving impact studies of teachers' professional development: Toward better conceptualizations and measures. *Educational Researcher, 38*(3), 181–199.

Desimone, L., Porter, A. C., Garet, M., Yoon, K. S., & Birman, B. (2002). Does professional development change teachers' instruction? Results from a three-year study. *Educational Evaluation and Policy Analysis, 24*(2), 81–112.

Fullan, M. (1993). *Change forces: Probing the depth of educational reform.* New York: Falmer.

Gonzales, P., Guzman, J. C., Partelow, L., Pahlke, E., Jocelyn, L., Kastberg, D., & Williams, T. (2003). *Highlights of the trend in International Mathematics and Science Study (TIMSS).* https://nces.ed.gov/timss/

Guskey, T. R. (1994). Results-oriented professional development: In search of an optimal mix of effective practices. *Journal of Staff Development, 15*(4), 42–50.

Guskey, T. R. (1997). Research needs to link professional development and student learning. *Journal of Staff Development, 18*, 36–40.

Hayes, L. L., & Robnolt, V. J. (2010). Data-driven professional development: The professional development plan for a reading excellence act school. *Reading Research and Instruction, 46*(2), 95–119.

Johnson, C. C. (2011). The road to culturally relevant science: Exploring how teachers navigate change in pedagogy. *Journal of Research in Science Teaching, 48*(2), 170–198.

Johnson, C. C. (2013). Educational turbulence: The influence of macro and micro policy on science education reform. *Journal of Science Teacher Education, 24*(4), 693–715.

Johnson, C. C., & Fargo, J. D. (2010). Urban school reform through transformative professional development: Impact on teacher change and student learning of science. *Urban Education, 45*(1), 4–29.

Johnson, C. C., Fargo, J. D., & Kahle, J. B. (2010). The cumulative and residual impact of a systemic reform program on teacher change and student learning of science. *School Science and Mathematics, 110*(3), 144–159.

Johnson, C. C., Kahle, J. B., & Fargo, J. (2007). A study of sustained, whole-school, professional development on student achievement in science. *Journal of Research in Science Teaching, 44*(6), 775–786.

Johnson, C. C., Sondergeld, T. A., & Walton, J. (2017). A statewide implementation of the critical features of professional development: Impact on teacher outcomes. *School Science and Mathematics, 117*(8), 350–351.

Joyce, B., & Showers, B. (2002). *Student achievement through staff development* (3rd ed.). Alexandria, VA: Association for Supervision and Curriculum Development.

Loucks-Horsley, S., Love, N., Stiles, K. E., Mundry, S., & Hewson, P. W. (2007). *Designing professional development for teachers of mathematics and science*. Thousand Oaks, CA: Corwin Press.

Putnam, R., & Borko, H. (1997). Teacher learning: Implications of new views of cognition. In B. J. Biddle, T. L. Good, & I. F. Goodston (Eds.), *The international handbook of teachers and teaching* (pp. 1223–1296). Dordrecht, The Netherlands: Kluwer.

Supovitz, J. A., & Turner, H. M. (2001). The effects of professional development on science teaching practices and classroom culture. *Journal of Research in Science Teaching, 37*(9), 963–980.

11 Frameworks and Advocacy for STEM

Shaun Yoder, Susan Bodary, and Carla C. Johnson

Ohio has experienced extraordinary progress in the realm of science, technology, engineering, and mathematics (STEM) education. In just 2007, the state's elected officials made an unprecedented commitment of more than $200,000,000 in state funding to support an array of cohesive STEM education policies, spanning the state's pre-K-20 education continuum (Ohio Business Alliance for Higher Education & the Economy, 2007). The Ohio investment has continued and continues today with policies to support the designation of STEM schools, STEM programs, and funding for higher education scholarships in STEM.

To understand how this major victory came about, we must look back to 2006, when STEM education in Ohio gained an unexpected champion. That year, Nationwide Insurance, headquartered in Columbus, Ohio, surveyed its 36,000 employees to measure its future workforce needs. The results of the survey were shocking to then-CEO Jerry Jurgensen. It revealed that Nationwide's largest employment sector was neither insurance nor any other aspect of the financial services industry. Instead, technology was its largest employment sector. The survey results – intended to help Nationwide plan for its future – came on the heels of the company's recent move to bring in a number of top-level computer scientists from India because it could not find the needed talent in Ohio (Kaplan, 2010).

The sobering survey results led Jurgensen to become a key voice in the campaign for increased STEM education in Ohio's schools. What followed was the development of a powerful, multi-partner STEM-education coalition never before seen in Ohio. With help from the Ohio Business Roundtable and Battelle's fledgling Ohio STEM Learning Network (OSLN), Nationwide joined forces with other Ohio-based businesses (e.g. Cincinnati Bell, GE Lighting, Marathon Petroleum, and Diebold) and leaders from K-12, higher education and philanthropy to advocate for the prioritization of STEM education in the state's upcoming biennial budget. The coalition was by no means assured of success, however. The state budget was being cobbled together in the midst of a crippling national recession that dried up state revenue and was sure to result in across-the-board budget cuts.

Despite the troubling economic environment, Ohio's elected officials heeded the call of the coalition and made STEM education a top priority. Of the $200M, they included funding to launch STEM schools and "Programs of Excellence" ($12.6M), support students in STEM schools with state education aid ($2.9M), expand supplemental STEM programs ($3.5M), increase the supply of STEM teachers ($26.9M), enhance STEM educator professional development ($9.3M), attract undergraduates into STEM disciplines through scholarships ($100M), and increase the supply of renowned STEM scientists and researchers across the state's institutions of higher education ($50M) (Kaplan, 2010).

Today, Ohio's 2008–2009 biennial budget serves as the exemplar for developing, informing, and advocating for STEM policies. It demonstrates the importance of focused coalitions of diverse partners from K-12, higher education, business, philanthropy, and community joining hands to develop policy and advocate for the funding and implementation of that policy. The Ohio General Assembly passed the budget by a near-unanimous margin in a historically challenging fiscal environment. Nearly, all members of the elected body viewed STEM as an investment worth making. This was a clear testament to the advocacy work pursued by partners.

Ohio represents just one success story in which key partners came together to create an aligned set of STEM policies and advocated in support of those policies. Other states have pursued similar approaches and have garnered similar results. From Texas to Tennessee, partners have united to develop, inform, and advocate for STEM education policies. Those same partners, in many cases, have stayed the course to ensure implementation, refinement, and follow-through at the local, regional, state, and national levels. And, while it may seem like common sense, the process of implementing, refining, following through, and sustaining a connected set of STEM policies stand as one of today's greatest innovations. Indeed, sticking to the long-term implementation of an identified set of policies is rare in education – where the field has become numb to revolving policies.

In this chapter, we explore why cohesive STEM education and talent policies are essential in today's fiercely competitive global society, where student success in STEM matters now more than ever. We outline four steps communities can take to develop and set in motion policies that enable, support, and in many cases, result in real change. We also showcase examples of strong STEM policies, successful advocacy approaches, and innovative tools that accelerate the work. Finally, the chapter highlights key characteristics of effective STEM programs, including processes used by a few states to transform traditional public and private schools into a STEM-focused school.

Four Steps to Building and Enacting Effective STEM Policies

Nationwide Insurance's dilemma is not limited to a single company or industry, and most certainly not to a single state. The shortage of workers skilled in STEM subjects is a well-documented crisis that must be addressed at the public policy level. The reality is that Nationwide's talent shortage – an experience shared by many other US companies – has its roots in the education system (Business-Higher Education Forum, 2007). Put simply, it is an education pipeline issue. Solving the pipeline problem requires the development and implementation of smart, transformative policies directed toward a clear set of local, regional, state, or national STEM-based goals and outcomes.

At its most basic form, Merriam-Webster defines *policy* as "a high-level overall plan embracing the general goals and acceptable procedures especially of a governmental body." From a STEM-perspective, policy and the making of that policy is far more complex. And, defining *STEM* for the purposes of developing education and talent policies can be just as complicated because of the multifaceted nature of the acronym. Sure, STEM stands for science, technology, engineering, and mathematics. But the acronym is greater than the sum of its parts. From a teaching and learning perspective, STEM is a verb that emphasizes the purposeful integration of the various disciplines in solving real-world problems (Breiner, Harkness, Johnson, & Koehler, 2012).

Pockets of transformative STEM programs, practices, and partnerships exist throughout the country. And they are marked by innovations that emerge when students, teachers, and partners engage in teaching and learning based on real-world work. When well executed, with critical input from practitioners and partners, STEM education and talent policies have the potential to scale change across change-resistant local education ecosystems.

To blast through the inertia, we identify four key steps that community leaders and stakeholders can use to enable and support the development and implementation of transformative STEM policies (see Figure 11.1).

#1: Collect and Use Relevant DATA to Rally Stakeholders, Inform Goals, and Push Policy

Experienced STEM advocates know how important data are in securing multi-sector partner support, making the case for policy action and whetting the appetites of policymakers to enact transformative change. Nationwide's workforce survey is but one example of a localized data collection effort that resulted in significant action. The same reaction could be triggered from the collection and analysis of STEM-specific regional, state, or national-level data.

230　*Shaun Yoder et al.*

```
┌─────────────────────────────┬─────────────────────────────┐
│  #1: Collect and use        │  #2: Build diverse, multi-  │
│  relevent DATA to rally     │  sector PARTNERSHIPS to     │
│  stakeholders, inform goals │  drive policy and advocacy  │
│  and push policy            │                             │
│              ┌──────────────────────────┐                 │
│              │     Transformative       │                 │
│              │     STEM Policies        │                 │
│              └──────────────────────────┘                 │
│  #3: Develop GOALS &        │  #4: FOLLOW THROUGH         │
│  POLICIES based on data     │  and seek transformative    │
│  and informed by multi-     │  action in a change-        │
│  sector partners            │  resistant ecosystem        │
└─────────────────────────────┴─────────────────────────────┘
```

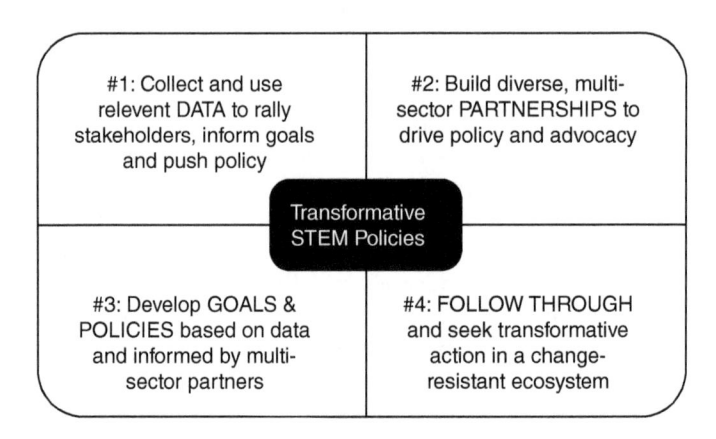

Figure 11.1 Transformative STEM practices.

Today, elected officials and policymakers at all levels are operating in challenging environments. Most seek a firm understanding of their community's future workforce needs, particularly in STEM. This creates an opportunity for STEM-vested partners to use data to identify what the community's future STEM-workforce and STEM-skills needs are, use those data to attract and secure a diverse array of power partners to join the cause, and develop transformative policies that will put the community on a trajectory for future success.

But knowing the STEM data is just one piece of the puzzle. Advocates and partners must also present the data in a clear and compelling way to stir discussion, promote questions, and ignite a collective search for policy solutions. Following are examples of the types of STEM-specific data that communities might consider collecting based on a national-level dashboard. The national-level data illustrate the powerful story that can be developed to pinpoint the problem and drive toward solutions. Each community and state has its own story to tell.

Where to Begin? Understand Your Community's STEM Workforce Needs

It is well documented that STEM knowledge and skills are in high demand. And regions, states, and the nation as a whole must be poised to capitalize on future opportunities where the advantage goes to companies that are first to invent and produce innovative products. From 2000 to 2010, the growth in STEM jobs was three times greater than that of non-STEM jobs (Economics and Statistics Administration, 2011). The US Department of Commerce estimates that in the coming years,

STEM occupations will grow 1.7 times faster than non-STEM jobs (US Department of Commerce, 2012). Currently, STEM jobs comprise 20% of all US jobs (Rothwell, 2013), and the share of STEM occupations is projected to increase by 26% between 2010 and 2020 (Carnevale, Smith, & Strohl, 2013). Within this decade, 95% of STEM jobs will require some postsecondary education and training, with approximately two-thirds requiring a bachelor's degree or better. Additionally, more than 75% of the top 25 jobs for 2014 identified by US News and World Report (2014) were in the STEM fields. According to a survey of Fortune 1000 companies, 89% continue to report "fierce" competition in finding candidates to fill jobs requiring four-year degrees in STEM-related fields (Bayer, 2013).

Despite a growing demand, the percentage of students earning STEM degrees has not substantially changed in recent years (National Center for Education Statistics, 2019). A new report by the labor-market analytics firm Burning Glass Technologies (2014) reveals a clearer picture of this STEM-skills gap for entry-level workers. Forty-eight percent of all entry-level jobs requiring at least a bachelor's degree are in STEM fields, while only 29% of students graduate with a STEM degree. What's more, demand for STEM skills stretches beyond the needs of STEM occupations to non-STEM fields, exacerbating shortages of STEM talent. In fact, almost 50% of students who graduate with a bachelor's degree in a STEM major do not enter a STEM occupation. Researchers ascribe this diversion from STEM fields to interests, values, and pay (Carnevale, Smith, & Strohl, 2011).

While these workforce data speak to what is happening at the national level, they also offer a template for how local, regional, and state-level data might be gathered to tell a more localized story. And many local and regional STEM advocates have become masters of working with data organizations to collect and mine reliable workforce data to galvanize action around a specific STEM initiative. For instance, long-time Long Island, New York residents Ken White (Brookhaven National Laboratory), Cheryl Davidson (Long Island Works Coalition), and Mark Grossman (New York State Department of Labor Commissioner's Regional Representative for Long Island) knew that STEM was an economic imperative for Long Island. Based on regional workforce employment trends, high-tech employers were having trouble identifying local talent to fill jobs. In fact, Long Island's largest employer, North Shore-LIJ Health System, struggled to fill more than 1,000 technical positions due to a gap in the local applicant pool's ability to do the work. This, while at least 100,000 Long Islanders remained unemployed, and at least 20% of the island lived in poverty exacerbated by the area's high cost of living. Based on local data, White, Davidson, and Grossman created urgency and opportunity for STEM and capitalized on regional economic development strategies already under development.

Just as Long Island partners did, community leaders should start by understanding how many STEM jobs are projected in the near future in their region or state. With those data in hand, communities must then gain an understanding of the condition of their "talent pipeline." In other words, is the pre-K-12 education system successfully preparing students that are STEM capable? Is the postsecondary system attracting and successfully graduating students with STEM degrees? These two segments of the STEM pipeline ultimately determine the yield of prospective STEM workers for a region or state.

Condition of Your Pre-K-12 STEM Education Pipeline: Is It Leaking?

Troubling STEM workforce data raise an important question: what is occurring in the pre-K-12 segment of the STEM education pipeline? Based on national student trends (Change the Equation, 2014), there is no question that pre-K-12 STEM education pipeline needs attention. State-administered assessments in the elementary grades – as early as fourth grade – indicate that students are not achieving at proficient levels in mathematics and science. This is cause for concern. A comprehensive STEM education pipeline analysis conducted in New Hampshire, widely considered a high achieving state, found that 51% of the state's fourth graders scored at a proficient or above level on the New England Common Assessment Program (NECAP) in science. The situation was worse for the state's eighth graders, with only 31% scoring at that level (New Hampshire Charitable Foundation, 2014). The narrowing of STEM skills begins early in many states and localities. New Hampshire pursued an impressive STEM education pipeline analysis to help drive its state-level STEM goal and policy development, which is currently being led by the governor in conjunction with a variety of public and private partners. The leaky pre-K-12 pipeline indicates that many students are simply underprepared to move from one grade to the next and unprepared for success in college and career. To effectively mitigate and stop leaks in the pre-K-12 STEM education pipeline, communities must have a handle on where they stand in the key benchmark areas. We use national-level data to illustrate how local, regional, and state communities might assess their benchmarks.

Formal and informal STEM learning opportunities: Nationally, elementary students are spending less time in formal science instruction. The average amount of time an elementary school student spent on science in 2009 was 2 hours per week (Blank, 2012). Informal STEM learning is just as important as formal STEM learning. It is proven to raise student confidence and classroom achievement in STEM and generate student interest in pursuing STEM studies and careers (Thomasian, n.d.). Types of informal STEM learning programs include those that provide pre-K-12 students afterschool, end, and summer activities over multiple years at institutions such as science museums, zoos, local universities, and research centers. Unfortunately, good, objective data that differentiate those programs having the greatest impact do not exist at the national level.

Student performance in STEM: What does student achievement look like on mathematics and science national assessments, particularly in grades 4 and 8? How are students performing on mathematics and science national assessments in grades 9–12? Fourth grade student performance on the National Assessment of Educational Progress (NAEP) increased in 2019 by one point to 241, compared to 240 in the most recent two NAEP assessments in 2015 and 2017. However, eighth grade student performance declined by one point overall in 2019. In twelfth-grade, students had no significant change in reading performance in 2019 (National Assessment of Educational Progress, 2019). Science has not been assessed with NAEP since 2015. Fourth and eighth graders scored four points higher in 2015 than in previous years (2009 and 2011 respectively). Twelfth-graders remained the same in 2015 as in 2009 with no significant change, performing just above the basic level. Internationally, students from the United States are performing above the OECD average on the Programme for International Student Assessment (PISA) in science (502) but below the average for mathematics (478). United States student achievement on the PISA remained flat for students in science and mathematics since 2006 and 2003 respectively (OECD, 2018).

Rigorous standards in STEM: The majority of states have adopted college- and career-ready standards in math and English language arts, with schools and districts being held accountable for student achievement in those subjects. Specifically, 43 states and the District of Columbia have adopted the Common Core State Standards (Common Core State Standards Initiative, 2014). Twelve states have committed to adopt *Next Generation Science Standards* (Camins, 2014). However, 25 states do not hold schools accountable for meeting student performance targets in science (Change the Equation, 2014).

Rigorous course completion in STEM: Are students completing rigorous mathematics, science, engineering, and technology courses in grades 9–12? Today, 35 states have established graduation requirements that require all high school graduates to complete college- and career-ready course requirements so that earning a diploma ensures that a student is prepared for postsecondary education.

Teacher effectiveness in STEM: Are STEM-specific teachers masters of their content, particularly in middle and high school? Do they know how to teach STEM methods? Are they supported with high-quality STEM professional development opportunities? Many states across the country have established teacher evaluation systems to determine who their best teachers are and ways to help support those teachers who struggle. While the systems are relatively new, and some have not yet produced or published data, they hold promise for targeting support and growing teacher effectiveness in STEM and other disciplines (Achieve, 2014).

Student success in STEM beyond high school: How many students go to college and do not require remediation in mathematics their freshman year? An August 2012 report notes that, nationally, too many first-time college freshmen require remediation in mathematics or reading. Nearly, 52% of students who entered a two-year college enrolled in remediation while almost 20% of those entering a four-year college required remediation (Complete College America, 2012). These pre-K-12 data trends are concerning. That said, these can be mitigated or even reversed over time through the development and enactment of smart STEM education and talent policies. We discuss such innovative policies later in the "Goals & Policies" section.

Condition of Your Local Postsecondary STEM Pipeline: Is it Yielding STEM Degrees?

The nation's postsecondary pipeline also experiences widespread leakage, with a range of factors hindering STEM degree production. For every 100 students who pursue a bachelor's degree, 28 choose to major in STEM, but only 15 earn a STEM degree, and even fewer actually enter a STEM occupation (Chen & Ho, 2012). Similarly, for every 100 students who pursue an associate's degree, only 31 earn a STEM degree (Chen & Soldner, 2013). To make matters worse, data reveal that STEM students are more likely to switch majors to a non-STEM major than non-STEM students are to change to a STEM major. Simply increasing the number of students entering STEM majors will not necessarily translate to higher STEM-degree production if postsecondary institutions do not also stem the tide of students out of STEM majors. Just as we suggested gathering benchmark data that speak to key transition points in the pre-K-12 segment of the pipeline, similar benchmarks exist for the postsecondary side of the pipeline. We suggest that communities have an understanding of the following postsecondary benchmarks:

Academic preparation and math proficiency: Are first-time freshman prepared for success? Again, remediation rates should be considered.

Design of developmental and gateway courses: If students require developmental or remedial coursework, then are those courses designed to promote continued success and persistence? Are gateway courses, or those first-year courses all students must pass to progress to the next level in a degree program (e.g. calculus in an engineering program), designed to support student learning or to "weed out" students?

Early immersion to STEM courses as freshman: Data from the National Center for Education Statistics (nd) suggest that students who dive into their STEM coursework as freshman are more likely to experience success and finish in STEM. Are students properly counseled and supported to take the threshold number of STEM course hours in their first year?

STEM transfer policies: Does your state have clearly understood and followed course transfer policies in place that compel four-year colleges to recognize and accept course credit from two-year colleges?

Strong connection to the STEM workforce: Does the postsecondary experience link students to actual work experience in the STEM field? Is industry satisfied with the skill and preparedness level of candidates?

Production of effective STEM teachers: Do STEM educators enter the classroom with an appropriate level of content knowledge? Do they have the pedagogical understanding to prepare interdisciplinary lessons and engage with real-world examples and partners?

Again, one might conclude that the challenges facing the STEM postsecondary pipeline are insurmountable. The reality, however, is that smart and targeted STEM policy development can help patch the leaky pipeline. We discuss such innovative policies later in the "Goals & Policy" section.

#2: Build Diverse, Multi-Sector PARTNERSHIPS to Drive STEM Policy and Advocacy

Diverse, multi-sector partnerships are invaluable to the advancement of STEM policy and advocacy. For starters, partners from pre-K-12, postsecondary, business, workforce, philanthropy, and the community significantly enhance the collection of sector-specific and localized STEM data. But, perhaps, the greatest benefit of partner engagement is the resultant robust policy development and powerful voice of advocacy. We have already referenced two examples – Ohio and Long Island – where multi-sector partnerships brought fortitude, focus, and follow-through to the STEM policy and advocacy table. In this section, we take a closer look at how partnerships can be built and organized to gain the most traction.

Developing policy collaboratively with an array of partners – as opposed to in isolation – brings together unique perspectives necessary for innovative, transformative, and sustainable STEM policies. It puts into action the "collective impact" model, which adheres to the premise that better cross-sector engagement and coordination lead to greater progress than the isolated intervention of individual organizations (Hanleybrown, Kania, & Kramer, 2012).

While the specific collective impact model calls for the creation of formal partnerships anchored in new types of nonprofit management organizations, the primary point, regardless of partnership structure (formal or informal), is that everyone has a role to play in ensuring the development of the right STEM policies: educators, business leaders, economic development advisors, workforce professionals, and higher education administrators, to name a few. We use tenets of the collective impact model to describe how to build lasting and impactful partnerships (Kania & Kramer, 2011).

Building Effective STEM Partnerships

STEM-focused partnerships can take on many forms. Generally, local partnerships have been more informal in nature, though that is changing as local initiatives become more sophisticated. State- and nationally focused partnerships tend to be more formally structured. Many are established networks. The structure of the partnership should be driven by its community needs and corresponding STEM policy goals, with form following function. However, there are some attributes that newly developed STEM-focused partnerships should include. First, leaders should focus on establishing a partnership that brings stakeholders together to prepare the strongest STEM learning and achievement policies that connect one segment of the education pipeline to the next. Whether the focus is local, regional, state, or national, the partnership should reflect the horizontal flow of education pipeline (as depicted in Figure 11.2). All sectors of the pipeline – from pre-K-12, postsecondary and workforce – should have a seat at the table to analyze data, develop goals, identify policy solutions, and advance those policies. Gathering these partners is critical, as national data indicate that some of the greatest leakages in the STEM pipeline occur at transition points where students are supposed to advance from one segment to the next.

Partners from across segments of the pipeline should be committed to the STEM cause and be willing to authentically participate in the initiative – from policy development to advocacy. Research has shown that through partnerships, which advocate for STEM, overall STEM community awareness is significantly increased (Sondergeld & Johnson, 2014). While targeted action may or may not focus on all segments at once, having the right representation allows the partners to develop the strongest possible solutions across the pipeline.

Partnerships should also be connected vertically (see Figure 11.3). This means that local-level partnerships, where possible, are linked to state-level partnerships, and state-level partnerships are at least aware of and, in some cases, tied to national-level efforts. These vertical linkages reinforce activities underway at each level, maximizing impact.

The STEM East partnership in Lenoir County, North Carolina offers a textbook example of vertical connectivity. This regional partnership knows that the most effective change begins at the local level. One of four local STEM communities in the state, Lenoir County has been

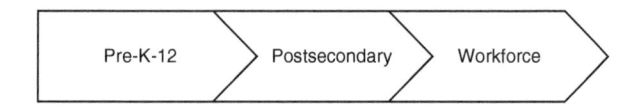

Figure 11.2 Horizontally aligned STEM partnerships.

Figure 11.3 Vertically aligned STEM partnerships.

battling an economic downturn for more than 15 years. In its heyday, the regional economy was driven by tobacco and textiles. But those jobs dried up as the local economy shifted from manufacturing to knowledge. For instance, DuPont, the Fortune 100 Company which patented Dacron polyester fibers, has operated a site in Lenoir County since 1953. At its prime, the factory employed nearly 4,000 workers. By 2005, it had fewer than 200 employees. Recognizing this regional workforce data, Lenoir community leaders knew they had to find a way to reshape the once textile-dependent workforce into a skills rich, STEM-literate community. That's what prompted local leaders to assemble a STEM leadership team. This partnership was intentionally designed to be horizontally aligned – with representatives from major local forces in education, economic development, government, and business – including the Director of Operations for aerospace industry giant Spirit Aero-Systems, which Lenoir County fought hard to recruit to the region. The STEM leadership team champions STEM community engagement and awareness building across the region. Its work is anchored in a community visioning process that included more than 200 people in the community, and the teachers, school leaders, and partners who now drive the region's thriving STEM strategies (Guillory & Quinterno, 2013).

Lenoir County's STEM East is fortunate to be linked to North Carolina (NC) STEM, North Carolina's state STEM network. The county's community-led effort was facilitated by NC STEM's Community Visioning & Design Process, a step-by-step plan for engaging all sectors of the community in visualizing, planning, and building education efforts that mirror the area's economic concerns (Guillory & Quinterno, 2013). Other tools developed by NC STEM include a list of STEM attributes or "hallmarks of programmatic quality in STEM education" and the NC STEM ScoreCard (North Carolina STEM Center, 2020) titled "Strategies that Engage Minds." The ScoreCard is aimed at helping the public and decision-makers chart a direction for the state's STEM-related economic future. It is designed around six domains that gauge the state's progress in (1) STEM workforce and economic impact, (2) informal education and STEM literacy, (3) strategic investments and innovation, (4) college and career readiness, (5) teacher quality and leadership, and (6) policy support. These domains target areas that will propel North Carolina forward in offering the best STEM learning opportunities in the nation.

NC STEM does not operate in a vacuum. The state-level network is connected to a multi-state STEM partnership known as STEMx. Created 'by states, for states' and anchored at Battelle headquarters in Columbus, OH, STEMx stands as a promising national-level partnership. Together, the 19 member states of STEMx established a shared vision, mission, and goals and articulated a value proposition for network membership. Member networks include those from Arizona, California, Colorado, Georgia, Idaho, Indiana, Kentucky, Michigan, New Mexico, New York, North Carolina, Ohio, Oklahoma, Oregon, Pennsylvania, Tennessee, Texas, Washington, and Washington DC. Similar to North Carolina's approach, each state-level network strives to maintain diverse partnerships that are horizontally aligned. As part of the network, states share and disseminate best practices aimed at accelerating STEM policies, practices, and partnerships. STEMx features the tagline: *Local Innovation. State Leadership. National Impact*, reinforcing the notion that vertically aligned partnerships matter.

Partnership Organizational Models

As mentioned earlier, local STEM partnerships tend to be organized informally. This enables initiatives to be nimble and responsive to emerging needs of the community. For instance, Dayton, Ohio's regional STEM partnership originally consisted of three primary partners: Dayton Regional STEM School, Dayton Regional STEM Center, and the lead convening partner, EDvention, which was intentionally established to be a simple, lean broker of opportunities, and convener of focused partners. The partnership, which had more than 21 members on its leadership council representing key horizontal partners, was purposely not created as a standalone 501(c)(3) organization. Rather, it was housed within a third-party organization so that it could focus efforts on partners and programs and not have to worry about funding its own existence. Over time, the effort morphed to become Learn to Earn Dayton, which maintains a broader focus on regional academic achievement and degree attainment, with specific goals and metrics along the education pipeline (Learn to Earn Dayton, n.d.). The Dayton Regional STEM Collaborative, a companion effort, focuses on the STEM-talent aspects of the Learn to Earn Dayton metrics and has attracted even more business and higher education leadership. Lean staffing is supported by contributions from partner organizations that have a vested interest in the region and joint grant opportunities. Other partners provide on-loan experts and talent-based resources aimed at advancing the work. Through nimble structures, the initiatives have harnessed the strongest champions, deepest supporters, and most influential leaders to advance the work in the region.

At the state level, STEM partnerships tend to be organized more formally, though only a few are incorporated 501(c)(3) organizations. The California STEM Learning Network benefits from generous donors including the S.D. Bechtel, Jr. Foundation, Chevron, Battelle, and the James Irvine Foundation, among others. Washington STEM is fueled by generous support from The Boeing Company, Bill & Melinda Gates Foundation, McKinstry Charitable Foundation, The Microsoft Foundation, and others. Leading partners in both states determined that the 501(c)(3) approach was best to meet their states' identified needs. That said, each has refined its approach over time to most effectively respond to the states' changing and evolving STEM needs.

Separate from establishing 501(c)(3) organizations, several states, including Ohio and Tennessee, have codified the establishment of public-private partnerships through the enactment of legislation or executive order that forges horizontal partnerships to jointly craft a statewide STEM agenda. The Ohio legislature established a STEM Committee consisting of the state Superintendent of Public Instruction, the Chancellor of the Board of Regents, the Director of Development, and four members of the public with STEM and/or business backgrounds. This state-level STEM committee was charged with distributing state funding for STEM schools and Programs of Excellence. The OSLN was available to provide technical assistance as needed. The Ohio statutory language essentially created a public-private partnership, where the OSLN works in conjunction with the STEM Committee to coordinate state-level private sector STEM partners and investments.

Similarly, the Tennessee STEM Advisory Council was enacted by an executive order signed by the Governor. It specifies that the Council, which serves as the leadership body of Tennessee's STEM Innovation Network, consists of the Commissioner of Education, Commissioner of Economic and Community Development, Chair of the Senate Education Committee, Chair of the House Education Committee, one representative from the State Board of Education, one representative from the Tennessee Board of Regents, one representative from the University of Tennessee, five representatives of STEM-related industries in the state, and two K-12 educators teaching in Tennessee public schools. The Council advises the Tennessee Department of Education and the Tennessee

STEM Innovation Network on promoting and expanding STEM teaching and learning. Other states, such as New York and Texas, have used public-private partnerships to make significant blended public-private STEM investments to support the development of new school models that directly connect to the world of work.

Helpful tools and resources are available to help communities build and support STEM networks. One such tool is the STEMx Sustainability Compass. A joint project of STEMx, Battelle, and Education First, the self-assessment tool is designed to help local, regional, and state-coalition leaders and partners gauge sustainability levels of their partnerships and offer materials and approaches to strengthen their work over time (STEMx, n.d.). The Sustainability Compass recognizes that better goal and policy development results from a blend of public and private horizontal partners that are connected, when possible, to vertical partners.

#3: Develop GOALS & POLICIES based on Data and Informed by Multi-Sector Partners

With a firm understanding of the data and strong partnerships in place, the table is now set for STEM goal and policy development. Together, partners must identify a set of clear, measurable goals and the supporting policy strategies or activities to accomplish them. The horizontally and vertically aligned partnerships should drive the goal and policy-setting process. In fact, goals and policies, in simplest form, should mirror the structure of the partnerships (Figure 11.4 illustrates this combined

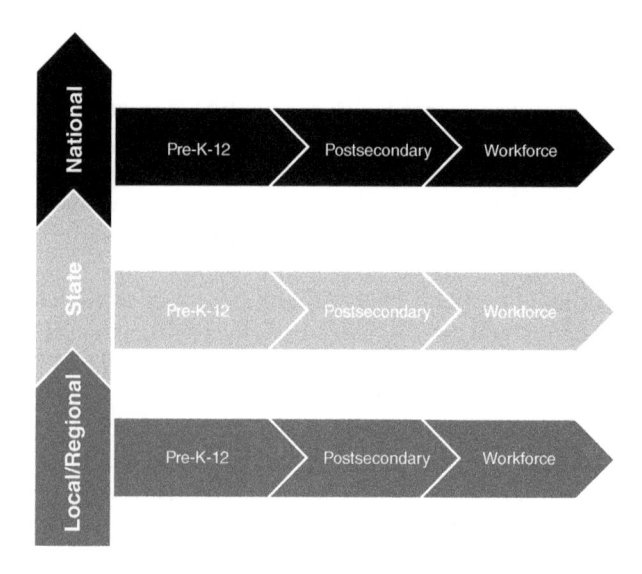

Figure 11.4 Horizontally and vertically aligned STEM partnerships.

approach). From a horizontal perspective, STEM policies should be strategically formulated to meet the needs of and close gaps between pre-K-12, postsecondary and workforce. The horizontal alignment reflects the continuous flow of students and achievement in the pre-K-20 education pipeline. Vertical policy alignment means that there are local connections to the state and national levels. This maximizes resources and results in stronger, more informed policies. Horizontal and vertical policy alignment reinforces the importance of building horizontal and vertical partnerships at the frontend of coalition development.

Establishing Goals and Policies

In the DATA section of this chapter, we identified key data benchmarks that speak to the condition of any pre-K-12 and postsecondary STEM education pipeline. These benchmarks help partners identify what is working well, illuminate leaky points in the pipeline, and determine where STEM goals and policy strategies are necessary to advance talent. Communities should consider these benchmarks as a first order of business when establishing goals and policies.

Next, partners must recognize the realities of the localized education ecosystem. *Lasting Impact: A Business Leadership's Playbook for Supporting America's Schools* (Allan et al., 2014) suggests that education in America is largely local, and each city or town has its own "education ecosystem." Understanding how this ecosystem functions, including which horizontal partners contribute to student achievement – not simply schools themselves, but non-profit organizations, teachers' unions, government agencies, businesses, faith-based organizations, etc. – is essential to designing goals and policies that have the greatest short-term and long-term impact.

Figure 11.5 illustrates the elements of a dynamic local pre-K-12 education ecosystem. Goals and policies at every level should be student-focused, informed by data, and use key "drivers" critical to advancing STEM progress. Pre-K-12 STEM drivers include effective teachers, high standards, quality curriculum, formal STEM learning, informal STEM learning, and embedded technology.

Certain enablers must be in place to support the use of the drivers to set goals and policies across the pipeline. The enablers create the right context for change and ensure the ongoing follow-through and long-term success of STEM policies. These include strong, committed leadership; horizontal and vertical partner engagement (as previously discussed); clear metrics; sophisticated data collection; and strong accountability.

In the end, STEM policy development should seek transformational action that results in improved student performance. STEM drivers must be used to fashion goals and policies that strategically push an ecosystem to a new configuration and level of performance (Allan et al., 2014).

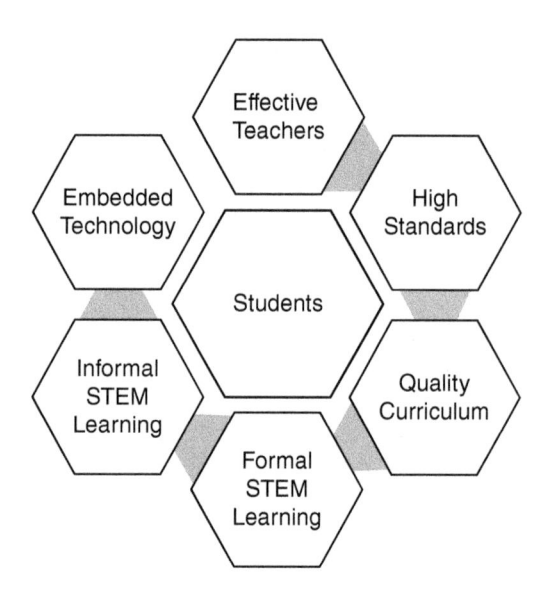

Figure 11.5 Pre-K-12 STEM drivers.

Such policies should help leaders and administrators fundamentally rethink how things are done.

Partners have often depended upon a single driver or enabling element within a single ecosystem to create impact. For instance, perhaps, a business partner developed a policy to provide supplies to a nearby school or to 'adopt' one school building. Rarely have such parochial efforts been enough to push an ecosystem to a new configuration and level of performance. Partners should consider what combination of drivers is likely to produce significant long-term impact.

#4: FOLLOW-THROUGH and Seek Transformative Action in a Change-Resistant Climate

We opened this chapter by discussing the success Ohio partners experienced in advocating for a set of comprehensive STEM policies and investments. We wind down the chapter by discussing how Ohio partners did it. The truth is that a great deal of time, collaboration, perseverance, planning, and follow-through drove the success of the initiative. That work continues today.

Ohio's state-level efforts to focus on STEM education date back to 2004, when the then-Governor Bob Taft commissioned a panel of horizontal partners to make recommendations on ways to maximize the use of higher education as a tool for economic growth. The panel recommended an intense focus on STEM higher education. This recommendation helped trigger a 10-year focus on STEM education.

Two years later, the OSLN was launched in 2006 by Battelle and the Ohio Business Roundtable. The OSLN led a process to assign specific roles to key partners. For instance, Battelle, anchor partner to the OSLN, was on point to handle member engagement and the day-to-day functions of the partnership. The Teaching Institute for Excellence in STEM (TIES) was charged with providing technical support to network programs across the state. The third partner, the Education Council, ensured that the best practices were captured and disseminated to partners across the network. Finally, the Ohio Business Roundtable led the network's policy and advocacy work. Its first order of business was to convene stakeholders to develop a set of goals, policy strategies, and metrics to drive STEM policy and advocacy in Ohio.

Once those goals were established, the OSLN gathered key stakeholders from around the state and facilitated a process to identify the critical few items all could rally around and advocate for at the state level. What followed was a coordinated set of advocacy activities that all partners engaged in to secure support from key state leaders, agency directors, and elected officials. Different partners were best positioned to do different things. And some partners had never engaged in advocacy work before. From visiting legislators to providing testimony to crunching the data and making the case, members of the OSLN focused on gaining policy change, securing financial support, and gathering state-level champions to fuel the drive toward reaching STEM education goals.

To guide partners, the Ohio Business Roundtable produced the Ohio STEM AdvoKit, intended as a one-stop tactical advocacy guide. In its original form, the AdvoKit contained an overview of the national and state STEM education landscape; tailored sets of talking points for STEM advocates, including students, parents, educators, employers, and community leaders; a STEM FAQ; clarification of what STEM education is and isn't; top-line messaging; and sample letters of support that could be used by stakeholders to frame letters to the legislature and the press. Today, the OSLN features the AdvoKit as one of its key tools, and others have used it as a platform to tailor for their own state work. To view the document in full, please visit: http://www.osln.org/wp-content/uploads/2013/03/Ohio-AdvoKit.pdf. This tool proved essential in coordinating stakeholder advocacy deployment and messaging.

The OSLN and its partners continue to follow through on implementing STEM practices and identifying enabling policies necessary for long-term goals. While 2008 might have been the beginning, it was certainly not the end. Successive state budgets continue to honor STEM as a critical investment to the state's future, including the most recent Mid-Biennial Review budget, which included several key provisions critical to the on-going success of STEM schools and the flexibility for innovation in STEM areas.

Other states have developed similar tactical tools to help partners advance STEM teaching and learning in coordinated and sophisticated fashion. North Carolina, for instance, used its "Do-It-Yourself Guide to STEM Community Engagement" to engage and dispatch a broad range of community leaders to support STEM education in a targeted way.

Similarly, the California STEM Learning Network maintains a strong focus on STEM policy and advocacy. The state-level network offers tips on what local community members can do to advance STEM education in schools and communities across the state. To assist its members, the network issues regular publications to guide including a strategy roadmap to transform STEM education in California, a policy brief that explained the importance of the Next Generation Science Standards, and brief aimed at rethinking teacher preparation and policy.

States across the nation are moving the STEM *policy needle* with the goal of moving the STEM *student achievement and talent needle*. This work is challenging and rests upon focused follow-through and long-term commitment. STEM goals and policies are tools for students and teachers to make increased STEM achievement and degrees a reality. The four steps discussed in this chapter are tools for communities and states to establish strong STEM goals and policies. If community and state leaders use data to inform the policy, organize the right people, recognize the right drivers to affect change, develop and advocate for the right policies, and make adjustments based on what works, then far more students will be prepared for college, careers, and engagement in a STEM-filled world.

Characteristics of Effective STEM Programs

There are clear characteristics of effective STEM programs emerging from the national STEM education work (e.g., Johnson, Mohr-Schroeder, Moore, & English, 2020; Johnson & Sondergeld, 2020; Johnson, Sondergeld, & Walton, 2017). These were articulated in Figure 11.5 as the PK-12 STEM drivers. Stakeholders involved in STEM education reform should carefully consider the use of each pre-K-12 success driver to set STEM goals and policies that promote innovation.

Effective Teachers

Research indicates that a classroom teacher's effectiveness is more important – and has more impact on student achievement – than any other factor controlled by school systems, including class size or the school a student attends (e.g. Darling-Hammond, 2010; Rivkin, Hanushek, & Kain, 2005). Thoughtful, skillful educators are the backbone to delivering innovative STEM instruction across elementary and secondary education. They understand the standards for what students should know and able to do. They know how to cleverly integrate those standards throughout curriculum and instruction. Teachers drive formal STEM learning and develop and deliver the hands-on, project-based instruction. Advancing policies that effectively prepare new teachers and sharpen the effectiveness of those already practicing, particularly in the STEM disciplines, will have a positive impact on student performance.

Cutting-edge, research-based professional development opportunities (e.g. Desimone, 2009; Johnson and Fargo, 2014) play a significant role in enhancing teacher effectiveness. Educators need exposure to high-quality professional development that sharpens their craft in the classroom. The engagement of horizontal partners, from business to higher education, often enhances innovative professional development offerings. For instance, a STEM-related business might open its laboratories to local teachers and give them an opportunity to work alongside laboratory technicians, strengthening content knowledge and offering real-world application, which can be transferred back to the classroom.

MC^2STEM, a Cleveland Metropolitan School District high school, maintains three campuses located on site at the Great Lakes Science Center, GE Lighting's Nela Park, and the Health Careers Center. The high school, established through a public-private partnership, leverages the expertise of professionals on each campus to deliver teacher professional development that focuses on cross-training experiences and trans-disciplinary instructional units. Industry partners and professionals from higher education not only enhance professional development opportunities but also they provide direct instruction to the students on many occasions (MC^2STEM High School, 2014). Ohio adopted a statewide policy to establish and invest in STEM training centers, enabling MC^2STEM to train educators in its district and across the state.

High Standards

Implementing rigorous STEM-related academic standards is a prominent, far-reaching driver that can impact every student. States that implement rigorous standards are setting expectations for what all students should know and be able to do, regardless of where students receive their education. As state policymakers consider adopting standards in mathematics, science, engineering, and technology – whether through the *Common Core State Standards* (for mathematics), the *Next Generation Science Standards* (NGSS), engineering standards, or other homegrown standards – they should carefully consider how the standards promote meaningful integrated STEM education opportunities. Integrating standards across the STEM disciplines can significantly enhance the student learning experience. The real world is integrated by nature, and an interdisciplinary approach provides authentic contexts for learning (Ronis, 2007, Roth, 1993).

"A Framework for K-12 Science Education" offers a prime example of what this disciplinary integration might look like for science standards. The Framework suggests that K-12 science standards be built around three dimensions: (1) science and engineering practices; (2) crosscutting concepts that unify the study of science and engineering through their common application across fields; and (3) core ideas in four disciplinary areas: physical sciences; life sciences; earth and space sciences; and engineering, technology, and the applications of science. The Framework goes on to indicate that these dimensions should be integrated into standards, curriculum, and instruction.

Integrated Curriculum

Standards, which are typically set at the state level, allow local educators to choose or design integrated curriculum and instruction tailored to the needs of their students. Integrated STEM curriculum, as discussed throughout this STEM Road Map as the pivotal component of STEM reform – brings together the content disciplines for the purposes of teaching and learning – makes learning more relevant for students. It helps them form deeper understandings and build connections among central concepts. Students become more interested and vested in school when instruction is based on integrated curriculum (Berlin, 1994; Berlin & White, 2012; George, 1996; Mason, 1996; Morrison & McDuffie, 2009). This is often done in elementary grades through project or problem-based learning units and in high school via hybrid courses, career, and technical education programs and focused STEM schools and programs. State policy, however, must allow for such programs and courses as well as provide appropriate waivers and approvals as needed. This may also include rethinking traditional assessments to ensure that more than just factual knowledge is being measured (Johnson, 2013).

Formal STEM Learning

Formal STEM learning most often occurs during the traditional school day. And while the school day might be traditional, the teaching and learning approach is anything but. An inquiry-based approach is prominent in many formal STEM learning opportunities. When professional engineers encounter a problem in the field, for instance, they implement a series of steps known as the engineering design process. The learner is given the opportunity to ask questions; define problems; model, plan, and conduct investigations; analyze and interpret data; apply mathematics and computations thinking; construct explanations and solutions; and communicate findings (Czerniak & Johnson, 2014). Inquiry-based instruction is maximized though the use of integrated curriculum. Project-based learning (PBL) is another approach to formal STEM learning. PBL experiences require students to uncover and address real-world problems and share findings with authentic audience (Riordian, 2013). PBL features curriculum and directly applies the engineering design process and inquiry-based learning (e.g. Czerniak & Johnson, 2104; Ronis, 2007; Roth, 1993).

Some regions and states have created and launched STEM schools to deliver this type of learning experience. For instance, Ohio, North Carolina, Tennessee, Texas, and others launched STEM schools, in partnership with horizontal and vertical partners, to completely transform the delivery of formal STEM learning.

Such learning experiences can also take place as units within traditional schools. Arizona, New Mexico, New York, North Carolina, and Texas have enacted policy changes that facilitate and promote new curricular and programmatic approaches aimed at transforming existing public schools. Arizona's approach is particularly unique. The Arizona STEM Network, in partnership with the Maricopa County Education Association (a key horizontal partner), established the STEM Immersion Guide to help schools and districts integrate STEM education into curriculum and instruction. The STEM Immersion Guide contains key design elements that support the development of project-based, interdisciplinary STEM instruction and provides practical tools and information to assist teachers, administrators, schools, and districts that want to improve student outcomes by integrating STEM (Science Foundation Arizona, 2013).

Informal STEM Learning

Informal STEM learning is perhaps one of the greatest examples as to why horizontal partners matter. If representatives from local museums, zoos, universities, business, and research centers are at the table as the policies are being considered, then there is greater chance that such partners would be committed to lending their time, talent, and resources to execute joint strategies where they are needed most.

Since most states do not consistently make informal science a partner in the STEM agenda, the role of vertical partners becomes even more important for regions. Consider that states provide only a small share of direct funding to informal science institutions, while the majority of the funding comes from the federal government, corporate, and private foundations and the general public. The National Science Foundation (NSF) is interested in growing the body of research that will help regions and states make the case for increased support for informal STEM learning. NSF has dedicated up to $14.4 million to advance new approaches to and evidence-based understanding of the design and development of

STEM learning in informal environments, provide multiple pathways for broadening access to STEM learning experiences, and advance assessment of informal STEM learning (National Science Foundation, 2014).

Local and state partners can maximize the use of the informal STEM learning driver by developing policies aimed at bringing formal and informal STEM learning experiences together. This type of merger has the potential to fuel high school internships for students, accelerate online course taking from third party partners, proliferate team teaching opportunities where teachers couple with STEM professionals, and increase student opportunity to earn credit from challenging out-of-school experiences. In some places, informal partners are also developing their own pre-schools and professional development for pre-K-12 teachers. This driver is ripe for policy innovation.

Embedded Technology

STEM education should employ the latest technologies as tools for teaching and learning and as the content for learning. The "technology" component of STEM seeks to prepare students to understand, deploy, and in many cases, develop technologies that are connected to real-world STEM applications. Technology is a part of the learning experience; embedded in it, not apart from it.

Many "blended learning" environments combine embedded technology, different pedagogical approaches, and unique classroom operations to help educators personalize learning for individual students. This often results in better student outcomes. Models should be structured so that every student has opportunities for individualized learning, and every teacher has the time and resources, including data, to differentiate small group or one-to-one instruction. Technology that is simply overlaid on an antiquated model of schooling increases the costs of education and the challenges to improving student achievement.

Example state-level policies might include efforts to ensure sufficient internet connectivity for schools or to provide competitive funds to districts that use technology and innovation to transform teaching and learning, such as Ohio's Straight "A" Fund (Ohio Department of Education, 2014).

These six drivers, if positioned and used properly, have the potential to significantly affect STEM student success in pre-K-12. They should command priority focus as communities determine STEM goals and policies. The drivers are inextricably linked to STEM student success beyond high school.

STEM School/Program Start-up Process

The STEM school start-up process is comprised of three main stages: strategic planning, development, and implementation (Johnson, 2014).

Frameworks and Advocacy for STEM 249

The strategic planning phase is focused on development of mission, vision, goals, objectives, and intended outcomes from the desired STEM approach. This should be planned collaboratively with a team that is representative of all stakeholders involved in the effort, including K-12 and community partners.

The development phase consists of developing a plan for necessary teacher professional development on new pedagogy, content, and technological skills. A second component is curriculum development, where teams of teachers work with expert STEM curriculum facilitators to develop integrated STEM curriculum through modification of existing resources or generation of new ideas and concepts that are designed to engage students in solving real-world problems. The third component is development of school climate, including scheduling, teacher collaboration time for planning, engagement of STEM experts and community partners in co-teaching, field trips, and planning of curriculum, and determining student STEM experiences that will take place outside of the school walls.

The implementation phase is the actual beginning of the implementation of the STEM school or program plan. During this phase, there should be considerable effort focused on providing teachers time to collaborate, refining and revising of curriculum, assessment of fidelity of implementation, assessment of teacher and student outcomes, and real-time professional development for teachers and partners engaged in the work.

STEM Innovations, LTD (www.steminnovations.com) has provided support for the STEM School/Program start-up process nationally for several years and will work with school districts and other agencies to collaboratively develop an individualized plan that will leverage existing resources. State-level STEM networks may also be a resource that would provide support to schools and programs that are interested in implementing a STEM approach.

Bringing It All Together: A Look at Tennessee and Texas

Across the country, STEM partnerships are developing comprehensive STEM goals and policies designed to transform the education ecosystem and achieve long-term impact. Tennessee and Texas, among others, are employing multiple drivers to accomplish their goals.

In August 2012, the Tennessee STEM Innovation Network, a state-level STEM partnership, released *Future-Ready Tennessee: Developing STEM Talent for 2018 and Beyond*. The strategic plan sets out to answer the question: "Will Tennessee have the competitive and skilled workforce it needs to prosper in a STEM-driven economy?" The plan (Tennessee STEM Innovation Network, 2012) used state STEM data to inform development of four goals aimed at accelerating STEM talent development:

1 Increase student interest, participation, and achievement in STEM;
2 Expand student access to effective STEM teachers and leaders;

3 Reduce the state's STEM talent and skills gaps; and
4 Build community awareness and support for STEM.

Each goal is supported by a set of strategies and progress metrics that track to the drivers discussed in this chapter. Goal #1, for instance, identifies four strategies including establishing regional STEM innovation hubs to bring horizontal partners together locally; launching STEM platform schools to change the teaching and learning model; ensuring all students have access to rigorous STEM courses; and identifying, developing, and sharing STEM curriculum tools.

This state-level STEM strategic plan is significant because it is anchored in data and builds upon broader, pre-existing state-level policy goals in K-12, higher education, and workforce. It also identifies and capitalizes on the state's STEM assets – from institutions of higher education to high-tech health and research organizations and global businesses – to enhance STEM teaching and learning for students across the state. Tennessee has been using its strategic plan as a guide for STEM policy and advocacy for the last two years.

Almost a decade ago, Texas launched T-STEM, an initiative squarely focused on using pre-K-12 STEM drivers to transform the delivery of teaching and learning to students and affect the state's broader education ecosystem. Then, the Texas High School Project (now Educate Texas) partnered with the governor, Texas Education Agency (TEA), and major philanthropists to start 36 STEM schools to serve students in greatest need. Today, T-STEM aims to empower teachers, inspire students, and advance studies in STEM. The public-private initiative includes academies, professional development centers and networks aimed at improving instruction, and academic performance in science and mathematics-related subjects at secondary schools.

The goals of the T-STEM Academies are clear: increase the number of students entering postsecondary studies and careers in STEM; promote quality school leadership that support school redesign efforts, quality teacher recruitment, and improved teacher preparation; and assist in long-term educational and economic development and alignment of the STEM fields (Communities Foundation of Texas, 2014a, 2014b).

Parallel to launching T-STEM Academies, the state launched T-STEM Centers, which are located at universities and regional education service centers, to create new STEM instructional materials and provide high-quality professional development. They coordinate with industry and business, which provide resources to T-STEM Academies. To connect the work of T-STEM Academies and T-STEM Centers, the state also launched the T-STEM Network, which offers professional development, exemplary profiles, and other STEM education resource.

To scale its T-STEM Academies with fidelity across the state, the initiative created a T-STEM Academy Blueprint and established a rigorous

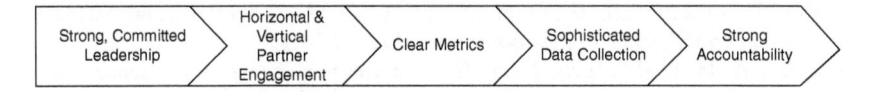

Figure 11.6 Five supporting enablers for PK-postsecondary success.

[Flow diagram: Strong, Committed Leadership → Horizontal & Vertical Partner Engagement → Clear Metrics → Sophisticated Data Collection → Strong Accountability]

process for existing schools to gain the STEM designation in concert with TEA. T-STEM Academies use the T-STEM Design Blueprint, Rubric, and Glossary as a guidepost to build and sustain STEM schools that focus on mission-driven leadership; school culture and design; student outreach, recruitment, and retention; curriculum, instruction, and assessment; strategic alliances; and academic advancement and sustainability. The T-STEM Rubric also addresses the five supporting enablers (Figure 11.6). To date, 70 T-STEM Academies and seven blended Early College High School/T-STEM Academies serve more than 40,000 students across the state. Impact is even more far reaching, thanks to the T-STEM Network, which ensures dissemination of best practices across the Lone Star state (Communities Foundation of Texas, 2014a, 2014b).

References

Achieve. (2014). *The ADP network*. Retrieved from http://www.achieve.org/adp-network

Allan, S., Grossman, A., Rivkin, J. W., Vaduganathan, N., Childress, S., Henry, T., ... Sommerfeld, M. (2014). *Lasting impact: A business leader's playbook for supporting America's schools*. Retrieved from http://www.hbs.edu/competitiveness/pdf/lasting-impact.pdf

Bayer. (2013). *Is there STEM workforce shortage?* Retrieved from http://www.bayerus.com/News/NewsDetail.aspx?ID=DCF6FA59-C9B5-2F14-A7AB69927FB04B0B

Berlin, D. (1994). The integration of science and mathematics education: Highlights from the NSF/SSMA Wingspread conference plenary papers. *School Science and Mathematics, 94*(1), 32–35.

Berlin, D., & White, A. (2012). A longitudinal look at attitudes and perceptions related to the integration of mathematics, science, and technology education. *School Science and Mathematics, 112*(1), 20–30.

Blank, R. K. (2012). *What is the impact of decline in science instructional time in elementary school?* Report to the Noyce Foundation.

Bureau of Labor Statistics. (n.d.). *Databases, tables, and calculators by subject*. Retrieved from http://bls.gov/timeseries/LNS14000000

Burning Glass Technologies. (2014). *Summary: Real-time insight into the market for entry-level STEM jobs*. Retrieved from http://www.burning-glass.com/media/3347/Real%20Time%20STEM%20Insight%20Summary.pdf

Business-Higher Education Forum. (2007). *An American imperative: Transforming the recruitment, retention, and renewal of our nation's mathematics and science teacher workforce*.

Breiner, J., Harkness, M., Johnson, C. C., & Koehler, C. (2012). What is STEM? A discussion about conceptions of STEM in education and partnerships. *School Science and Mathematics, 112*(1), 3–11.

Camins, A. H. (2014, July). Next steps for the next generation science standards. *Education Week.* Retrieved from http://www.edweek.org/ew/articles/2014/07/22/37camins.h33.html

Carnevale, A. P., Smith, N., & Strohl, J. (2011). *STEM: Science, technology, engineering, mathematics.* Georgetown University Center on Education and the Workforce. Retrieved from https://georgetown.app.box.com/s/cyrrqbjyirjy64uw91f6

Carnevale, A. P., Smith, N., & Strohl, J. (2013). *Recovery: Job growth and education requirements through 2020.* Georgetown University Center on Education and the Workforce. Retrieved from http://cew.georgetown.edu/recovery2020

Change the Equation. (2014). *STEMtistics – science.* Retrieved from http://changetheequation.org/stemtistics/science?page=1

Chen, X., & Ho, P. (2012). *STEM in postsecondary education: Entrance, attrition, and coursetaking among 2003–2004 beginning postsecondary students* (NCES Report 2013152). Retrieved from http://nces.ed.gov/pubs2013/2013152.pdf

Chen, X., & Soldner, M. (2013). *STEM attrition: College students' paths into and out of STEM fields* (NCES Report 2014001). Retrieved from http://nces.ed.gov/pubs2014/2014001rev.pdf

Common Core State Standards Initiative. (2014). *Standards in your state.* Retrieved from http://www.corestandards.org/standards-in-your-state/

Complete College America. (2012). *Remediation: Higher educations' bridge to nowhere.* Retrieved from http://completecollege.org/docs/CCA-Remediation-final.pdf

Communities Foundation of Texas. (2014a). *Educate Texas.* Retrieved from http://www.edtx.org/college-ready-standards-and-practices/t-stem/

Communities Foundation of Texas. (2014b). T-STEM academy design blueprint. *Educate Texas.* Retrieved from http://www.edtx.org/college-ready-standards-and-practices/t-stem/

Czerniak, C.M. & Johnson, C.C. (2014). Interdisciplinary Science and STEM Teaching. Invited handbook chapter to appear in Lederman, N.G. & Abell, S.K. (Eds.), *Handbook of Research on Science Education 2nd Edition,* Lawrence Erlbaum Associates, Inc., 395-412.

Darling-Hammond, L. (2010). *The flat world and education: How America's commitment to equity will determine our future.* New York: Teachers College Press.

Desimone, L.M. (2009). Improving impact studies of teachers' professional development: Toward better conceptualization and measures. *Educational Researcher, 38*(3), 181-199.

Economics and Statistics Administration. (2011). *STEM: Good jobs now and for the future.* Washington, DC: United States Department of Commerce.

George, P. S. (1996). The integrated curriculum: A reality check. *Middle School Journal, 28,* 12–19.

Guillory, F., & Quinterno, J. (2013). *Strategies that engage minds: Empowering North Carolina's economic future.* Retrieved from http://ncsmt.org/wp-content/uploads/2014/03/NCSTEMScorecard.pdf

Hanleybrown, F., Kania, J., & Kramer, M. (2012). Channeling change: Making collective impact work. *Stanford Social Innovation Review*. Retrieved from http://www.ssireview.org/blog/entry/channeling_change_making_collective_impact_work

Johnson, C.C., Mohr-Schroeder, M., Moore, T. & English, L. (2020). *Handbook of Research on STEM Education*. Routledge.

Johnson, C.C., & Sondergeld, T.A. (2020). Outcomes of an integrated STEM high school: Enabling access and achievement for all students. *Urban Education*, 1–27, https://doi.org/10.1177/0042085920914368

Johnson, C.C., Sondergeld, T.A., & Walton, J. (2017). A statewide implementation of the critical features of professional development: Impact on teacher outcomes. *School Science and Mathematics*, 117 (8), 350-351.

Johnson, C. C., & Fargo, J. D. (2014). A study of the impact of Transformative Professional Development (TPD) on Hispanic student performance on state-mandated assessments of science. *Journal of Science Teacher Education*, 25, 845–859.

Johnson, C. C. (2013). Educational turbulence: The influence of macro and micro policy on science education reform. *Journal of Science Teacher Education*, 24(4), 693–715.

Johnson, C. C. (2014). *Indiana STEM School Summit*. Retrieved from www.education.purdue.edu/news/STEM Summit.pdf

Kania, J., & Kramer, M. (2011). Collective impact. *Stanford Social Innovation Review*. Retrieved from http://www.ssireview.org/articles/entry/collective_impact

Kaplan, D. A. (2010, June). The STEM challenge. *Fortune Magazine*, p. 25.

Learn to Earn Dayton. (n.d.). Retrieved from http://learntoearndayton.org/

Mason, T. C. (1996). Integrated curricula: Potential and problems. *Journal of Teacher Education*, 47(4), 263–270.

MC^{2}STEM High School. (2014). STEM connection: From classroom to workplace. Retrieved from http://www.hbs.edu/competitiveness/pdf/lasting-impact.pdf

Morrison, J., & McDuffie, A. R. (2009). Connecting science and mathematics: Using inquiry investigations to learn about data collection, analysis, and display. *School Science and Mathematics*, 109(1), 31–44.

National Assessment of Educational Progress (2019). NAEP Report Card: Mathematics. Retrieved from https://www.nationsreportcard.gov/mathematics/nation/scores/?grade=4 https://nces.ed.gov/programs/raceindicators/indicator_reg.asp

National Assessment of Educational Progress (2015). NAEP Report Card: Science. Retrieved from https://www.nationsreportcard.gov/science_2015/#scores/chart_loc_1?grade=12

National Science Foundation. (2014). Advancing informal STEM learning (AISL). Retrieved from http://www.nsf.gov/funding/pgm_summ.jsp?pims_id=504793

New Hampshire Charitable Foundation (2014). Building a STEM Workforce. Retrieved from https://www.nhcf.org/what-were-up-to/smarter-pathways-4/

North Carolina STEM Center (2020). NC STEM Scorecard. Retrieved from https://www.ncstemcenter.org/learn/stem-scorecard/

Ohio Business Alliance for Higher Education & the Economy. (2007). KidsOhio.org. Retrieved from http://www.kidsohio.org/wp-content/uploads/2008/02/7-12-07-stem-analysis-h-b-119-doc-sy-_3_-3.pdf

Ohio Department of Education. (2014). *Straight A fund*. Retrieved from http://education.ohio.gov/Topics/Straight-A-Fund

Organisation for Economic Co-operation and Development (OECD). (2018). Programme for International Student Assessment (PISA) Results. Retreived from https://www.oecd.org/pisa/publications/pisa-2018-results.htm

Riordan, R. (2013, January). Change the subject: Making the case for project-based learning. *Edutopia*. Retrieved from http://www.edutopia.org/blog/21st-century-skills-changing-subjects-larry-rosenstock-rob-riordan

Rivkin, S. G., Hanushek, E. A., & Kain, J. F. (2005). Teachers, schools and academic achievement. *Econometrica, 73*(2), 417–458.

Ronis, D. L. (2007). *Problem-based learning for math and science: Integrating inquiry and the internet*. Thousand Oaks, CA: Corwin.

Roth, W. M. (1993). Problem-centered learning for the integration of mathematics and science in a constructvist laboratory: A case study. *School Science and Mathematics, 93*(3), 113–122.

Rothwell, J. (2013, June). The hidden STEM economy: Key findings. *Brookings*. Retrieved from http://www.brookings.edu/research/interactives/2013/the-hidden-stem-economy

Science Foundation Arizona. (2013). *Welcome to the STEM immersion guide*. Retrieved from http://stemguide.sfaz.org/

Sondergeld, T., & Johnson, C. C. (2014). Using the Rasch Model for the development of affective measures in science education research. *Science Education, 98*(4), 581–613.

STEMx. (n.d.). *Sustainability compass*. Retrieved from http://www.stemx.us/sustainability-compass/destination-sustainability/

Tennessee STEM Innovation Network. (2012). *Future read Tennessee: Developing STEM talent for 2018 and beyond*. Retrieved from http://thetisin.org/wp-content/uploads/2013/04/FINAL-TN-STEM-Strategic-Plan-Aug2012.pdf

Thomasian, J. (n.d.). *The role of informal science in the state education agenda*. National Governors Association Issue Brief. Retrieved from http://www.nga.org/files/live/sites/NGA/files/pdf/1203INFORMALSCIENCEBRIEF.PD

US Department of Commerce. (2012). *The competitiveness and innovative capacity of the United States*. Washington, DC.

US News and World Report: Money and Careers. (2014). *The best 100 jobs*. Retrieved from http://money.usnews.com/careers/best-jobs/rankings/the-100-best-jobs

Appendix A
Sample STEM Module One: Grade 7

Janet B. Walton and James M. Caruthers

STEM Road Map Curriculum Module Overview

STEM Road Map Module Theme and Grade Level: Cause and Effect, Grade 7

STEM Road Map Module Topic: Transportation – Motorsports

Lead Discipline: Science

Module Summary:

Students will take on the role of design engineers as they work in teams to design, within a set of design constraints, an innovative prototype vehicle with a new safety aspect and powered by energy transformations. As they move through the module, students will investigate types of energy, energy transformations, the law of conservation of energy, the concepts of speed, friction, aerodynamic drag, and the engineering design process (EDP). The module will culminate in the design project, The Automotive X-Challenge. Students will participate in a race day event in which cars will compete for speed and will present their design to industry professionals to be judged for design, innovation, teamwork, and presentation quality.

Established Goals/Objectives:

The goals for this module are for students to be able to:

- Understand the physics concepts of motion, force, and energy (Science)
- Utilize the EDP in an authentic, real-world problem situation (Engineering)
- Make connections to the historical, economic, geographic, and cultural aspects of motorsports (Social Studies)
- Utilize appropriate mathematics practices and content to complete authentic tasks (Mathematics)

- Communicate learning and experiences through various forms of writing and speaking while effectively using relevant vocabulary and grammar (English/Language Arts)
- Build mastery of relevant 21st Century Themes and Skills

Science (NGSS Performance Objectives):

Motion and Stability (MS-PS2); Energy (MS-PS3); Engineering Design (MS-ETS1)

Students will understand that there are different forms of energy with unique characteristics; understand the concepts of speed, friction, and aerodynamic drag; and understand the elements of the EDP. Students will apply their understanding of energy, energy transformations, speed, friction, aerodynamic drag, materials, and the EDP by working in teams to design a vehicle powered by energy transformations and will be able to discuss the design process and the energy transformations that power their vehicle. Student teams will investigate a topic of their choice related to motorsports and will present their designs and topical research projects through oral and visual presentations.

Driving Question/Problem for Students to Solve:

How can we design and build a mode of transportation that is powered by energy transformations?

Launch: Introduce unit by showing video of automotive X-prize (https://www.youtube.com/watch?v=car1X_YElxk). Discuss automotive innovations and concept of engineering design. Present invitation/flyer for a class X-challenge to design a prototype car.

Lesson Plan #1 – Transportation – Motorsports

Lesson Title: Start Your Engines

Lesson Summary: This lesson introduces the X-Challenge unit. The concept of innovative car design will be introduced via a video and discussion. Engineering as a profession will be introduced using racecar designers as an exemplar. Students will be introduced to the driving question for the unit and receive an invitation to participate in the Automotive X-Challenge. The elements of the EDP will be introduced, and students will use the EDP in a mini design challenge.

Essential Question(s):

- What skills will we use to design a solution for the unit's challenge?
- What do engineers do and how do they do their work?

Table A.1.1 Content Standards Addressed in STEM Road Map Module – Transportation – Motorsports

NGSS Performance Outcomes Standards	Common Core Mathematics Standards	Common Core English/Language Arts Standards
MS-PS2-1. Apply Newton's Third Law to design a solution to a problem involving the motion of two colliding objects.	**CCSS.Math.Practice.1.** Make sense of problems and persevere in solving them.	**CCSS.ELA-Literacy.RL.7.1** Cite several pieces of textual evidence to support analysis of what the text says explicitly as well as inferences drawn from the text.
MS-PS2-2. Plan an investigation to provide evidence that the change in an object's motion depends on the sum of the forces on the object and the mass of the object.	**CCSS.Math.Practice.2.** Reason abstractly and quantitatively.	**CCSS.ELA-Literacy.RL.7.7.** Compare and contrast a written story, drama, or poem to its audio, filmed, staged, or multimedia version, analyzing the effects of techniques unique to each medium (e.g. lighting, sound, color, or camera focus and angles in a film).
	CCSS.Math.Practice.3. Construct viable arguments and critique the reasoning of others.	**CCSS.ELA-Literacy.W.7.1.** Write arguments to support claims with clear reasons and relevant evidence.
MS-PS2-3. Ask questions about data to determine the factors that affect the strength of electric and magnetic forces.	**CCSS.Math.Practice.4.** Model with mathematics.	**CCSS.ELA-Literacy.W.7.1a.** Introduce claim(s), acknowledge alternate or opposing claims, and organize the reasons and evidence logically.
	CCSS.Math.Practice.5. Use appropriate tools strategically.	
MS-PS2-5. Conduct an investigation and evaluate the experimental design to provide evidence that fields exist between objects exerting forces on each other even though the objects are not in contact.	**CCSS.Math.Practice.6.** Attend to precision.	**CCSS.ELA-Literacy.W.7.2.** Write informative/explanatory texts to examine a topic and convey ideas, concepts, and information through the selection, organization, and analysis of relevant content.
	CCSS.Math.Practice.7. Look for and make use of structure.	**CCSS.ELA-Literacy.W.7.2a.** Introduce a topic clearly, previewing what is to follow; organize ideas, concepts, and information, using strategies such as definition, classification, comparison/contrast, and cause/effect; include formatting (e.g. headings), graphics (e.g. charts, tables), and multimedia when useful to aiding comprehension.
MS-PS3-1. Construct and interpret graphical displays of data to describe the relationship of kinetic energy to the mass of an object and to the speed of an object.	**CCSS.Math.Practice.8.** Look for and express regularity in repeated reasoning.	
	CCSS.Math.Content.7.RP.A.1. Compute unit rates associated with ratios of fractions, including ratios of lengths, areas, and other quantities measured in like or different units.	**CCSS.ELA-Literacy.W.7.2b.** Develop the topic with relevant facts, definitions, concrete details, quotations, or other information and examples.

(*Continued*)

Table A.1.1 (Continued)

NGSS Performance Outcomes Standards	Common Core Mathematics Standards	Common Core English/Language Arts Standards
MS-PS3-2. Develop a model to describe that when the arrangement of objects interacting at a distance changes, different amounts of potential energy are stored in the system.	**CCSS.Math.Content.7.RP.A.2.** Recognize and represent proportional relationships between quantities.	**CCSS.ELA-Literacy.W.7.6.** Use technology, including the internet, to produce and publish writing and link to and cite sources as well as to interact and collaborate with others, including linking to and citing sources.
MS-PS3-5. Construct, use, and present arguments to support the claim that when the kinetic energy of an object changes, energy is transferred to or from the object.	**CCSS.Math.Content.7.NS.A.3.** Solve real-world and mathematical problems involving the four operations with rational numbers.	**CCSS.ELA-Literacy.W.7.7.** Conduct short research projects to answer a question, drawing on several sources and generating additional related, focused questions for further research and investigation.
MS-ETS1-1. Define the criteria and constraints of a design problem with sufficient precision to ensure a successful conclusion, taking into account relevant scientific principles and potential impacts on people and the natural environment that may limit possible solutions.	**CCSS.Math.Content.7.EE.B.3.** Solve multi-step real-life and mathematical problems posed with positive and negative rational numbers in any form (whole numbers, fractions, and decimals), using tools strategically. Apply properties of operations to calculate with numbers in any form; convert between forms as appropriate; and assess the reasonableness of answers using mental computation and estimation strategies.	**CCSS.ELA-Literacy.W.7.8.** Gather relevant information from multiple print and digital sources, using search terms effectively; assess the credibility and accuracy of each source; and quote or paraphrase the data and conclusions of others while avoiding plagiarism and following a standard format for citation.
		CCSS.ELA-Literacy.W.7.9. Draw evidence from literary or informational texts to support analysis, reflection, and research.
MS-ETS1-2. Evaluate competing design solutions using a systematic process to determine how well they meet the criteria and constraints of the problem.		**CCSS.ELA-Literacy.SL.7.1.** Engage effectively in a range of collaborative discussions (one-on-one, in groups, and teacher-led) with diverse partners on grade 7 topics, texts, and issues, building on others' ideas, and expressing their own clearly.
		CCSS.ELA-Literacy.SL.7.1a. Come to discussions prepared, having read, or researched material under study; explicitly draw on that preparation by referring to evidence on the topic, text, or issue to probe and reflect on ideas under discussion.

MS-ETS1-3. Analyze data from tests to determine similarities and differences among several design solutions to identify the best characteristics of each that can be combined into a new solution to better meet the criteria for success.	CCSS.Math.Content.7.EE.B.4. Use variables to represent quantities in a real-world or mathematical problem and construct simple equations and inequalities to solve problems by reasoning about the quantities.	CCSS.ELA-Literacy.SL.7.1b. Follow rules for collegial discussions, track progress toward specific goals and deadlines, and define individual roles as needed.
		CCSS.ELA-Literacy.SL.7.1c. Pose questions that illicit elaboration and respond to others' questions and comments with relevant observations and ideas that bring the discussion back on topic as needed.
MS-ETS1-4. Develop a model to generate data for iterative testing and modification of a proposed object, tool, or process such that an optimal design can be achieved.	CCSS.Math.Content.7.SP.A.1. Understand that statistics can be used to gain information about a population by examining a sample of the population; generalizations about a population from a sample are valid only if the sample is representative of that population. Understand that random sampling tends to produce representative samples and support valid inferences.	CCSS.ELA-Literacy.SL.7.1d. Acknowledge new information expressed by others and, when warranted, modify their own views.
		CCSS.ELA-Literacy.SL.7.3. Delineate a speaker's argument and specific claims, evaluating the soundness of the reasoning and the relevance and sufficiency of the evidence.
		CCSS.ELA-Literacy.SL.7.4. Present claims and findings, emphasizing salient points in a focused, coherent manner with pertinent descriptions, facts, details, and examples; use appropriate eye contact, adequate volume, and clear pronunciation.
		CCSS.ELA-Literacy.SL.7.5. Include multimedia components and visual displays in presentations to clarify claims and findings and emphasize salient points.

Table A.1.2 21st Century Skills Addressed in the STEM Road Map Module

21st Century Skills	Learning Skills and Technology Tools (from P21 Framework)	Teaching Strategies	Evidence of Success
21st century interdisciplinary themes	Financial, economic, business, and entrepreneurial literacy	• Draw connections between academic content and a variety of career pathways using a variety of resources including videos and classroom guests. • Highlight the importance of motorsports and manufacturing to the US economy. • Provide students with budget constraints for their prototype designs.	• Students can discuss the variety of jobs that support the motorsports industry. • Students can discuss the role of motorsports in the US economy. • Students create a cost-effective prototype. • Students can create a compelling presentation to a diverse audience highlighting the benefits of their prototype design.
Learning and innovation skills	Creativity and innovation; critical thinking and problem-solving; communication and collaboration	• Teach and facilitate student use of Engineering Design Process (EDP) throughout unit and design challenge. • Facilitate group work and instruct students on use of design journals.	• Students can implement the EDP in a group setting to create and present a prototype within budgetary and engineering constraints with evidence of collaboration. • Design journals reflect students' critical thinking and are used to draw connections between ideas and concepts presented in class and the prototype designs.

Information, media and technology skills	Information, communications, and technology (ICT) literacy	• Have students use technology to research their group motorsports topic. • Provide students with opportunities to incorporate multimedia elements into presentations. • Support appropriate use of technology resources and proper use of sources (i.e. citing sources appropriately).
		• Student presentations include information from internet research and/or multimedia presentation techniques. • References are acknowledged and cited where appropriate.
Life and career skills	Flexibility and adaptability; initiative and self-direction; social and cross-cultural skills; productivity and accountability	• Scaffold student group work through a series of investigations/lab activities to support team prototype design and topical research efforts. • Use EDP to teach flexibility (through redesign), time management and goal management. • Provide guidelines and practice opportunities for student presentations emphasizing professionalism and inclusivity of all team members.
		• Team projects are completed on time with evidence of participation by all team members. • Students present to peers, industry professionals, and teachers using appropriate language and professional demeanor. • Students are able to respond to questions regarding the design process and teamwork.

Table A.1.3 Desired Outcomes and Monitoring Success

Desired Outcome	Evidence of Success in Achieving Identified Outcome	
	Performance Tasks	Other Measures
Students will understand the elements of the engineering design process and use the process to apply the concepts introduced in the unit and the findings from their inquiry activities to the final design challenge.	• Students will maintain design journals (individual activity worksheets, Engineer It! worksheets, and reflections) • Students will design a working prototype car (team) • Student teams will research and present on a topic of their choice (team) • Students will be able to discuss how they applied their understanding of energy transformations and other concepts introduced in the unit to their designs (individual and team)	• Collaboration rubric

Table A.1.4 Assessment Plan

Major group products	• **Prototype vehicle** (evidence of incorporation of science concepts; evidence of innovation and creativity; evidence of use of EDP principles; evidence of collaboration – see prototype design rubric) • **Topical research project** (evidence of use of multiple sources of information) • **Presentation** (use of appropriate information and language; use of good presentation skills; response to audience questions; use of audiovisual aids; evidence of collaboration – see presentation rubric)
Major individual products/ deliverables	• Lab reports • **Design journal reflections** • **Topical quizzes** • **Participation** in prototype design and presentation

Table A.1.5 STEM Road Map Module Timeline – Weeks 1–3

Day 1 (Lesson 1)	Day 2 (Lesson 1)	Day 3 (Lesson 2)	Day 4 (Lesson 2)	Day 5 (Lesson 3)
Start Your Engines Launch the module. Introduce challenge, engineering design process (EDP)	Start Your Engines Students use EDP in a mini design challenge. Introduce design journals.	Let's Get Energetic! Introduce potential/kinetic energy as the two major categories of energy. Energy flow game.	Let's Get Energetic! Introduce Law of Conservation of Energy. Explore energy transformations and energy sources for cars.	Materials matter Introduce concept of materials science and gravitational potential energy. Ball drop lab activity.

Day 6 (Lesson 3)	Day 7 (Lesson 4)	Day 8 (Lesson 4)	Day 9 (Lesson 5)	Day 10 (Lesson 5)
Materials Matter Students investigate materials used in race cars and safety aspects of current technologies.	Stretching It Elastic Potential Energy. Introduce/demonstrate concept. Rubber Band Shooters activity.	Stretching It. Data analysis for Rubber Band Shooters. Effect of heat on elastomers.	Rubber Band Racers Introduce speed and begin Rubber Band Racers.	Rubber Band Racers Reflect on Rubber Band Racers design and draft ideal materials list for Challenge.

Day 11 (Lesson 6)	Day 12 (Lesson 6)	Day 13 (Lesson 7)	Day 14 (Lesson 7)	Day 15 (Lesson 7)
Fact or Friction? Introduce Friction. Demonstrations and Roll Down Test inquiry activity.	Fact or Friction? Students research race track materials and reflect on the role of friction in their X-Challenge design.	Ready, Set Race: The X-Challenge Team planning, Identify & Research Problem, Brainstorm.	Ready, Set Race: The X-Challenge Brainstorm, sketch designs, make budget.	Ready, Set Race: The X-Challenge Continue design sketches, purchase materials.

Table A.1.6 STEM Road Map Module Timeline – Weeks 4 and 5

Day 16 (Lesson 7)	Day 17 (Lesson 7)	Day 18 (Lesson 7)	Day 19 (Lesson 7)	Day 20 (Lesson 7)
Ready, Set Race: The X-Challenge Build, test, evaluate, and redesign.	Ready, Set Race: The X-Challenge Build, test, evaluate, and redesign.	Ready, Set Race: The X-Challenge Build, test, evaluate, and redesign.	Ready, Set Race: The X-Challenge. Build, test, evaluate, and redesign.	Topical research Ready, Set Race: The X-Challenge Topical research. Create presentation materials.

Day 21 (Lesson 7)	Day 22 (Lesson 7)	Day 23 (Lesson 7)	Day 24 (Lesson 7)	
Ready, Set Race: The X-Challenge Create presentation materials and practice presentation.	Ready, Set, Race: The X-Challenge Design presentation day.	Ready, Set, Race: The X-Challenge Design presentation day.	Ready, Set, Race: The X-Challenge Reflecting on designs, review feedback from industry "judges."	

Appendix A   265

Established Goals/Objectives:

- Students will understand that engineers work in teams to design products.
- Students will understand that engineers use a process (the EDP) to do their work.
- Students will understand that engineers work within design constraints and that they must consider multiple objectives when designing products.
- Students will understand that engineers and other manufacturing industry professionals must be able to present their work to multiple audiences.
- Students will understand and be able to use design journals as a reflective tool to prepare for their design challenge.
- Students will be able to apply the EDP to a group design challenge.
- Student teams will present their designs to the class.

Time Required: Two days

Necessary Materials:

- Audiovisual equipment (internet access) to show videos
- Design journals – three ring binders with dividers
- Snow Proof School materials – 50 index cards and one roll of office tape per three to four students; metal washers for weights to test designs.
- EDP graphic handouts
- Engineer It! Worksheet handouts

Teacher Background Information

This lesson provides an introduction to the unit using the Automotive X-Prize as a "hook" for the unit's culminating design challenge and driving question. The original Automotive X-Prize, sponsored by Progressive Insurance, awarded $10 million in 2010 to three teams that designed safe, affordable, fuel-efficient vehicles that could be marketed to consumers (see http://auto.xprize.org/ for more information). The X-Prize principles of innovative design and marketability will be used throughout the unit to scaffold students' understanding of science principles and to provide a gateway connection to the motorsports and manufacturing industries.

Engineering is introduced as a career connection in this unit with a particular emphasis on design engineers in the motorsports industry. Students will be challenged to act as engineers, using engineering thinking (the EDP) to ultimately work in teams to build their vehicles.

Table A.1.7 Key Vocabulary – Lesson One

Key Vocabulary	Definition
Engineering	Applying math and science skills to solve real-world problems by designing solutions or products.
Engineering Design Process	A series of steps engineers use to come up with solutions to problems (identify problem; brainstorm; design; build; test and evaluate; redesign; share solution)
Innovate	To do something in a new way or have new ideas about how something can be done.
Collaborate	To work with another person or group to achieve something
Prototype	An early model of a product or process used for testing and from which other forms are developed.

An in-depth knowledge of auto racing is not necessary to teach this unit; however, some background will be helpful. A summary of each car series' specifications can be found at http://www.indycar.com/Fan-Info/INDYCAR-101/The-Car-Dallara/Car-Comparisons.

Students may be familiar with the concept of race engineers in motorsports and may not understand the difference between race engineers and design engineers. Race engineers are individuals who act as a conduit between the race driver and race mechanics. They provide drivers with critical information about strategy and respond to drivers' feedback about the car's performance. This information is used to make adjustments to improve the car's handling and performance. Race engineers typically have a design engineering background (see http://www.formula1.com/news/features/2009/9/9885.html for a discussion of the role of Formula One race engineers). In contrast, design engineers work primarily in a manufacturing setting.

Your students may be familiar with the scientific method but may not have experience with the EDP. Students should understand that the processes are similar but are used in different situations. The scientific method is used to test predictions and explanations about the world. The EDP, on the other hand, is used to create a solution to a problem. In reality, engineers use both processes and your students' experience will reflect this. They will use the scientific method within the research and knowledge building phase of the EDP as they engage in their inquiry activities and will use the EDP during their final X-Challenge design challenge. A good summary of the similarities and differences in the process can be found at http://www.sciencebuddies.org/engineering-design-process/engineering-design-compare-scientific-method.shtml. A graphic representation of the EDP is provided at the end of this lesson. It may be useful to post this in your classroom.

Appendix A   267

The X-Challenge design challenge is a team-based challenge. You may wish to assign teams now or you may choose to observe student group work for the first two lessons before assigning teams. These teams should be composed of four to six students each. Research suggests that teacher designated teams comprised of students with ability levels are preferable for project and problem-based learning units (Belland, Glazewski, & Ertmer, 2009; Oakley, Felder, Brent, & Elhaji, 2004).

Lesson Preparation

This unit will culminate with a race day event during which students will present their designs and a related research project. Inviting outside guests to assess projects and talk to students about their design process and what they have learned adds real-life context to students' work and requires that they prepare presentations that are engaging and professional.

Have available:

- Access to audiovisual equipment with internet access to show videos.
- Copies of X-Challenge Invitation (option: this can be used as a cover for student design journals)
- Copies of EDP graphic
- Copies of Snow-Proof School building challenge
- Supplies for Snow-Proof School building challenge
- Copies of Engineer It! Design process sheets for each student
- Students should have design journals/lab notebooks (3-ringer binder with dividers)

Learning Plan Components

Introductory Activity/Engagement

Science class:

- Show video of Automotive X-prize: https://www.youtube.com/watch?v=car1X_YElxk
- Discuss what the problem was and the relevance of the problem to society and the industry (limited resources, cost efficiency, safety, etc.).
- Tell students that they will be challenged to create a vehicle that uses no traditional fuel in the X-Challenge and that they will use the same processes that the X-Prize teams did.
- Have students brainstorm about what the X-Prize teams needed to consider (fuel efficiency, weight, parts withstanding rapid acceleration, etc.).
- Have students brainstorm about what will they need to consider in creating a fuel-free car. Ask students to recall who the winners in

the mainstream category ended up competing against? Discuss the concept of constraints and that people who design things often work within a set of constraints or goals they need to meet.
- Introduce the concept of engineering:
  - Ask students who designs things like cars (engineers).
  - Show "What is Engineering?" video: https://www.youtube.com/watch?v=bipTWWHya8A
  - Discuss that students will be assuming the role of engineers for this unit.
  - Ask students how many people were involved in designing each of the X-prize cars? Point out that engineers work in teams.
  - Ask students to name different sorts of engineers (civil, mechanical, nuclear, materials, chemical, electrical). Ask students to consider what sorts of engineers design cars (mechanical)
  - Have students brainstorm about what makes an individual a good team member/create a class list. Make sure that students understand that they will be assessed on their teamwork for this unit.
- Introduce the concept that engineers use a process:
  - Introduce EDP
  - Discuss similarities/differences with scientific method
  - Introduce the concept of prototypes as a preliminary or first model that is often on a small scale
  - Option: Show video about EDP, http://www.pbslearningmedia.org/resource/phy03.sci.engin.design.desprocess/what-is-the-design-process/.
- Introduce Design Journals (three sections: Engineer It!, Lab/Activity Worksheets, and Reflections):
  - Explain that they will be used:
  - As lab notebooks to store lab/activity worksheets
  - To reflect on activities and make connections with the X-challenge
  - To track their use of the EDP using Engineer It! Worksheets
  - Note: these journals can be kept electronically using the worksheets as templates for students to set up their own journals in Word or other word processing software.
- Distribute Team Member Expectations page and review expectations and have each student sign and place in the front of their Design Journal notebooks

Activity/Investigation

Science class: Use EDP in the Snow-Proof School Challenge

Introduce the *Snow-Proof School Challenge*: Ask students to recall a recent heavy winter (i.e. 2014). Heavy snows were a concern for schools and other public buildings since there was a possibility that they could collapse under the weight of the snow. This was especially a problem in states that usually do not experience heavy snows and where buildings are not designed with the expectation that roofs will need to support this extra weight. Many school principals and maintenance workers actually shoveled snow off roofs to ensure that they didn't collapse (show pictures). Introduce the idea that students will act as engineers to design a snow-proof school building (discuss what sort of engineers design buildings – architects and civil engineers).

- Have students form teams of 3–4
- Distribute Snow-Proof School Challenge description
- Distribute Engineer It! Sheets

Teams should work collaboratively to solve the challenge. Each student should complete an entry in their Design Journals and present their designs to the class.

Mathematics connections: math practices (constructing buildings), geometry (what shapes were best design for function), and measurement (using precision in constructing buildings)

ELA connections: writing (design journals), speaking and listening skills (group discussion and group work), and reading (current events related to racecars)

Social Studies connections: economics (budget constraints, resource limitations) and geography (examining regions with high snowfall)

Explain

Science class

- Introduce EDP
- Emphasize teamwork components of EDP
- Prototypes

Extend/Apply Knowledge: What opportunities will students have to apply what they have learned through their work in this lesson explicitly, if any?

Social studies: Student groups will conduct research on the locations in the United States where the racing industry has a presence and explore how this has impacted local culture and quality of life.

Mathematics: Students will gather data on the impact on the local economy for one region with a racing industry presence.

English/Language Arts: Students will utilize writing skills to develop a one-page overview of their selected region in the United States and the racing industry presence outlining the pros and cons of living in a region with this type of sports industry.

Assessment

Performance Tasks

- Completion of Snow-Proof School Challenge
- Engineer It! Sheets (design journal)
- One-page overview paper

Other Measures

- Assessment of collaboration skills

Internet Resources

Race car information/series comparisons: http://sports.yahoo.com/irl/news?slug=txindycarseriesprimer and http://www.indycar.com/Fan-Info/INDYCAR-101/The-Car-Dallara/Car-Comparisons.

What is Engineering? http://www.youtube.com/watch?v=bipTWWHya8A.

IndyCar Factory information: http://www.indycarfactory.com/about.html.

Interview with Luca Pignacca, chief designer at Dallara (2012): http://www.youtube.com/watch?v=8-eMjny_PJ.

Progressive Insurance: X-prize: http://auto.xprize.org/.

X-prize video: https://www.youtube.com/watch?v=car1X_YElxk

EDP video: http://www.pbslearningmedia.org/resource/phy03.sci.engin.design.desprocess/what-is-the-design-process.

EDP versus scientific process, summary:
http://www.sciencebuddies.org/engineering-design-process/engineering-design-compare-scientific-method.shtml.

Engineer It!

Name: _____   Date: _____

Project/Activity: _____

Step 1: (Identify the problem)

- State the problem:

- Identify the conditions that must be met to solve the problem:

- Identify anything that might limit the solution (cost, availability of materials, safety):

Step 2: Research & Brainstorm

- If you did any research, summarize your findings here:

- Brainstorm! What solutions do you and your team imagine?

Step 3: Plan Design & Sketch

Include a sketch or sketches here (you may include additional sheets). Label and include materials you need:

Why did you choose the design?

List the materials you will need for your prototype.

Step 4: Build

Did your building process go as expected? What turned out differently than you thought it would when you designed and sketched your ideas?

Did you need additional materials? If so, list them here:

Step 5: Test & Evaluate

How did you test your prototype?

What were the results of your tests?

What are the strengths of your design?

What are the weaknesses of your design?

Did your design solve the problem?

Step 6: Improve the design

What changes would help your design perform better?

Step 7: *(Present Solutions)*

How will you share your design?

Decide who will present various aspects of your design and the design process. List team member responsibilities here:

The Snow Proof Challenge

Your Challenge:

Design and build a prototype building that has at least three surface levels (basement, mid-floor, and roof), that is at least 20 cm high, and can support as much weight as possible.

Design rules:

- Materials are 50 index cards and one roll of office tape
- Cards can be folded but not torn
- No piece of tape can be longer than 2 inches
- Building cannot be taped to the floor or table
- Building must have a roof surface on which to put the test weights (washers)
- Time to design and build: 40 minutes
- Height is measured from the ground to the roof level
- Tower must support the weight for at least 10 seconds

Lesson Plan #2 – Transportation – Motorsports

Lesson Title: *Let's Get Energetic!*

Lesson Summary: This lesson introduces the concept of energy as the ability to do work, the idea that all forms of energy fall into the categories of potential and kinetic energy, the concept of energy transformations, and the Law of Conservation of Energy. Students explore energy and energy transformations through teacher demonstrations, an interactive game, and energy inquiry stations. Students will work in their design teams to brainstorm and research energy sources for cars, the implications of each of these energy sources, and discuss what energy sources they could use to power their X-Challenge vehicle.

Essential Question(s):

- What is energy?
- What are the different ways things can have energy?
- How is energy transferred from one thing to another?
- In what ways could your X-Challenge car be powered?

Established Goals/Objectives:

- Students will understand that energy is the ability to do work
- Students will be able to identify various types of energy (sound, light, heat, chemical)
- Students will be able to differentiate between potential and kinetic energy.
- Students will understand that one form of energy can be converted to another.
- Students will be able to discuss and identify energy transformations and trace the conversion of one form of energy to another.
- Students will understand and discuss the Law of Conservation of Energy.
- Students will create a database of energy sources for cars.

Time Required: two classes

Necessary Materials:

Introduction and Demonstration:

- Flashlight (battery-operated)
- Jump rope
- Ball (to bounce)
- Toy car
- 1 plastic container of sand, about 2/3 full
- 1 thermometer
- Energy Flows worksheets (one per student)

Transformation Stations:

- Transformation Station Instructions (1–2 per station)
- Transformation Station Student Record sheets (1 per student)
- Transformation Station materials (see materials lists in station instructions)

Teacher Background Information

The classic definition of energy, the ability to do work, may be difficult for students to conceptualize. Energy can be introduced as a physical property of objects. Unlike color, mass, etc., however, energy is better understood by what it can do, rather than how it looks or feels, so that we define energy as the ability to do work.

Table A.1.8 Key Vocabulary – Lesson Two

Key Vocabulary	Definition
Energy	The ability to do work; work is done whenever something moves.
Work	Using a force to move an object a distance.
Potential Energy	Energy stored in an object because of its state or position.
Kinetic Energy	Energy of motion.
Energy Transformations	When one type of energy is converted to another type.
Temperature	A measure of the amount of heat in an object or substance.
Thermal Energy	Heat. The internal energy in substances caused by the movement of atoms.
Mechanical Energy	The energy of a moving object.
Electrical Energy	The energy of electrons moving through a substance.
Chemical Energy	Energy stored in the bonds in atoms and molecules; released when chemical compounds change or react.
Nuclear Energy	The energy that holds the nucleus of an atom together.
Gravitational Potential Energy	The energy of a place or position.
Elastic Potential Energy	Energy stored in objects by applying force.
Sound Energy	The movement of energy through substances in longitudinal waves.

All energy falls into two major categories: potential and kinetic (see Energy page at the end of this lesson). All other varieties of energy fall into one of these two categories. There are many different forms of energy. The Energy handout may be a useful way to organize this visually for students.

Although students will design their X-Challenge cars with a limited set of materials, this is a useful time for them to brainstorm about how the cars could be powered. As an option, this activity may lead to a discussion of renewable versus non-renewable energy sources and energy conservation. The National Energy Education Project (NEED) provides a curriculum guide containing information and student resources for renewable and non-renewable energy sources. It can be accessed online at http://www.need.org/files/curriculum/guides/Energy%20Flows.pdf.

Lesson Preparation

Have materials available for introductory activity at the start of class
Assemble sand demonstration materials
Prepare copies of Energy organizer chart to share with class in a discussion

Set up Transformation Stations
Prepare copies of Transformation Station instruction sheets
Prepare copies of Transformation Station student record sheets

Learning Plan Components

Introductory Activity/Engagement

Science class:
Begin the class with asking one student to turn on a flashlight, one to jump rope, one to bounce a ball, one to roll a toy car down a ramp. Ask students what it took for each of these actions to occur? What else happens when we do these things?

Ask students to brainstorm about the question "What is Energy?" Lead students to an understanding that any kind of work or change requires energy, so energy is the ability to do work or change something.

Activity/Investigation

Science class:

1 Sand Energy Demonstration:

  1 Place the bulb of a thermometer about halfway to the bottom in the middle of a plastic container filled about 2/3 with sand.
  2 After about 30 seconds, read the temperature of the sand.
  3 Remove the thermometer and place the lid on the plastic container. Make sure the lid is sealed all the way around the container.
  4 Have each student shake the container vigorously for a few seconds (a total of about 2.5 minutes.)
  5 Remove the lid and submerge the thermometer bulb under the sand for about 30 seconds. Read the temperature of the sand.

Ask students:

- Why did the temperature change? (kinetic energy of particles – particles bouncing off one another produced heat and therefore increase in temperature)
- Where did the energy for the temperature change come from?

Lead students to an understanding that one kind of energy can be transformed to another kind of energy (i.e. kinetic energy of shaking to heat energy).

2 Energy Flow Activity

Discuss as a class the flow of energy from the students' breakfast that morning through the energy transformations that ultimately result in an increase in temperature of the sand.

Tracing this energy flow can be an introduction to the Law of Conservation of Energy. Students should understand that energy is not created or destroyed, but can be transformed to another kind of energy.

Ask students to consider what the source is of most of the energy on earth (the sun).

Have students write in their journals regarding where their energy comes from and where it goes on a daily basis. Point out to students that they should consider the flow of energy to them, but that the ways they expend energy may not necessarily be expressed as a flow (for instance, they may conclude that they conclude energy by breathing, thinking, growing, and digesting food). Have students share their ideas.

2 Transformation Stations

Divide students into four groups. Leave a copy of the station instructions at each of the four stations and give each student a record sheet. Students should have about 8 minutes at each station to investigate what types of energy are involved in energy conversions.

Station#1: Wind up flashlight. Mechanical to electrical to light energy
Station #2: Music box. Mechanical to sound.
Station #3: Baking soda and vinegar balloons. Chemical to mechanical energy.
Station #4: Sand jars. Mechanical to thermal.

Mathematics connections: Students will utilize conversion formulas to calculate temperature conversions. Students will also create an energy audit and will project how much money a household could save in one month using fluorescent bulbs rather than incandescent bulbs.

ELA connections: Student teams will develop a blog advocating for their selected choice of alternative energy.

Social Studies connections: Students will explore the use of alternative energy sources in the United States for a variety of purposes. Further, students will learn what types of alternative energy sources are in use in their community.

Explain

Science class:

- Energy is the ability to do work; a physical property that is observed by its effects.
- All energy falls into two categories: potential and kinetic.
- Potential energy is energy stored in an object because of its state or position; kinetic energy is the energy of motion

- One type of energy can be converted to another.
- There are numerous forms of energy that fall within the larger two categories of energy.
- Law of conservation of energy.

Extend/Apply Knowledge

Science class: Students work in groups to brainstorm ideas for energy sources for design challenge cars. Create a class database of energy sources for the design challenge and discuss the feasibility of each.

Mathematics connections: Introduce kinetic energy calculations (k = 1/2 mv^2) and calculate the kinetic energy of various items; conduct a home energy audit and calculate cost of energy used in a week.

ELA connections: Research alternative energy sources and write position papers about their usefulness or create a brochure to advertise energy alternatives.

Social Studies connections: Discuss government role in energy conservation (for example, tax credits for energy-efficient appliances)

Assessment

Performance Tasks

- Completion of Transformation Stations
- Transformation Station record sheets

Other Measures

- Observation of participation/collaboration in brainstorming session.

Internet Resources

NEED Energy Flows resources: http://www.need.org/files/curriculum/guides/Energy%20Flows.pdf.

Transformation Stations Overview

Station #1: Light Up My Life

Students will be provided with two Wind-up LED flashlights at this station. They will investigate how long the flashlight remains illuminated relative to how many times they turn the crank in order to investigate the relationship between the input of mechanical energy and the output of light energy.

Materials: two crank flashlights, two timers, station instructions

Station #2: Making Music

At this station students will be provided with two transparent manually operated music boxes. Students will observe what happens inside the music box as they wind it up and think about how the sound is generated and it reaches their ears.

Materials: two transparent music boxes, station instructions, plastic drinking straws (six per student), scissors (one per student), masking tape (two rolls)

Station #3: Sand Shakers

This station is a variation on the sand temperature demonstration. Two containers of sand: one completely full (so little movement is possible when shaken) and one about 1/3 full. Students will measure and record initial temperatures and make hypotheses about what will happen when they shake each container for 2 minutes.

Materials: two containers of sand (1 full, the other 1/3 full), two thermometers, timer, station instructions

Station #4: Blow It Up!

At this station students will use baking soda and vinegar to create a chemical reaction to blow up a balloon. They will be provided with two flasks, each with 100 mL of vinegar and two quantities of baking soda (1½ tsp and ½ tsp). They will put the baking soda into balloons using a funnel and then attach the balloons to the vinegar flasks. Students should see that the balloon with the smaller amount of baking soda is smaller and be able to conclude that more chemical reagents result in more chemical energy, which is transformed to mechanical energy to blow up the balloons.

Materials: eight flasks of vinegar (two per student group), baking soda (pre-measured for each group), balloons, safety glasses, station instructions

Transformation Station Instruction Sheets

Station #1: Light Up My Life

You will investigate how long the flashlight remains illuminated in relation to how many times they turn the crank and record your observations on your record sheet.

Materials:

- Two Wind-up LED flashlights
- Two timers

Procedure:

1. Turn the crank once. What happens?
2. Now, have one person timing and another turning the crank. Turn the crank exactly one cycle. Time how long the light stays lit.
3. Next, turn the crank one more cycle. How long did the flashlight stay lit? Is it brighter?
4. Repeat step 2, cranking one additional cycle each time and timing how much longer the light stays on.

Station #2: *Making Music*

Have you ever wondered how a music box works? At this station you will be able to see what happens within the music box when you wind it up and make your own musical instrument. Record your observations on your record sheet.

Materials:

- Two transparent music boxes
- Drinking straws
- Scissors
- Masking tape

Wind up the music box. What do you see? What do you hear? Record your observations.

Now, try to make a musical instrument that will play different notes using six plastic drinking straws per person (hint: you will need to blow across the top of the drinking straws to make a sound!).

Station #3: *Sand Shakers*

Think about what happened to the sand in the demonstration at the beginning of class. What do you think will happen with these two containers? Will the amount of sand in them make a difference?

Materials:

- Two containers of sand: one completely full and one about 1/3 full
- Two thermometers
- One timer

1. Make a hypothesis about what will happen to the temperature of the sand in each of the two containers when you shake them. Record this on your record sheet.
2. Use the thermometer to find the temperature of the sand. Record it on your record sheet.
3. Put lids on the containers. Choose two people to shake the containers and one person to time.
4. Shake the containers for 2 minutes.
5. Take the lid off and measure the final temperatures for each container. Record this on your record sheet.
6. Repeat steps 3 and 4 with different people shaking the containers.

Station #4: Blow it Up!

You will see a chemical reaction at this station and you will capture the products of the reaction inside a balloon. Vinegar reacts with baking soda and turns into carbon dioxide and water. What do you think will happen to the balloon?

Materials:

- Two flasks, each with 100 mL of vinegar
- Two pre-measured quantities of baking soda (1½ tsp and ½ tsp)
- Two balloons
- Funnel
- Safety glasses (one per student)

Procedure:

1. Be sure that everyone in the group is wearing safety glasses.
2. Attach the balloon opening to the funnel and use the funnel to add the smaller amount (1/2 tsp) of baking soda to the balloon.
3. Without allowing the baking soda to fall into the flask, attach the balloon to the flask with the vinegar.
4. Once the balloon is firmly attached to the flask, lift the balloon so the baking soda empties in to the flask.
5. Observe what happens and touch the balloon to see if it feels warm or cold. Record your observations on your record sheet.
6. Repeat the procedure with a new balloon, vinegar flask, and the larger amount of baking soda (1½ tsp).

Transformation Station Student Record Sheets

Name _____

Light Up My Life

Number of Cranks	Amount of Time Light Remains Lit (seconds)
1	
2	
3	
4	
5	
6	

What happens when you crank the flashlight more? Why do you think this is?

What energy transformations do you think are happening here (hint: the flashlight has a battery inside it)?

Transformation Station Student Record Sheets

Name _____

Making Music

1 What did you see when you wound the music box?

2 How do you think that what you see inside the music box creates sound (hint: think about what happens to the surface of a drum when you hit it)?

3 How do you think that your straw instrument makes sound? How is that the same as the way the music box makes sound?

4 What energy transformations do you think are happening?

Transformation Station Student Record Sheets

Name _____

Sand Shakers

1. State your hypothesis – what do you think will happen when you shake the two containers. Will it be different or the same for the two containers?

2. Record your data:

Quantity of Sand in Container	Beginning Temperature Trial 1	End Temperature Trial 1	Change in Temperature Trial 1	Beginning Temperature Trial 2	End Temperature Trial 2	Change in Temperature Trial 2
Full						
Not Full						

3. Was your hypothesis correct?

4. What energy transformations do you think are happening?

Transformation Station Student Record Sheets

Name _____

Blow It Up!

1. What happened when you put the first amount of baking soda into the balloon?

2. Touch the balloon. Does it feel warm or cold? Why do you think this might be?

3. What happened when you put the second amount of baking soda into the balloon? Was it different than the first amount of baking soda?

4. What energy transformations do you think are happening in the reaction?

Lesson Plan #3 – Transportation – Motorsports

Lesson Title: *Materials Matter*

Lesson Summary: This lesson introduces the role of materials in energy transformations and in car design. Students will investigate the Law of Conservation of Energy and the effect of materials in energy transformations in the Ball Drop activity and will calculate gravitational potential energy (GPE). A discussion of the different materials and their performance in the Ball Drop activity serves to segue into a discussion of materials used in car design. If student design teams for the X-Challenge have not already been formed, teams should be chosen during this lesson. Student design teams will investigate the various materials used in IndyCar race cars and the effect of those materials on car performance. Design teams will present their findings to the class. The design team research project will be an opportunity to discuss roles of team members, using the various members of a racing team to illustrate the division of duties and collaboration that occurs in successful teams.

Essential Question(s):

- What effect do position and weight have on GPE?
- What energy transformations can we observe and how can we account for the Law of Conservation of Energy?
- How can we work effectively as a team to accomplish a goal?

Established Goals/Objectives:

- Students will understand and observe the Law of Conservation of Energy.
- Students will understand the relationship of position and weight in GPE and make appropriate calculations.
- Students will understand qualitatively the role of materials in elastic potential energy.
- Students will construct bar graphs using the results from their Ball Drop investigation.
- Students will understand that materials have different properties and observe the effect of materials on energy transformations.
- Students will understand the various roles of race team members and apply that understanding to their own teamwork.
- Student teams will investigate racecar materials and their effect on car performance.
- Student teams will present findings.

Time Required: Two classes

Table A.1.9 Key Vocabulary – Lesson Three

Key Vocabulary	Definition
Law of Conservation of Energy	Energy can be neither created nor destroyed; instead it is transformed from one form to another.
Elastic Potential Energy	Energy stored as the deformation of an elastic object such as a spring or an elastomer.
Gravitational Potential Energy	The energy an object possesses because of its position in the gravitational field (i.e. its height). Calculated as GPE = m × g × h where m = mass in kg, g = acceleration of gravity (9.8 N/kg), and h = height in m.
Materials Science	A field that deals with the discovery and design of material.
Rebound	To bounce or spring back from force of impact.

Necessary Materials:

Introductory Activity:

- One rubber ball (basketball)
- One foam ball
- Audiovisual equipment (internet access)

Driving in a Material World group research project:

- Internet access
- Audiovisual equipment for presentations

Ball Drop Activity:

- Three balls per student group (golf ball, tennis ball, rubber ball)
- 1 scale or balance per student group
- 2 meter sticks per student group
- 1 chair per student group
- Calculators (1 per student)
- Student internet access (energy calculator and graphing)

Teacher Background Information

The law of conservation of energy states that energy cannot be created or destroyed, but can be transformed. In the case of dropping a ball, you transfer energy from your muscles to the ball when you lift it, giving it GPE, or the energy gained by an object as its height increases. After you drop the ball, its GPE is converted to kinetic energy, which will continue to increase until the ball hits a surface, at which point the kinetic energy is transformed into other forms of energy (some into sound, some thermal from friction, some elastic potential energy from the deformation of the ball when it hits the ground). The elastic potential energy is the reason that the ball bounces or rebounds. This is an example of an inelastic collision, in which part of the kinetic energy changes to another form of energy after a collision. A car crash is an example of an inelastic collision since when cars collide, the kinetic energy transfers to sound, thermal energy and the mechanical energy that causes the cars to change shape.

Body materials for racecars are chosen with weight and safety considerations. In their materials research, students may find references to carbon fibers, aluminum, and reinforcing materials such as Zylon. They should make the connection that the weight of body materials affects car performance (speed) and safety.

Lesson Preparation

- Assemble materials for introductory activity (one rubber ball, one foam ball)
- Prepare ball drop activity materials
- Ball drop worksheets (one per student)
- Design journal reflection sheets (one per student)
- Collaboration rubrics (one per student, optional)

Learning Plan Components

Introductory Activity/Engagement

Science class:

Introduce the class with the video of the race car tire bouncing: http://www.youtube.com/watch?v=3tMJ8U-2ZMU

Ask students what energy transformations they see. Ask them what happens to the energy in the tire. Introduce the Law of Conservation of Energy. Refer to the energy flow worksheet from the last lesson and ask if there was more or less energy in the energy inputs (right side) than the energy outputs (left side) (if all energy outputs are accounted for they should be equal). If students feel that the two sides don't balance, what do they think happened to the extra energy?

Have two students bounce a rubber ball (basketball, etc.) to each other. Have students diagram the trajectory of the ball they see and work as a class to label the energy transformations they see including GPE, elastic potential energy, thermal energy, sound energy, and kinetic energy.

Now repeat the ball bouncing activity with a foam ball. Ask students what they observe about the differences in how the two balls behave. Introduce the idea that different materials have different properties and that this is important in designing products, including cars.

Appendix A   287

Activity/Investigation

Science class:

1 Ball Drop Activity

Students will investigate the law of conservation of energy and calculate GPE of various balls to investigate the effect of the ball's weight and position on its GPE and the effect of its material on its elastic energy (see Ball Drop Activity student worksheets at the end of this lesson).

Materials: Three balls per student group (about four students): golf, tennis, and rubber ball; scales on which to measure ball weights in grams, two meter sticks per group. One chair per group (to stand on for 200 cm drop) and Ball Drop activity lab sheet (attached at the end of this lesson).

Give each group three balls: a golf ball, a tennis ball, and a rubber ball. Allow students to "experiment" with bouncing each ball for one minute. Remind students that this is a ball drop activity and that they should drop rather than throw the ball.

Each student group will set up a testing station to investigate rebound heights. This will require attaching a meter stick to a vertical surface (i.e. a wall) at the approximate point of rebound for the balls.

Students should first weigh each of the three balls and record their weights on their lab record sheet (they will use these to calculate GPE for each ball at each height).

Students will then drop each type of ball four times from 100 cm and then 200 cm and measure the rebound heights (for first rebound) using a meter stick. Students should also observe how many times the ball bounces. These results should be recorded on the lab record sheet.

After the trials for each ball are completed, students should calculate the average heights of the first rebound for each ball and the average number of rebounds.

Students should graph their results on a bar graph (one bar graph for each drop height).

To extend this activity, students may also calculate GPE for each ball from each height (GPE = mass (in kg) × gravity (9.8 N/kg) × height (in meters)).

Hold a class discussion about how the different materials behaved and why they might act differently. Use this as a transition to have students think about why racecars are made of specific materials.

2 Driving in a Material World

Show students a picture of an IndyCar. Ask them to observe what kinds of materials they see in the car. Tell students that their first design team task will be to research materials used in racecars. Each team will be assigned a materials topic to research (body materials, tire materials, materials for safety, engine materials).

Teams will make a 5-minute multimedia presentation for the class at the end of the lesson. This should include:

- Factual information
- History
- Pictures/videos
- Oral narration

Ask them how they think that their team of (4–6) will work together to finish this project and their design challenge.

Team members should decide on roles (i.e. research facts, research history, find pictures/videos, create PowerPoint, and act as narrator).

After Ball Drop data analysis and materials presentations are complete, students can make a design journal entry using the "Design Journal Reflection" sheet to relate the findings to their X-Challenge design.

Mathematics connections: Calculate fuel usage in a typical IndyCar race and compare this to data on public transportation and personal transportation. Students will also work on unit conversions. Finally, students can determine if the purchase price of a hybrid vehicle is warranted – meaning will the consumer come out ahead on gasoline savings after the initial cost of ownership difference is negated.

ELA connections: In Language Arts, students will conduct research on the pros and cons on the use of seatbelts and airbags in personal vehicles and will develop a public service announcement (PSA) targeting adolescents with their desired messaging regarding the use.

Social Studies connections: Civics – discuss government role in safety (seatbelt, child seat laws) and safety innovations that come from racecars.

Explain

Science class:

- Law of Conservation of Energy
- Gravitational Potential Energy
- Elastic Potential Energy
- Unit conversions (cm to m; g to kg)

Extend/Apply Knowledge

Science class: Apply findings from Ball Drop and Material World activities to design challenge via design journal reflection.

Assessment

Performance Tasks

- Completion of Ball Drop activity
- Ball Drop Lab worksheet (make three copies for the three trials per group)
- Design journal reflection entries

Other Measures

- Observation of participation/collaboration in materials research project (collaboration rubric attached at the end of this lesson)

Internet Resources

Race car tire bouncing video: http://www.youtube.com/watch?v=3tMJ8U-2ZMU

Energy Calculator web tool: http://easycalculation.com/physics/classical-physics/potential-energy.php

Graphing web tool: http://nces.ed.gov/nceskids/createagraph/default.aspx

Ball Drop Worksheet

Name _____

Procedure:

1. Weigh each of the three balls and record their weights (in grams).
2. Designate one person to drop the ball, one person to hold the meter stick vertically to measure rebound height, one person to observe the rebound height, and one person to observe the number of bounces.
3. Measure 100 cm from the floor.
4. Hold ball #1 at 100 cm and drop it (NOTE: be sure to drop, not throw the ball).
5. Observe the height of the first rebound (or how high it bounces) in centimeters and the number of bounces before the ball comes to a rest. Record this.
6. Repeat for four trials.
7. Switch roles (choose a new person to drop the ball, a new person to hold the meter stick, etc.)
8. Measure 200 cm from the floor.
9. Repeat procedure for four trials from this height.
10. Repeat procedure for the other ball materials.
11. Calculate average rebound heights and numbers of bounces.
12. Construct a bar graph for each ball material at each drop height.

BALL #_____

Material/type of ball: Mass of ball:

Trial #	100 cm Drop Height		200 cm Drop Height	
	Rebound Height (cm)	Number of Bounces	Rebound Height (cm)	Number of Bounces
1				
2				
3				
4				
Total of all trials				
Average (Total / 4)				

Design Journal Reflection Template

Name _____   Date _____

Name of activity/idea

_____

Summary of findings

Reflection (for example, "I was surprised when…." Or, "I wonder what would happen if….")

Connection to design challenge (for example, "Car tires are made of a material similar to rubber bands. The findings from this activity make me think that we should think about _____ when we design our prototype.")

Any other thoughts, ideas, or sketches (this is a space for you to include anything else you might be thinking about that will relate to your prototype design).

COLLABORATION RUBRIC (30 points)

Student Name _____ Team Name _____

Individual Performance	Below Standard (0–3)	Approaching Standard (4–7)	Meets or Exceeds Standard (9–10)	Student Score
Individual Accountability	Student is unprepared Student does not communicate with team members and does not manage tasks as agreed upon by the team Student does not complete or participate in project tasks Student does not complete tasks on time Student does not use feedback from others to improve work	Student is usually prepared Student sometimes communicates with team members and manages tasks as agreed upon by the team, but not consistently Student completes or participates in some project tasks but needs to be reminded Student completes most tasks on time Student sometime uses feedback from others to improve work.	Student is consistently prepared Student consistently communicates with team members and manages tasks as agreed upon by the team. Student discusses and reflects on ideas with the team. Student completes or participates in project tasks without being reminded Student completes tasks on time Student uses feedback from others to improve work.	
Team Participation	Student does not help the team solve problems; may interfere with team work Student does not express ideas clearly, pose relevant questions, or participate in group discussions Student does not give useful feedback to other team members Student does not volunteer to help others when needed	Student cooperates with the team but may not actively help solve problems Student sometimes expresses ideas, poses relevant questions, elaborates in response to questions, and participates in group discussions Student provides some feedback to team members Student sometimes volunteers to help others	Student helps the team solve problems and manage conflicts Student makes discussions effective by clearly expressing ideas, posing questions, responding thoughtfully to team members' questions and perspectives Student gives useful feedback to others so they can improve their work Student volunteers to help other if needed	
Professionalism & Respect for team members	Student is impolite or disrespectful to other team members Student does now acknowledge or respect others' ideas and perspectives	Student is usually polite and respectful to other team members Student usually acknowledges and respects others' ideas and perspectives	Student is consistently polite and respectful to other team members Student consistently acknowledges and respects others' ideas and perspectives.	

Lesson Plan #4 – Transportation – Motorsports

Lesson Title: Stretching It

Lesson Summary: This lesson will focus on elastic potential energy and will build on students' understanding of the role of materials in design by focusing on the role of elastomers in racecar design and performance. A demonstration with a rubber band testing stand will introduce the concept that the amount of energy stored in an elastomer changes as force is applied to it. Students will conduct an inquiry, Rubber Band Shooters, into how the amount of stretch and width of rubber bands affect potential and kinetic energy and will graph results. Connections to race car tires and tire manufacturing will be made through video clips and discussion.

Essential Question(s): What effects the amount of energy stored in an elastomer? How are properties of elastomers used in racecar tire design and manufacturing and how do these properties affect performance?

Established Goals/Objectives:

- Students will observe and investigate the properties of elastomers and elastic potential energy.
- Students will understand the concept of thermal energy.
- Students will be able to use their understanding of elastomers to discuss the properties and performance of racecar tires.
- Students will discuss the technology and careers associated with tire manufacturing in Indiana.
- Students will understand and discuss how the properties of elastomers affect tire design and race car performance.
- Students will create line graphs using the results of their inquiry.

Time Required: Two classes

Necessary Materials:

Introductory Activity:

- Audiovisual Equipment (internet access)
- Slinky
- Spring
- Snake-in-a-can

Rubber Band Testing:

- Testing stand
- Milk jug
- Sand
- Paper funnel
- Various width rubber bands
- Calculators
- Internet access (optional for graphing)

Rubber Band Shooters:
- 1 ruler per group
- 1 meter stick per group
- 3 (6.5 – 7.5 cm long) rubber bands per group, 1 each of the following widths: 1mm, 3mm, 6mm
- Masking tape or chalk to mark ground
- Calculators
- Internet access (optional for graphing

Heat It Up (optional):
- Rubber band testing stand
- Heat lamp
- Thermometer
- Masking tape
- Duct tape
- Push pin
- Ice cube

Teacher Background Information

Elastomer is simply an umbrella term for the family of materials commonly referred to as rubbers. The word elastomer is derived from "elastic polymers," reflecting that they are composed of long chainlike molecules, or polymers, that can recover their shape after being stretched. Under normal conditions the chains of molecules are coiled, but straighten out when the material is stretched. When releasing the stretch, the molecules spontaneously return to their coiled shape (the "snap" of a rubber band).

Most rubber bands are manufactured using natural rubber because of its superior elasticity. Natural rubber is obtained by tapping the bark of the rubber tree to extrude latex, which hardens and becomes elastic when exposed to air.

Temperature changes affect elastomers in an unexpected way. When a rubber band is heated, it contracts and expands when cooled. This

Table A.1.10 Key Vocabulary – Lesson Four

Key Vocabulary	Definition
Elastomer	A natural or synthetic material that has elastic properties
Elastic Potential Energy	Potential energy stored as a result of deforming an elastic object, such as stretching a spring; equal to the work done to deform the object.
Thermal Energy	Energy possessed by an object or a system due to the movement of particles within the object or system; a type of kinetic energy
Force	Any influence that changes the motion of an object

property has to do with the properties of entropy and the fact that the molecules are coiled in the "resting" state of a rubber band. For an explanation of these properties, see http://www.physlink.com/education/askexperts/ae478.cfm.

Racecar tires are the most obvious use of elastomers in their design. These tires typically contain more synthetic elastomers than natural rubber (approximately 65% synthetic on average). Reinforcing materials such as carbon black and silica are also added to the elastomeric makeup of tires. The major difference between racecar and passenger car tires is that they are made with efficiency – moving as quickly as possible without sticking too much to the road – as the primary goal. Therefore, the tires are soft so that they grip the road, but have no treads. Because the tires are soft, the material needs support from the rubber around it, which is part of the reason that racecar tires have no treads. If there were treads, the grooves would allow the soft rubber to move too much and it would overheat. When tires overheat, the properties of the rubber changes and becomes oily resulting in potential slippage.

Tire manufacturing is an example of advanced manufacturing – manufacturing that uses highly technical processes and employs technically skilled people in a variety of roles. For an overview of the racecar tire manufacturing process, visit Hoosier Racing Tire's description of their manufacturing process at https://www.hoosiertire.com/index.htm.

Students may be familiar with the image of racecars weaving when they are in warm-up laps. This weaving action allows the tires to warm up for maximum grip, since the tire becomes literally sticky when it warms. In drag racing on the other hand, there is not typically time or space for warm-up laps, so a solvent is poured on the asphalt, and the drivers spin the rear wheels in it to heat their tires for maximum grip at the start of the race.

Elastomers are used in other facets of automotive manufacturing as well. They are increasingly being used to make lightweight auto body components and are used in various seals and gaskets, engine and transmission mounts, and brakes. Many of these are manufactured using an injection molding process in which the heated material is injected into a mold where it cools and hardens into the shape of the mold.

This lesson will connect to product (tire) quality control in the Rubber Band Testing stand demonstration. Details about how Bridgestone Tires quality tests its products can be found at http://www.bridgestonetrucktires.com/us_eng/real/magazines/bestof3/speced3_quality_control.asp.

Lesson Preparation

- Assemble Rubber Band Testing stand
  - Cut a 2" × 4" to between 3 and 4 feet in length, making sure both ends are square
  - Make base that is approximately 10" × 12" from ¾ plywood or from 2" × 10"
  - Screw 2" × 4" to base
  - Cut top piece from 1" × 4" that is approximately 8" long and screw to top of 2" × 4"
  - Drill hole in top piece at least 5" from the front of the 2" × 4"
  - Get an empty ½ gal. milk jug and large rubber band
  - Insert rubber band through milk jug handle and then loop rubber band through itself and pull tight
  - Attach free end of rubber band to the hook on the stand
  - Take a sheet of paper and make a filling cone. Make sure the end of the cone can fit into the opening on the ½ gal. milk jug. Tape or staple the cone into its final shape.
- Make paper funnel
- Prepare Rubber Band Shooter activity supplies
- Prepare Testing It worksheets
- Prepare Rubber Band Shooters worksheets

Learning Plan Components

Introductory Activity/Engagement

Remind students of the importance of materials they investigated in the last lesson. Show Anatomy of a Pit Stop graphic (from lesson 3) and ask what part of the car gets the most attention during a pit stop (tires).

Ask students to recall what tires are made of (rubber/elastomers). How do they think tires are made?

Show video of tire manufacturing http://www.youtube.com/watch?v=0BSgWKLkv9o.

Ask students if this was what they expected a tire factory to look like? What kinds of jobs did they see people doing? Introduce the idea that advanced manufacturing requires people with all kinds of technical skills, including computer programming, robotics, and the skills to operate high-tech equipment.

Ask students what properties they think are important for manufacturers to consider when producing racecar tires (speed, safety, wear, etc.)

Ask the students to recall the definition of elastic potential energy is (potential energy stored by deforming an elastic object). Ask students to brainstorm some items that have elastic potential energy (rubber band, balls, tires, springs).

Demonstrate with a slinky, a spring, and a "snake-in-a-can."

Ask what happens to the elastic potential energy in each case. Elastic potential is converted to kinetic energy. Ask students to indicate what type of kinetic energy (mechanical, sound, small amount of heat).

Tell students that they will investigate the relationship between elastic potential energy and kinetic energy using rubber bands.

Tell students that they will be considering the properties of elastomers in this lesson – the same properties that tire manufacturers and racecar teams need to consider when manufacturing and using tires.

Activity/Investigation

Students will observe and investigate the properties of elastomers and potential to kinetic energy transformations through a series of observations and activities.

1 Rubber Band Testing Stand

Introduce the idea of quality control – which products need to be tested to ensure that they meet quality standards. Ask where this might be a part of the EDP (test/evaluate and redesign).

Ask students why this is important (safety, quality of product, manufacturer reputation). Ask them why this might be important for a product such as a tire (potential for blow-outs, accidents, etc.).

Tire manufacturers as well as other types of manufacturers test a sample of their products – that means that they choose some tires at random as they come off the assembly line and test them to be sure that they are of good quality. They do this in two ways: non-destructive and destructive tests. Ask students what they think the difference is. Ask them to name some non-destructive test for tires (X-rays, weight, visual inspections). And some destructive tests (tire is cut into pieces so they can be looked at microscopically to make sure the components are the correct size, shape, and position within the tire; punctured to determine resistance to damage; performance testing – tires are mounted on wheels and they spin against a surface for hours).

Tell students that the class as a group will be testing rubber bands to see how much work can be done to a rubber band.

Show students the testing stand. Tell students how much one-cup of sand weighs (356 grams). Start with a thin rubber band. Ask them how many cups they think they can put into the milk jug before the rubber band will break. Ask them if this is destructive or non-destructive testing. Attach the rubber band onto the testing stand and onto the handle of the milk jug (photos attached at the end of this lesson). Use the filling cone to add sand by 1 cup (or increments of 1 cup with students calculating weights). Measure how much the band stretches with each added increment of weight.

Record the length and change in weights in a chart visible to the class and have students enter into their Testing It worksheets.

Ask students what kind of energy does the rubber band have (potential). When the band breaks, ask what kind of energy the band had (mechanical). Point out that before the band broke it was storing mechanical energy – that was its potential energy.

Ask students how this would be useful to a manufacturer who wanted to know if the rubber band was strong enough for a certain task. How would she express that?

Remind them of their calculations for GPE in the last lesson. Ask how they might calculate the energy that was stored in the rubber band before it broke?

Stored Energy = force × change in length where the force is the weight applied to the rubber band. So:

Elastic Potential Energy = change in length × weight applied

Calculate the stored energy for the first band (using the highest weight before the band broke) as a class. Students will record this on their Test It worksheet.

Repeat with various width rubber bands – ask students to predict what they think will happen with thicker rubber bands and to guess what weights they will hold. Record the weights and length changes on a class chart as you go.

Have students work in pairs to calculate the Stored Energy for each band and graph this data (using stored energy on the x-axis and change in length on the y-axis). This will be most effective if all band sizes are plotted on one graph. An option is to use a graphing web tool such as Graphing web tool: http://nces.ed.gov/nceskids/createagraph/default.aspx.

Ask students what they observe from their graphs (most energy is stored when the rubber band is highly stretched; wider bands can store more energy without breaking, etc.).

Students will reflect on this demonstration along with the following activities in their design journals at the end of the lesson.

2 Rubber band shooters

(This activity requires an open space for students to shoot rubber bands approximately 10 m; this space should be marked in increments of 1 m for each group using sidewalk chalk if outside, masking tape if inside).

Remind students that the vehicles they design will be powered by energy transformations. In the last activity observed, mostly energy is stored as potential energy. Now, they will observe that energy being converted into kinetic energy.

Remind students that they observed GPE (related to the height of the object off the ground) in the Ball Drop activity. Now they will observe elastic potential energy.

Ask students if they have shot a rubber band. How does the band generate energy? Point out that they are doing work on the band, providing the energy to the band that is then converted to kinetic energy when released.

Explain the procedure to students and ask them what they think will happen with thicker rubber bands – will they travel farther, not as far, or the same length? Why do they think that?

Procedure (see Rubber Band Shooters procedure sheet and worksheet):
- Students should work in groups of 4.
- Each group will have three rubber bands of similar length (6½–7½ cm) and varying widths (1 mm, 3 mm, 6 mm).
- Each group will have a ruler with centimeter markings, a meter stick, and a level surface from which to shoot their rubber bands (optional if outside, but without a surface students should take care to hold their ruler shooter level and at about the same height for each trial).
- Groups will mark off 10 meters in increments of 1 meter (marking each meter with sidewalk chalk or masking tape).
- Students will pull each of their rubber bands back to three lengths (10 cm, 15 cm, 20 cm) and "shoot" their rubber bands, conducting four trials for each of the three pull lengths. After each trial they will measure distance traveled and record on their Stretching It worksheet. They will repeat this for each of the three widths of rubber bands (students should take care to keep their ruler shooters level and at the same distance from the ground for each trial).
- After they have completed their trials, students will compute average distance traveled for each of the trials.
- Students will construct either a line graph (preferably on one graph) for the amount of stretch (on x-axis) and distance traveled (on y-axis) for each rubber band. Option: students can use a web graphing tool such as the NCES kids' graphing tool at http://nces.ed.gov/nceskids/createagraph/default.aspx

Explain

Science class:

- Elastomers
- Elastic Potential Energy
- Thermal Energy

Mathematics connections:

- Mean/average
- Types of graphs and graphical representations of data

Social Studies connections: Biodegradability of plastics and recycling; History: space shuttle challenger and O-rings

Extend/Apply Knowledge

Science class: Students may complete design journal reflections based upon their activities in this lesson. They will consider what implications the properties of elastomers and the energy conversions they observed have for their car designs.

Mathematics connection: Students will calculate standard deviations for their Rubber Band Shooters data.

ELA connections: Student teams will research the use of plastics in toothpaste (or other unusual uses of materials for the benefit of society) and discuss in the larger groups. Student groups will develop position papers that will be shared with the community based upon their findings.

Social Studies connections: Students can apply their understanding of the thermal properties of elastomers to understanding the cause of the Space Shuttle Challenger disaster (historical event).

Assessment

Performance Tasks

- Test It Chart – project on overhead for students to copy in their journal.
- Rubber Band Shooters Worksheet
- Design Journal Reflection Entries

Internet Resources

Explanation of rubber band properties: http://www.physlink.com/education/askexperts/ae478.cfm

Tire manufacturing video: http://www.youtube.com/watch?v=0BSgWKLkv9o

Hoosier Tire Manufacturing Process: https://www.hoosiertire.com/index.htm

Graphing web tool: http://nces.ed.gov/nceskids/createagraph/default.aspx

Testing It Worksheet

Name _____

Rubber Band #_____, Width _____ mm

# Cups of Sand	Weight of Sand (# cups × 367 grams)	Length of Rubber Band	Change in Length (Final Length – Original Length)	Energy (Weight × Change in Length)

Rubber Band Shooters Worksheet

Name _____

Procedure:

1. Mark your starting spot for shooting your rubber bands.
2. Place a mark (chalk or masking tape) every 1 meter from your starting spot for 10 meters.
3. Record the width (in mm) of your first rubber band.
4. Place your ruler on a flat surface if possible or be sure that your ruler is parallel to the ground.
5. Place the rubber band around the end of your ruler and pull it back to a stretch of 10 cm.
6. Release the rubber band.
7. Measure the distance traveled using your marks and your meter stick. Record the distance in the table.

8. Repeat for three more trials (NOTE: be sure to keep your shooter level and at the same distance from the ground for each trial).
9. Pull the rubber band back to a distance of 15 cm and repeat for four trials.
10. Pull the rubber band back to a distance of 20 cm and repeat for four trials.
11. Repeat procedure for the other two widths of rubber bands.
12. After all trials are complete, compute the average distance the rubber bands traveled for each trial.
13. Construct a line graph for the amount of stretch (on *x*-axis) and distance traveled (on *y*-axis) for each rubber band (plot on same graph).

Rubber Band #_____,   Width:_____mm

Trial	Distance Traveled		
	10 cm Stretch	15 cm Stretch	20 cm Stretch
1			
2			
3			
4			
Total			
Average (Total ÷ 4)			

Lesson Plan #5 – Transportation – Motorsports

Lesson Title: *Rubber Band Racers*

Lesson Summary: This lesson will build upon students' understanding of the effect of stretching an elastomer on its potential energy and will introduce the concept of speed. Student design teams will construct a rubber band racer using a set of simple materials and the EDP. Students will calculate speed of their racers, and teams will participate in a race. Students will reflect on design features that enhanced or detracted from the performance of the various teams' racers. Students will reflect on the role of the materials in their racer and consider what materials might have made their car perform better. Student design teams will begin to draft an ideal materials list for their X-Challenge car designs.

Essential Question(s): How does car design affect speed?

Established Goals/Objectives:

- Students will be able to state the definition of speed.
- Students will be able to calculate speed.
- Students will understand the relationship between potential energy and speed.
- Student teams will design and build a vehicle powered by rubber bands.
- Students will be able to relate their findings from building a rubber band car to their X-Challenge design challenge.

Time Required: two classes

Necessary Materials:

Introductory Activity:
- Audiovisual equipment (Internet access)

Speedy Olympics:
- Stopwatch (1 per each group of 4 students)

Rubber Band Racers:
- Engineer It! Worksheets (1 per student)
- Rubber Band Racers challenge description (1 per student)
- Extra washers to add weight to cars

For each design team:
- 4 CDs
- 4 plastic plates (small)
- 6 rubber bands (various widths)
- 4 unsharpened pencils (or wooden dowels)*
- 4 drinking straws
- 5 metal paper clips
- 1 piece of corrugated cardboard (about 8 x 8)
- 1 piece of foam board (about 8 x 8)
- 4 metal washers (1/4 inch)
- 10 craft sticks
- Scissors
- 1 role masking tape
- meterstick
- stopwatch

* Be sure that the pencils or dowels fit inside of the straws and can turn freely

Table A.1.11 Key Vocabulary – Lesson Five

Key Vocabulary	Definition
Speed	The path covered by an object over an amount of time. Speed = distance/time. Does not depend on the direction of travel.
Velocity	The change in position (or displacement) of an object over an amount of time. Depends on the direction of travel. Velocity = displacement/time

Teacher Background Information

In this lesson you will introduce students to the concept of speed. Speed is a scalar quantity described as a magnitude regardless of direction that represents how much distance was covered during a specified amount of time.

Average speed = total distance/time.

Students will understand the concept of speed based upon a car speedometer.

Be sure not to interchange the terms speed and velocity, however. Velocity is a vector quantity (depends on magnitude and direction) that measures total displacement.

Average velocity = displacement/time.

This distinction is important in considering motorsports since races are often conducted on a circular track, meaning that if a car begins and ends at the same spot, its average velocity is 0 although its speed may be nearing 200 mph! You may wish to introduce the concept of velocity to your students.

For a more complete description of this distinction (along with a video), see http://education-portal.com/academy/lesson/speed-and-velocity-difference-and-examples.html#lesson.

You may wish to review some designs for rubber band cars before the students create their Rubber Band Racers. There are numerous videos online, including http://www.youtube.com/watch?v=v3pbVAYkGf0.

Lesson Preparation

- Prepare Rubber Band Racers Challenge description
- Prepare Engineer It! Worksheets
- Assemble Rubber Band Racer "kits" with sets of supplies for each design team

Learning Plan Components

Introductory Activity/Engagement

Science class: Begin the lesson with asking students how the winner of an IndyCar race is determined (the fastest). Ask them how much time difference they think there usually is between the first and second finishers (can be seconds or tenths of seconds). Ask them to guess what the time difference was in the closest IndyCar race ever. To answer, show video of top 10 closest IndyCar races in history http://www.youtube.com/watch?v=HI8MnBrUdhE.

Ask students what speed is. Guide students to an understanding that speed is distance traveled in a certain amount of time and can be calculated as Speed = Distance/Time. Emphasize the importance of units in speed.

Introduce the technology behind race timing using the diagram from the Indy Car Fan Info page (diagram is attached at the end of this lesson) to emphasize how important it is to measure speed to very precise standards.

Tell students that in this lesson they are going to create rubber band vehicles that can go as fast as possible and measure the speed of rubber band vehicles they create.

Activity/Investigation

Science class:

1 Speedy Olympics

Divide students into groups and assign each an activity to do "the fastest" (e.g. running, doing jumping jacks, pushing an eraser with their noses, doing pushups, bouncing a ball, and crawling). Provide each team with a stopwatch. Have the teams practice with some members doing the activity and some timing/measuring (or counting repetitions in a given time). After they've practiced for 5–10 minutes have each group present their "Speedy Olympic" skill and time or measure the results (i.e. distance/time or repetitions/time). Enter this into a class data table.

Ask students for what activities they can calculate speed and for which they cannot (they cannot calculate speed for jumping jacks, pushups because there is no distance involved). Ask student groups to calculate speed for the activities involving distance and share answers as a class. Discuss the importance of units (i.e. steps per minute or feet traveled per 5 seconds).

2 Rubber Band Racers

Have students brainstorm about what factors affect speed in a racecar. Compile a class list.

Introduce the activity by telling students that their design teams are going to have a chance to practice designing a car. They will be given only limited materials to use and only the remainder of this class period and the first half of the next class meeting to complete their cars. This means that they will need to use the EDP strategically.

Ask students to name the steps in the EDP. Distribute Engineer It! Worksheets and the Rubber Band Racers project description.

Each design team should receive the following materials:

- Four CDs
- Four plastic plates (small)
- Six rubber bands (various widths)
- Four unsharpened pencils or wooden dowels
- Four drinking straws
- Five metal paper clips
- One piece of corrugated cardboard (about 8 × 8)
- One piece of foam board (about 8 × 8)
- Four metal washers (1/4 inch)
- Ten craft sticks
- Scissors
- One role masking tape
- Meterstick
- Stopwatch

Students should use Engineer It! Worksheets to organize their work.

The constraints are as follows:

- The car can be powered by no more than three rubber bands.
- It must travel at least 3 meters.
- The car may not be propelled by human inputs (pushing).
- Rubber bands may not be used to "slingshot" the car.
- Only the materials provided can be used in the construction of the car.
- The car must have at least three wheels.
- They have only 50 minutes to design, build, and test their vehicles.

Watch for teams that are having difficulty in designing their car and be prepared to ask them guiding questions such as, "how could you attach your axle (pencil) to the car so that your axle can still turn freely?" (bend paperclips around the axle or put the axle inside the straw).

After the designated amount of time, each team should post its average speed and distance traveled (based upon three trials). Students may need a reminder of how to convert meters to centimeters. You may have students "compete" in a race to determine which car is the fastest.

After the fastest car has been identified, tape four metal washers to the back of the car and retime the trial. Add four washers and time again. Ask students to reflect on how the extra weight affected the speed.

Mathematics connections: Unit conversions, meters to centimeters, algebra, and speed calculations.

ELA connections: Student groups will research state mandated speed limits in at least four states across the US and develop a risk/benefit analysis related to higher speed limits and travel.

Social Studies connections: Students will examine the geography, industry, population, accident, and commerce in their four assigned states and use this data to make conclusions about the speed limits related to quality of life.

Explain

Science class:

Introduce the concepts of speed and velocity and the difference between the two.

You may need to introduce the concept of axles before the rubber band racer activity. Students should understand that axles in cars are steel rods that connect the tires to the car and that turn the wheels when the driver accelerates. Axles hold the majority of the weight of the car and are a critical component in its design.

Extend/Apply Knowledge

Science class: Based upon their experiences building the rubber band racer, design teams should begin to compile a list of materials they think would be useful in constructing their prototype car for the X-challenge. This can be done in individual design teams or as a whole-class brainstorming session. Compile a list of student ideas and ask students to discuss the rationale behind material choices.

Students should complete a design journal reflection based upon this lesson.

Mathematics connection: Extend understanding of unit conversion to English measurement system/metric system conversions (for example miles per hour to kilometers per hour).

ELA connections: Students research the history of the English and metric measurement systems and write a composition comparing the two systems, their histories, current-day use, and their opinion about which is the more efficient system to use.

Social Studies connections: Discuss the development of the automobile and how the innovations associated with automotive technology (safety, efficiency, speed, development of interstate highway systems, etc.) have influenced society.

Assessment

Performance Tasks

- Completion of rubber band racers
- Engineer It! Worksheets
- Design journal reflections

Internet Resources

Speed versus velocity: http://education-portal.com/academy/lesson/speed-and-velocity-difference-and-examples.html#lesson (for teacher reference, portions of this explanation/video contain discussions of scalar versus vector measurements).

Information on IndyCar timing/speed calculations: http://www.indycar.com/Fan-Info/INDYCAR-101/Understanding-The-Sport/Timing-and-Scoring.

Sample design for rubber band car (for teacher reference – note that students may create alternative designs – if the car works, there is no right or wrong design!): http://www.youtube.com/watch?v=v3pbVAYkGf0.

Video of top 10 closest IndyCar finishes: http://www.youtube.com/watch?v=HI8MnBrUdhE.

Rubber Band Racers Design Challenge

In this activity, your design team will be challenged to create a rubber band powered car using the provided set of materials. Your objective is to create a car that travels the greatest distance in the shortest time.

The rules for this challenge are:

- The car can be powered by no more than three rubber bands.
- It must travel at least 3 meters (How many centimeters is that?).
- The car may not be propelled by human inputs (pushing).
- Rubber bands may not be used to "slingshot" the car.
- Only the materials provided can be used in the construction of the car. The meter stick and stopwatch may not be used in the construction.
- The car must have at least three wheels.
- You have only 50 minutes to design, build, and test your cars. You must complete three trial runs with your completed design.

Your team should conduct at least three timed trials in the "Test and Evaluate" step of the EDP. You will be racing your car against the other teams' cars, so your goal is to make your car as fast as possible!

Record the times for your three trials on your Engineer It! Worksheets using a chart like this one:

Trial #	Distance Traveled (in centimeters)	Time Elapsed (in seconds)	Speed (cm/second)
1			
2			
3			
Average Speed			

Lesson Plan #6 – The Automotive X-Challenge

Lesson Title: *Fact or Friction?*

Lesson Summary: This lesson introduces the concept of friction through demonstrations and inquiry activities. The overarching objective is for students to understand friction and the effect of various materials on the amount of friction. A discussion of whether friction is "good or bad" will be followed by an inquiry activity, Frictional Forces, in which students investigate the effects of various "roadway" materials on friction. Connections will be made to racecar tires and race track materials (optional). Students will reflect on the role of friction in their X-Challenge car design.

Essential Question(s):
- What is friction?
- What effect do various materials have on the amount of friction?

Established Goals/Objectives:
- Students will have a conceptual understanding of friction.
- Students will observe the effect of surface materials on friction.
- Students will research racetrack materials and their effect on friction.

Time Required: 2 classes

Necessary Materials:

Introductory Activity:

- Audiovisual equipment (internet access)
- 20 ounce plastic bottle (empty)
- 1 pound uncooked rice
- pencil
- tennis ball

Frictional Force:

- Frictional Forces worksheet – 1 per student

For each student group:

- Small box (about 5" x 5")
- Washers or pennies (100)
- String (3 feet)
- Masking tape (1 role)
- Plastic sandwich bag
- 6 unsharpened pencils
- 5 marbles
- 5 rubber bands
- Scale/balance (to measure in grams)
- Surfaces of different roughness:
- Sandpaper, aluminum Foil, wax paper, plastic wrap, non-skid drawer liner.

Teacher Background Information

This lesson introduces the concept of friction. Atoms and molecules sliding over each other cause friction. A rough surface produces more friction than a smooth surface, but no matter how smooth a surface appears, it is still "rough" at the atomic level. Friction causes kinetic energy to be converted to thermal energy, and therefore some amount of heat is always generated through friction.

There are several types of friction, but these fall into two major categories:

1. Static friction is friction between two items that are not moving (adhesion or electrical friction/static electricity).
2. Kinetic friction is friction between two moving objects (this encompasses rolling friction, sliding friction, and fluid friction).

Friction does not depend on the amount of contact surface area of the two bodies or on the relative speed of the two bodies in contact. The major consideration for friction is the type of surface.

Students will investigate the effect of various surfaces on kinetic friction in order to gain a qualitative understanding of friction.

Table A.1.12 Key Vocabulary – Lesson Six

Key Vocabulary	Definition
Forces	Pushes or pulls on an object.
Friction	The resistance that one surface or object encounters when moving over another; the force that opposes sliding motion.

Friction is an important factor in racecar design. Tires need enough friction to stay on the track without slipping. However, too much friction will slow the car down. Brake pads need a high amount of friction to stop a car effectively while automakers seek to produce engine pistons with very low friction (see http://www.caranddriver.com/features/everything-you-ever-wanted-to-know-about-pistons-feature for an explanation of pistons).

Student X-challenge designs will consider friction primarily in their choice of wheel materials and axle rotation. This lesson is geared toward introducing friction in a qualitative manner and will not introduce the mathematical formulas associated with calculating frictional forces.

Although students will not have a choice of road surfaces in their X-challenge design, they may be interested in the development of the course materials at the Indianapolis Speedway. The original 1909 racing surface was crushed stone-sprayed with tar. After the first automobile race (in August, 1909) on the track, management realized that a paved surface was necessary for safety. Later that year, over 3 million paving bricks, each weighing 9.5 pounds, were laid to create a new surface. This led to the track's nickname, "The Brickyard." In 1961, the track was resurfaced with asphalt. In 2004 the track was resurfaced with a Steel Slag Stone Mastic Asphalt formula that research showed to be very smooth and very durable (see http://www.acs.org/content/acs/en/pressroom/newsreleases/2013/september/indy-500-track-continues-to-foster-better-technology-for-everyday-driving.html for more detailed information).

Lesson Preparation

- Prepare materials for introductory demonstrations
- Prepare Snow Day Friction! Worksheets if student group option is chosen
- Prepare Fictional Forces worksheets
- Prepare Fictional Forces activity materials
- Prepare design journal reflection worksheets (one per student)

Learning Plan Components

Introductory Activity/Engagement

Science class: Begin class by asking students if they think that you can pick up 20,000 grains of rice with a pencil.

Conduct the following demonstration (Floating Rice):

- Use a funnel to fill a 20-ounce water/soda bottle with rice (you'll need almost a pound to fill it nearly to the top).
- Have student volunteers put a full size pencil into the rice bottle and stab it into the bottle continuously. Ask them what happens as they continue to stab the pencil into the rice (gets more difficult to move).

- Stab the pencil into the rice a few more times (using quick stabs – the goal is for the pencil to get "stuck"). At some point you should be able to carefully lift the bottle by just holding the pencil.

Ask the students what they think is happening.

Now hold a piece of paper parallel to the ground and drop it. Ask students what is happening.

Tell students that what they are seeing is friction at work. Ask students what friction is and develop a definition as a group (friction is a force that works in the opposite direction of motion; it is the force two surfaces exert when they rub against each other).

Ask students if friction is good or bad?

Roll a tennis ball across the floor and allow it to stop at a point in the middle of the floor (without hitting a wall). Ask students to predict what would happen to the ball if there were no friction.

Show video, A World Without Friction http://www.youtube.com/watch?v=7EPwwMU94OA to introduce the concept that there are different types of friction (static and kinetic). Have students brainstorm some examples of each.

Activity/Investigation

Science class:

Begin the investigation by asking students how friction is important for racecars (tires need to grip the track, but not too much). Ask students what they think is different about racecar tires than regular car tires (race car tires are "slicks" – have no treads). Lead students to an understanding that friction depends on the types of surfaces touching.

Introduce the Frictional Forces activity by telling students that they are going to investigate the effects of surfaces on friction. Students are challenged to design a device to measure friction using their understanding of the EDP. The second option will require a longer amount of time and a greater amount of student autonomy.

Challenge student design teams to design a friction-testing device. Students will use the Engineer It! Worksheets and the EDP to design and build their device. The amount of time you allow for this challenge is up to you, but make sure that students know how much time is available to them.

Mathematics connections: Students will construct graphs and make calculations related to friction.

ELA connections: Students write a short story based in a world in which there is no friction.

Social Studies connections: Students will conduct research on how necessary commodities are delivered to Alaska on roads that are frequently covered in ice.

Explain

Science class:

Forces: Students should understand the concept that forces are pushes or pulls on an object and those forces have direction (for instance, gravity has a downward pull).

Difference between gravity and friction. Students should understand that gravity is always a downward pull; friction always pushes or pulls in the direction opposite of the direction that the object is sliding (or would slide with no friction). Friction always acts parallel to the surfaces in contact.

Friction: Students will need to understand the concept of friction qualitatively and that friction is a force that is exactly large enough to prevent sliding; another force is applied that is large enough to overcome friction, the object will slide.

Energy transformations and friction: As the forces on an object overcome friction, potential energy is converted to kinetic energy. The force of friction works to convert kinetic energy into thermal energy.

Extend/Apply Knowledge

Science class:

Students will complete a design journal reflection for the X-challenge based upon their findings from the Frictional Forces activity. Students should concentrate on identifying areas in their car design that might be affected by friction (wheels, axles) and the implications for materials in their car designs.

Prompt students to think about the exterior and mechanics of the car (brakes, tires, axles, other rotating parts).

As an extension, students may include research on racetrack materials and their effect on friction in their design journal reflections.

Extend the activity by prompting students to consider the interior of the car and where friction is important (anywhere where driver grip is needed such as steering wheel, gear shifter, accelerator, brake pedal, seat, etc.)

Assessment

Performance Tasks

- Frictional Forces worksheet
- Design Journal Reflection

Other Measures

Internet Resources

A World Without Friction: http://www.youtube.com/watch?v=VUfqjSeeZng.

Explanation of engine pistons and friction: http://www.caranddriver.com/features/everything-you-ever-wanted-to-know-about-pistons-feature.
History of Indianapolis Speedway track material: http://www.acs.org/content/acs/en/pressroom/newsreleases/2013/september/indy-500-track-continues-to-foster-better-technology-for-everyday-driving.html.

Frictional Forces

Your design team will take on the role of product designers in auto manufacturing. Car brakes work through friction so it is important to be able to measure friction of different types of brake pads since when the driver presses the brake pedal, it is these brake pads that make contact with the brake rotors to make the tires stop turning and stop the car.

Your team is responsible for devising a way to measure the kinetic friction of various materials your company is considering using on brake pads.

Your design must follow these rules:

- You may use only the materials provided (see Materials list).
- You must have a way to measure the friction of the various materials.
- You must be able to compare friction between the materials.
- You must complete your design and record data for each surface material within the time your teacher designates.

Materials list:

- Small box (about 5" × 5")
- Washers or pennies (100)
- String (3 feet)
- Masking tape (one role)
- Plastic sandwich bag
- Six unsharpened pencils

Lesson Plan #7 – Transportation – Motorsports

Lesson Title: *Ready, Set, Race: The X-Challenge*

Lesson Summary: This lesson is comprised of the design challenge that students have been working toward in the previous lessons. Using their understanding of the scientific concepts and EDP incorporated in the unit, students will use the EDP to design, build, test, redesign, and present their car designs. Teams will be presented with specific goals for each class period to support their teamwork and use of the EDP. Student teams will choose one motorsports-related topic to research as a group. The lesson will culminate with presentations and a race event.

Essential Question(s): How can we use our understanding of energy and forces to design a prototype car powered by energy transformations?

Established Goals/Objectives:

- Students will be able to apply their understanding of science concepts to design and build a prototype car within the specifications and constraints they are given.
- Students will be able to use the EDP to design and build their design.
- Student teams will identify one topic of interest related to racing to investigate.
- Students will create presentation materials based upon their designs and topical research.

Time Required: 12 classes

Necessary Materials:

- X-Challenge Student Packets
- Engineer It! X-Challenge Worksheets
- Parts Warehouse
- Access to audiovisual equipment with internet access for video
- Student technology access for project research and presentation preparation
- Materials for student research project presentations

Teacher Background Information

This lesson represents the culminating design challenge for the unit. Student teams will have some autonomy in the design process, but you should give students an overview of what they should accomplish each day during the process. X-Challenge Engineer It! Worksheets are attached to this lesson and provide an outline of what students should accomplish each day.

1 Design and build their prototype car using all steps of the EDP.
2 Complete a team research project and presentation on one of the racing industry-related topics provided.

The challenge culminates with a 'Race Day.' Inviting guest judges to assess projects and talk to students about their design process and what they have learned adds real-life context to their work and requires that they prepare presentations that are engaging and professional. You should begin to think at least a week ahead about whom you might invite to judge projects. It is preferable to ask industry representatives at least two to three weeks ahead of time in order for them to plan appropriately.

Students should be prepared to give a brief team overview of their car design and design process to judges and to present their research projects to the class and to the guest judges.

Students should be reminded throughout the design and building process to refer to what they know about various types of energy, energy transformations, materials, friction, and aerodynamics and be prepared to talk about these concepts with the guest judges.

You will act as the manager of the Parts Warehouse. Guidelines for the Parts Warehouse are included in the student packet. You may incorporate these materials at your discretion; blank spaces were left on the materials list so that you may add materials and prices if you wish. You may choose to create a scarcity of some items (i.e. if there are six design groups, provide only four of each body style). You may have teams visit the Parts Warehouse to purchase their supplies all at one time or you may create a sequence in which each team can purchase one item (or one lot of the same item) at each visit.

Lesson Preparation

- Prepare the "Parts Warehouse" including any additional items from student parts requests lists
- Prepare student X-challenge packets
- Prepare copies of daily X-Challenge Engineer It! Worksheets
- Prepare collaboration rubrics

Learning Plan Components

Introductory Activity/Engagement

Science class:

Remind students that the X-Challenge "officially" begins today and that their teams will use the next 12 class periods to create a prototype vehicle and also to investigate a topic of their team's choice about motorsports.

Show the Dallara Italy video: http://www.youtube.com/watch?v=FgTRNJ32fWA.

Ask students to name the jobs they saw people doing in the video. Connect this with the teamwork students will participate in during their challenge.

Activity/Investigation

Science class: Student teams will use the EDP to create a prototype vehicle for the X-Challenge.

Mathematics connections: Students calculate the speeds of their vehicles and calculate averages over a number of trials.

ELA connections: Students utilize technology to research a motorsports-related topic and create a presentation on that topic.

Explain

Science class:

Remind students about the steps of the EDP.
Remind students about science concepts from the unit:

Types of energy

Energy transformations

- Friction
- Aerodynamics

Mathematics connections: Students remind students about speed calculations, speed versus velocity, and calculating averages.

ELA connections: Discuss research skills, citing references, and presentation skills.

Social Studies connections: Students can explore the economics of the racing industry as well as examine resource scarcity around the globe.

Extend/Apply Knowledge

Science class: Students use their understanding of motorsports and manufacturing careers to research a topic related to the racing industry.

Assessment

Performance Tasks

The X-challenge will be assessed in four ways:

1. *Collaboration* (assessed during lesson 9) – each student will be assessed on collaboration during the module and will use the collaboration rubric provided earlier in the module (individual grading – 30 points, rubric attached).
2. *Engineering Design documentation* (assessed during lesson 10) – each team member will submit a completed set of three Engineer It! Design worksheets (attached at the end of this lesson) for the project (individual grading – 30 points).
3. *Final Design* (assessed during lesson 10) – judges will use the design rubric to assess the team's car design (team grading – 30 points, rubric attached).
4. *Research Presentation* (assessed during lesson 10) – judges will use the presentation rubric to assess the team's presentation and research project (team grading – 30 points, rubric attached).

NOTE: The design judging and research presentations are included in Lesson 10. Rubrics are attached to this lesson for reference and are also attached to Lesson 10.

Internet Resources

Dallara Italy video: http://www.youtube.com/watch?v=FgTRNJ32fWA.

Appendix A 317

PROTOTYPE DESIGN RUBRIC (30 POINTS)

Team Name _____

Team Performance	Below Standard (0–2)	Approaching Standard (3–4)	Meets or Exceeds Standard (5–6)	Team Score
Creativity and Innovation	• Design reflects little creativity with use of materials, lack of understanding of project purpose, and no innovative design features • Design is impractical • Design has several elements that do not fit	• Design reflects some creativity with the use of materials, a basic understanding of project purpose, and limited innovative design features • Design is limited in practicality and function • Design has some interesting elements, but may be excessive or inappropriate	• Design reflects creative use of materials, a sound understanding of project purpose, and distinct innovative design features. • Design is practical and functional • Design is well-crafted, includes interesting elements that are appropriate for the purpose	
Conceptual understanding	• Design incorporates no or few features that reflect conceptual understanding of science concepts (energy types, energy transformations, frictions, and aerodynamics)	• Design incorporates some features that reflect a limited conceptual understanding of science concepts (energy types, energy transformations, frictions, and aerodynamics)	• Design incorporates several features that reflect a sound conceptual understanding of science concepts (energy types, energy transformations, frictions, or aerodynamics)	
Designed within specified requirements	• Design violates challenge rules and/or specifications, and design is not finished • Design team exceeded budget by more than 10% ($30)	• Design meets most challenge rules and/or specifications, and design is finished on time • Design team exceeded budget by less than 10% ($30)	• Design meets all challenge rules and/or specifications, and design is finished on time • Design team stayed within budget	

(Continued)

Team Performance	Below Standard (0–2)	Approaching Standard (3–4)	Meets or Exceeds Standard (5–6)	Team Score
Performance	• Vehicle does not function or faces substantial problems (more than one pit stop) in traveling the required distance	• Vehicle functions, but does not travel the required distance • At least one pit stop is required	• Vehicle travels the required distance • Team requires one or no pit stops	
Design Presentation	• Team members are unable to articulate their design process • Team members are unable to identify or justify design features in terms of science concepts • Team members speak in a manner inappropriate to the audience (slang, poor grammar, mumbling)	• Team members articulate their design process, but not clearly or coherently • Team members make some reference to science concepts when discussing design features • Team members mostly speak in a manner appropriate to the audience but presentation may be confusing or not engaging to audience	• Team members articulate their design process clearly and coherently • Team members clearly refer to science concepts when discussing design features • Team members clearly outline the advantages of their design • Team members speak in a manner appropriate to the audience and are engaging and concise	

Appendix A 319

PROTOTYPE DESIGN RUBRIC (30 POINTS POSSIBLE)

Team Name _____

Team Performance	Below Standard (0–2)	Approaching Standard (3–4)	Meets or Exceeds Standard (5–6)	Team Score
Sources of information	• Team uses only one source for research • Team does not include references to information sources	• Team includes more than one source for research • Team includes some references to sources of information	• Team includes multiple sources for research • Team includes complete references for each source of information	
Ideas and Organization	• Team does not have a main idea or organizational strategy. • Presentation does not include an introduction and/or conclusion • Presentation is confusing and uninformative • Team uses presentation time poorly	• Team has a main idea or organizational strategy, but it is not clear or coherent. • Presentation includes either an introduction or conclusion, but not both • Presentation is somewhat coherent, but not well organized, and is somewhat informative. • Team uses presentation time adequately, but presentation may be somewhat too long or too short	• Team has a clear main idea and organizational strategy. • Presentation include both an introduction and conclusion • Presentation is coherent, well organized, and informative • Team uses presentation time well and presentation is neither too short nor too long.	

(Continued)

Team Performance	Below Standard (0–2)	Approaching Standard (3–4)	Meets or Exceeds Standard (5–6)	Team Score
Presentation Style	• Only one or two team members participate in the presentation • Presenters do not look at audience, reads notes • Presenters are difficult to understand • Presenters use language inappropriate for audience (uses slang, poor grammar, and frequent filler words such as "uh" and "mm")	• Some, but not all, team members participate in the presentation • Presenters make some eye contact with audience, but rely on notes • Most presenters are understandable, but volume may be too low or some presenters may mumble • Presenters use some language inappropriate for audience (slang, poor grammar, and some use of filler words such as "uh," um")	• All team members participate in the presentation • Presenters make eye contact with the audience and refer to notes only occasionally • Presenters are easy to understand • Presenters use appropriate language for audience (no slang, poor grammar, and infrequent use of filler words such as "uh" and um")	
Visual Aids	• Team does not use any visual aids to presentation • Visual aids are used but do not add to the presentation	• Team uses some visual aids to presentation, but they may be poorly executed or distract from the presentation	• Team uses well-produced visual aids or media that clarifies and enhances presentation.	
Response to audience questions	• Team fails to respond to questions from audience or responds inappropriately	• Team responds appropriately to audience questions but responses may be brief, incomplete, or unclear	• Team responds clearly and in detail to audience questions and seeks clarification of questions.	

The Automotive X-Challenge Overview

Your team is challenged to design and build an innovative prototype car powered by energy transformations. You will use the science concepts you have learned during the unit and your design journal reflections to help you with this process. Your team will have a budget of $300 X-bucks to spend. Your team will present their design in an oral presentation to judges on Race Day and must be able to answer questions about the design, team process, and science concepts associated with your car.

Your team will have seven class periods to complete the X-challenge and create a presentation for the challenge judges.

There are two parts to the X-challenge:

1. Team car design and design presentation
2. Team research project add presentation

Your team will choose one racing industry-related topic from the list below to research. You will include your findings as part of your X-challenge presentation. Your team will create a presentation based upon your research project. This can be a display board, a media presentation, a creative oral presentation (for example a mock debate or mock trial), a brochure, or another creative method of presenting your research findings.

1. Race Guide – create a race-watching guide that will attract new fans to watch IndyCar races. The guide should be informational but also should highlight reasons why non-fans should watch IndyCar races.
2. Women in Racing – Women are involved in car racing at all levels from race car designers to the pit crew. Choose three women involved in racing and highlight their accomplishments and what they did to achieve their successes.
3. Sport or Show? There is a debate about whether car racing should be classified as a sport or simply as entertainment. Research opinions about this and the justifications for each and present both sides of the argument.
4. Safety – Research the various safety innovations that the racing industry uses to keep drivers, crews, and fans safe. What safety features in passenger cars come from race car design?
5. Fashion & Design at the Track – Investigate the fashion elements of racing and what the racing industry does to make cars and drivers make a statement.
6. Racing: Then and Now – Investigate the history of car racing. What has changed over the years? How has the history influenced the development of modern-day racing?
7. Ways to Race – Research the different types of auto racing. Provide an overview of each one and highlight similarities and differences of the different types of racing.

Constraints

Your team will be given a budget of $300 X-bucks to build your prototype. Your X-bucks will be accepted at the parts warehouse.

- You may return or exchange items at the warehouse, but keep in mind that there may be limited availability of some items.
- You can trade items with other design teams, but you may not buy and sell items with other groups – only the parts warehouse accepts X-bucks!
- If you run out of money before your design is over, you may apply for a loan from the parts warehouse manager. Keep in mind that your design will be judged on cost-effectiveness and you should make every effort to stay within budget.
- You should record all of your transactions in your financial ledger (journal).

Each team will be provided with a no-cost start-up kit that contains *scissors, masking tape, safety glasses, a meterstick, and a timer*. All items used in the team's design must be purchased from the parts warehouse.

The following are the items, and their prices (in X-bucks) are available in the warehouse:

Styrofoam block – $50
Wood block – $35
Cardboard box – $40
Tires/wheels (black plastic tire material) – $40 each
CDs – $10 each
Wooden disks – $25 each
Spindles – $25 each
Drinking straws – $10 for 4
Wooden skewers – $20 for 2
Pipe cleaners – $10 for 4
Pencils – $10 for 2
Rubber bands, 6 mm – $40 for 2
Rubber bands, 3 mm – $30 for 2
Rubber bands, 1 mm – $20 for 2
Balloons – $30 for 2
Waxed paper – $10 per linear foot
Aluminum foil – $10 per linear foot
Poster board – $20 per half sheet
Craft sticks – $10 for 5
Paper cups – $30 for 4
Baking soda – $30 for ½ cup
Vinegar – $20 for ¼ cup

Plastic spoons – $10 for 2
Paper clips – $10 for 5
Wire hanger – $15 each
Cotton balls – $10 for 10
Index cards – $10 for 10
String – $10 for 3 feet
Toothpicks – $10 for 10
Wooden clothespins – $10 for 4
Set of poster paints and brushes – $20
Stickers – $10 per sheet

Your car must be designed according to the following rules:

- Your car must be able to travel at least 3 meters.
- Your car must travel in a straight line.
- You may use only the materials in your start-up kit and materials purchased from the parts warehouse.
- Your vehicle must use at least one energy transformation to power it, and it may not be powered by a force applied by a person (you can't push your car to make it go!).
- Once your energy conversion has started, you may not touch your device.
- All team members must participate in both the car design and the research project.
- On race day, teams will have one minute to prepare their cars at the starting line.
- If your car breaks down during a race, you will be given one pit stop of 3 minutes to repair your vehicle and then begin the race again.

The evaluation of your X-Challenge project is composed of four parts:

1 **Collaboration** – your teacher will assess you on collaboration, or how well you work with your team, during the unit – 30 points
2 **Engineering Design Process documentation** – each team member will submit completed Engineer It! Design worksheet for the project – 30 points
3 **Final Design** – judges will use a design rubric to assess the team's car design – 30 points
4 **Project Presentation** – judges will use the presentation rubric to assess the team's presentation and research project – 30 points

Your team will create a presentation for your research topic to present to your classmates, teacher, and guests. You will have **5 minutes** to make your presentation and **5 minutes** to answer questions.

Your Engineer It X-Challenge sheets are attached at the end of this packet. Here are a few pointers:

- Remember to refer to your design journal reflections when you are creating your design – it may include some useful information about energy transformations, friction, and aerodynamic drag that could be useful.
- The X-Challenge judges will ask you about your design process on Race Day, so be sure that all team members are familiar with all stages of the design. Be sure to be able to talk about your design decisions, the science concepts you considered, and testing and re-designing work.

X-Challenge Engineer It!

Name_____

1 **Name Your Team!**

Our team name is _____

2 **Identify the problem and constraints**

- State the problem:

- Identify the conditions that must be met to solve the problem:

- Identify anything that might limit the solution (cost, availability of materials, safety)

3 **Ideas**

- Is there anything from your design journal reflections that might be helpful? Summarize that here:

- Brainstorm! What solutions do you and your team imagine?

4 **Team Planning**

Record your team's plan here. Remember, everyone needs to be involved in both your car's design and your research project, but if some of your team members want to take on special tasks, you can record that here.

(For instance, you may wish to have an "accountant" to keep track of the budget; do you have someone who is very good at drawing who will make sketches? Is someone great at putting together multi-media presentations? Are there any other tasks team members want to lead?)

5 Design Features

- Based on your brainstorming in the last class, what features do you want to include in your car?

- Include a sketch or sketches here. Label your sketches:

6 Why did you choose this design?

7 Materials

- What materials do you think you will need (see the Parts Warehouse list)? List them here.

- How much will those parts cost in X-bucks?

8 Decide on your research project topic with your team. Record the topic here.

9 Test and Evaluate

- How did you test your prototype?

- What were the results of your tests?

- What are the strengths of your design?

- What are the weaknesses of your design?

10 Improve Your Design

What changes would help your design perform better? Record your ideas and make changes to your design!

11. **Present/Share your car prototype design**

    Decide who will present various aspects of your design and the design process. List team member responsibilities here:

12. **Present your research project**

    How will your team present your research topic?

    Decide who will present various aspects of your presentation. List team member responsibilities here:

Appendix B
Sample STEM Module Two: Grade K
Jennifer Suh

STEM Road Map Curriculum Module Overview

STEM Road Map Module Theme and Grade Level: The Represented World, Grade K

STEM Road Map Module Topic: Patterns on Earth and in the Sky

Module Summary

In this investigation, the students will begin to see patterns as they emerge during the year from a solar, weather perspective in the sky and the adaptability of animals on Earth. The problem/challenge for this unit is: *A petting zoo needs you to investigate how the patterns of the sky and the animals on Earth adapt to changes over one year and create a year-long calendar to demonstrate what you have observed throughout the year. Create a presentation for the petting zoo to explain to their customers the changes that animals experience over a year.*

Much of the observations the students will make can be recorded in their class STEM notebooks and used as talking points during this unit. The lead discipline of this unit's theme revolves around mathematics so much of the observations will take the form of quantitative relationships backed by qualitative observations and aligning the different patterns of the sky and animals. Data can be collected using illustrations of the cycles of the Sun, the Moon, the seasons, and how animals adapt to these changing conditions. Weather observations can also be collected and analyzed based on the seasons. This unit can span the entire school year so that the students can understand how patterns of the sky and the Earth change. If there was a pond or river/stream located close to the schoolyard, it would be an excellent area to make these observations and note the patterns that change throughout the year. The capstone project at the end of this unit might include a year-long calendar that illustrates the changes in the Sun, Moon, seasons, and animals that the students observed. The mapping of content standards associated with this theme/

328 Jennifer Suh

topic can be found in Table 4.4. Potential careers to explore: meteorologist, astronomer, ecologist, and animal husbandry.

Established Goals/Objectives:

The goal for this PBL is for student to learn and demonstrate their knowledge about weather patterns and how animals on Earth adapt to their changing environment.

- Understand change and observable patterns of weather that occur from day to day and throughout the year.
- Make connections that change is something that happens to many things in the environment based on observations made using one or more of their senses.
- Summarize daily weather conditions noting changes that occur from day to day and compare weather patterns that occur from season to season.
- Learn about animal characteristics and how they adapt to their environment.

The specific Next Generation Science Standards (NGSS) addressed are:

- **K-ESS2-1** Use and share observations of local weather conditions to describe patterns over time.
- **K-ESS3-1** Use a model to represent the relationship between the needs of different plants and animals (including humans) and the places they live.
- **K-PS3-1** Make observations to determine the effect of sunlight on Earth's surface.
- **K-LS1-1** Use observations to describe patterns of what plants and animals (including humans) need to survive.

Challenge and/or Problem for Students to Solve: *A petting zoo needs you to investigate how the patterns of the sky and the animals on Earth adapt to changes over one year and create a year-long calendar to demonstrate what you have observed throughout the year. Create a presentation for the petting zoo to explain to their customers the changes that animals experience over a year.* Compelling question for this unit include: *How do the patterns on earth including cycles of the Sun, the Moon, and the seasons' impact animals on Earth?*

Table A.2.1 Content Standards Addressed in STEM Road Map Module – Patterns on Earth and in the Sky

NGSS Performance Objectives	Mathematics	Common Core ELA
K-ESS2-1 Use and share observations of local weather conditions to describe patterns over time.	CCSS.Math.Practice.MP7. Look for and make use of structure	CCSS. ELA-Literacy. RI.K.1. With prompting and support, ask, and answer questions about key details in a text.
K-ESS3-1 Use a model to represent the relationship between the needs of different plants and animals (including humans) and the places they live.	CCSS.Math.Practice.MP8. Look for and express regularity in repeated reasoning	CCSS. ELA-Literacy. RI.K.3. With prompting and support, describe the connection between two individuals, events, ideas, or pieces of information in a text.
K-PS3-1 Make observations to determine the effect of sunlight on Earth's surface.	CCSS.Math.Content.K.CC.B.4 Understand the relationship between numbers and quantities; connect counting to cardinality.	CCSS. ELA-Literacy.W.K.2.Use a combination of drawing, dictating, and writing to compose informative/explanatory texts in which they name what they are writing about and supply some information about the topic.
Discuss and describe different types of weather.	CCSS.MATH.CONTENT.K.MD.A.1 Describe measurable attributes of objects, such as length or weight. Describe several measurable attributes of a single object.	CCSS. ELA-Literacy. W.K.5. With guidance and support from adults, respond to questions and suggestions from peers and add details to strengthen writing as needed.
K-LS1-1 Use observations to describe patterns of what plants and animals (including humans) need to survive.	CCSS.MATH.CONTENT.K.MD.A.2 Directly compare two objects with a measurable attribute in common, to see which object has "more of"/"less of" the attribute, and describe the difference. Classify objects and count the number of objects in each category.	CCSS.ELA-Literacy.W.K.7 Participate in shared research and writing projects.
	CCSS.MATH.CONTENT.K.MD.B.3 Classify objects into given categories; count the numbers of objects in each category and sort the categories by count.[1]	CCSS. ELA-Literacy.SL.K.1. Participate in collaborative conversations with diverse partners about kindergarten topics and texts with peers and adults in small and larger groups.
		CCSS. ELA-Literacy. SL.K.5. Add drawings or other visual displays to descriptions as desired to provide additional detail.
		CCSS.ELA-Literacy.SL.K.3 Participate in collaborative conversations with diverse partners about kindergarten topics and texts with peers and adults in small and larger groups.
		CCSS.ELA-Literacy.RL.K.1 With prompting and support, ask and answer questions about key details in a text.

Table A.2.2 21st Century Skills Addressed in the STEM Road Map Module

21st Century Skills	Learning Skills and Technology Tools (from P21 framework)	Teaching Strategies	Evidence of Success
21st century interdisciplinary themes	Global awareness Civic literacy	Teachers will allow students to explore weather patterns in other parts of the world.	Students will tell if they have travel to different locations around the Earth and if they have experienced different weather patterns.
Learning and innovation skills	Creativity and innovation Critical thinking and problem-solving Communication and collaboration	Using the 4Cs, teachers will launch a challenge to create an improved habitat for animals at the Petting Zoo	Students will collaboratively think about the needs of a newborn farm animal. They will be creating a presentation called the Petting Zoo infomercial and will be able to use their creativity.
Information, media, and technology skills	Information literacy Media literacy ICT literacy	Teachers will use several different web resources and books to build students' background knowledge for this project.	Students will learn more background knowledge using the website resources and books to design their final product.
Life and career skills	Flexibility and adaptability Initiative and self-direction Social and cross-cultural skills Productivity and accountability Leadership and responsibility	Teachers will monitor students engaged in collaborative projects to access their group cooperation skills and their leadership skills.	Students will work together to make a plan for their projects throughout the unit. Students will work effectively in collaborative groups and be clear about roles of each member.

Launch

Plan a field trip to the Petting Zoo: [Ideally, it should be a field trip to the Petting Zoo but if that is not possible, arrange for virtual field trip (via internet) or a guest speaker from the petting zoo, farm, or zoo with at least one animal].

The Petting Zoo owner/visitor can make the project more authentic by sharing that they need help attracting more visitors to the Petting Zoo by making people more aware of the events that happen at their Petting Zoo year round. The job is for the Kindergarten students to learn more about the seasonal change and how animals adapt to that change to be able to share with the visitors what to look for when they visit the Petting Zoo.

Prerequisite Key Knowledge: What are the key concepts that are most important for students to know in each discipline for the unit?

Table A.2.3 Prerequisite Key Knowledge

Prerequisite Key Knowledge	Application of Knowledge	Differentiation for Students Needing Knowledge
Students should know about zoos and petting zoos.	This is important because they will need to learn about how people care for the animals and how they adapt to their environment. They will need to think about the habitats, pens, and enclosures that the pets live and play in during the year.	There may be students with varied exposure and familiarity with zoos or farms. Provide many books, websites, and video clips of animals at the zoo or the farm. Provide farm animal toys and other zoo animals so they can play pretend zoo.
Students should know about weather and how it affects our daily lives.	This is important because they will build on the prior knowledge of being familiar with different weather patterns.	There may be students who have lived in areas with four seasons and others who have other regions of the United States without the four seasons. Use this difference as a teachable moment for students to share about different places with different climates.
Students should know that tools and technology helps us in our daily lives and that is result of engineering design.	This is important because they will be building ways to improve the current habitat for one of the petting zoo animal. In addition, they will see how wool or down can be used to make warmer coats for people.	Provide many everyday tools and technology that is an example of an engineering design.

Desired Outcomes and Monitoring Success: Identify student outcomes to be met through the unit. Some students may be successful given only the desired outcomes, while other students may need scaffolding by providing benchmark goals along the way. Students can use these desired outcomes to self-monitor and check that they are progressing in a positive direction.

Assessment Plan: Define the products and artifacts for the project. Be sure to include a variety of assessments for learning that are closely tied to the content, learning skills, and technology tools outcomes. The products

Table A.2.4 Desired Outcomes and Monitoring Success

Desired Outcome	Evidence of Success in Achieving Identified Outcome	
Students will explain how people and animals adapt to the weather and seasonal change through their Petting Zoo's Calendar of Events project.	*Performance Tasks* Students will create the Petting Zoo Calendar of events and Infomercial that feature appropriate activities for different seasonal visits to the Petting Zoo that will demonstrate their understanding of changes in season and the different ways animals adapt to the change.	*Other Measures* Students will have formative assessment through each lesson that assesses their understanding of changes in weather, season, and animal adaptability to the environment. There are multiple formative assessments that teachers can collect from Science, Literacy, Math, and Social Studies mini lesson activities.

Table A.2.5 Assessment Plan

Major group products	• Petting Zoo-Calendar of Events created by the class • Photo calendar with student drawings representing the seasonal change and how animals will adapt to the changes • Infomericial to advertise the calendar of events at the Petting Zoo-video presentation • Graph of weather patterns throughout the unit. • Model of an improved habitat for one of the Petting Zoo animal and an engineering design of an enhancement to the habitat.
Major individual products/ deliverables	• Graph daily weather and use it to compare pattern. Explore ways we can use materials to decrease the temperature of an area from the sun. Graph weather over a period of time several times throughout the year. Compare weather patterns using graphing data. • Choose a favorite animal and build a model habitat for it. • Create a calendar representing the seasonal change and how animals will adapt to the change

and criteria must align with the objectives and outcomes for the project. State the criteria for exemplary performance for each product. Plan for assessments that provide student feedback as the project progresses and provide for a culminating appraisal of performance or product with an accompanying rubric that clearly assesses the learning targets.

Resources:

School-based Individuals

Art teachers can provide lessons on drawing farm animals and discuss how to draw a landscape of a petting zoo. The media specialist can provide books, websites, and videos about farm animals and how animals adapt to their environment and seasons.

Technology

Age-appropriate website resource collection can be created for kindergarten students to access as they are doing research on their animal and about the local Petting Zoo.

Community

Contacting a Local Petting Zoo will be important to the authenticity of the project. If there is no local Petting Zoo, a local farm or an animal zoo will be helpful to contact to get field trip arranged and/or have a guest speaker from the zoo or farm visit the class.

Materials

Modeling Materials: Clay, toy animals, construction paper, children magazine.

Lesson Plan – A Glance at Week 1–2 – Patterns in our World Kindergarten STEM Unit

Lesson Title: *Patterns in Our World and How It Impacts Living Things*

Lesson Summary

This lesson is part of a unit that focuses on student understanding of patterns in nature, natural cycles, and changes that occur both quickly and slowly over time. An important idea represented in this unit is the relationship among Earth patterns, cycles, and change and their effects on living things. The topics developed include noting and measuring

Table A.2.6 STEM Road Map Module Schedule Weeks 1 and 2

Day 1	Day 2	Day 3	Day 4	Day 5
Students will be introduced to the "Patterns on Earth" as they play a Scavenger Hunt for Patterns in our World (Go outdoors) *Science and Language arts:* List all the patterns in our world	Students will observe local weather and record their observation and pretend to be weather reporters. *Science and Language arts:* Use weather vocabulary to tell about the weather.	Students will learn how people adapt to their environment by playing "Dressing for the Weather." *Science and Engineering:* Brainstorm technology to keep people warm or cool.	Students will learn how people participate in different weather activities. *Science Math and Language Arts:* Vote and create a table of their favorite things to do in different weather/season.	Students will observe local weather and record their observation. *Science and Math:* Graph the weekly weather pattern and compare the results.

Day 6	Day 7	Day 8	Day 9	Day 10
Students will learn about the center of our solar system, the Sun, and read and sing about the sun. *Language Arts:* Read the fable The North Wind and the Sun.	Students will learn about what makes day and night. *Science and Art Integration-* "Dance" the rotation on earth's axis with a light source to model day and night.	Students will learn about the moon and its phases and learn about the reflection of the sunlight on the moon. *Science and Language arts:* Read Papa, Please get the Moon for Me by Eric Carle.	Students will learn about reason for seasons. *Science and Art Integration:* "Dance" the revolution around the sun. Read the book The Reason for Seasons by Gail Gibbons	Students will display the landscape including changes in animals and plants throughout the seasons. *Science, Language & the Arts:* Make a season wheel with photos or magazine cut outs to show the landscape.

Appendix B   335

Table A.2.7 STEM Road Map Module Schedule Weeks 3 and 4

Day 11	Day 12	Day 13	Day 14	Day 15
Plan a Visit to the Petting Zoo				
Students will read about a true story called *Beatrice's Goat* and retell how one goat can sustain a family's need.	Students will identify animal characteristics of goats that help human needs and what they need to survive. *Science and Math: how much milk does the goat produce?*	Students will identify characteristics of horse that help human needs and what they need to survive. *Social Studies: Learning about how Horses Providing transportation.*	Students will identify animal characteristics of chicken that help human needs and what they need to survive. *Science and Social Studies: Learn about the life cycle of a chicken and about the goods it produces.*	Students will identify animal characteristics of sheep and geese that help human needs and what they need to survive. *Science and Engineering: Learn about the wool that the sheep produces and the down that geese produce that warms people. Weatherproof technology.*

Day 16	Day 17	Day 18	Day 19	Day 20
Students will create a calendar for Spring. *Science, Math Language arts: Baby animals are born in the spring. Learn names to describe baby animals and how many babies they give birth to at one time.*	Students will create a visual display of events at the petting zoo in the Spring. *Science, Math Language arts: Read the book, Is Your Mama, a Llama? Matching Activity- parent and young. How do adult animals care for their babies?*	Students will create a calendar for Summer: Learn about life cycles of animals. *Science and Language arts: Create a life cycle chart of a given animal.*	Students will learn how animals adapt to the summer. *Science and Language arts: Students will create a visual display of events at the petting zoo in the Summer.*	Students will learn how animals behave in the fall. *Science and Language Arts: Read a blog post from "Ask a Naturalist" (post 84) from the Charlotte Nature Museum about what animals do in the fall.*

Table A.2.8 STEM Road Map Module Schedule Week 5

Day 21	Day 22	Day 23	Day 24	Day 25
Students will learn how cold blooded animals behave in the fall. *Science and language arts: Read a blog post from "Ask a Naturalist" from the Charlotte Nature Museum and learn about what cold-blooded animals do in the fall.*	Students learn about how animals not only prepare for winter, but use coats, fat deposition, to survive in the cold. *Science and Language arts: Research to learn how animals use coats and fat deposits to keep warm.*	Students learn about how animals not only prepare for winter, but use hibernation, and migration to survive in the cold. *Science and Language arts: Teacher models how one can research to learn on how animals hibernate and migrate.*	Students will create participate in shared writing to explain how the four seasons impact animals' adaptability to the environment. *Science and math: Students put together their drawings of the different behaviors' of animals and creates a calendar of events.*	**Petting Zoo infomercial** Students will present their calendar and make an infomercial for visitors about the calendar of events and highlights to look forward to at the petting zoo.

changes, weather and seasonal changes which will connect to later lessons on how these impact animals' behaviors.

Essential Question(s)

what questions will guide student learning in this lesson?

- What patterns do we have in our world?
- How do we respond to these patterns daily and throughout the year?

Established Goals/Objectives

Students will understand (big ideas/key knowledge), know, and be able to do what (key skills)?

- Observe and identify daily weather conditions – sunny, rainy, cloudy, snowy, windy, warm, hot, cool, and cold.
- Predict daily weather based on basic observable conditions.
- Chart daily weather conditions.
- Identify characteristics of the different seasons.
- Recognize that plants, animals, and people adapt to the changing seasons in different ways.

Time Required: 10 days

Necessary Materials

Outdoor access and window from the classroom, calendar, weather graph, internet access

Table A.2.9 Key Vocabulary

Key Vocabulary	Definition
Weather	The state of the atmosphere at a place and time as regards heat, dryness, sunshine, wind, rain, etc.
Season	One of the four periods of the year (spring, summer, autumn, and winter)
Sun (solar)	The star that is the center of the solar system, around which the planets revolve and from which they receive light and heat
Moon (lunar)	The earth's natural satellite, orbiting the earth
Temperature	A measure of the warmth or coldness. Use words like hot, warm, cool, cold, and freezing
Precipitation	Falling products of condensation in the atmosphere, as rain, snow, or hail
Cycle	Cycle is a repeated pattern. A sequence is a series of events that occur in a natural order

Teacher Background Information

What content does the teacher need to know about to deliver this lesson? Brief summary.

Studying the weather is a great way to learn about observations and patterns and to help us prepare for upcoming days based on the weather predictions. For kindergarteners, weather pattern is easily seen and recorded because weather happens daily and can be easily seen and recorded. There are lots of daily and seasonal patterns that occur in weather. Use words to describe the sky, sunny, mostly sunny, partly cloudy, and cloudy and words for precipitation such as, dry, rain, snow, hail, and sleet. Measuring temperature may be difficult for the kindergarteners but exposing them to the temperature will make them learn that it is measureable and that there are tools like a thermometer (technology) that help measure how hot or cold it is outside. Using words like hot, cold, warm, and freezing will help students associate temperature with the weather and seasons.

Lesson Preparation

What will the teacher need to plan ahead of time for this lesson?

The series of lessons will take place over two weeks. This provides a glance at the two-week unit on weather and seasons and cycle on Earth.

Learning Plan Components

Introductory Activity/Engagement

Science class

Discuss the patterns in our world. Play "Scavenger Hunt for Patterns" in our World. They will go outside and look for patterns. Some may look for visual pattern with colors, shapes, and plants. Connect their observations with patterns we have like day and night, life cycle of a plant, moon phases, seasons, and weather patterns. As a class, students can pretend to be a weather reporter – can they predict the weather?

Mathematics class

Use the calendar to record the weather and use a graph to chart the pattern each day and throughout the unit. Students will collect data using observations. The class can represent data using different displays such as tables and graphs (bar graphs or picture graphs).

ELA class

Use communication skills to discuss patterns in our world.

Read aloud book

Fable: *The North Wind and the Sun* by Gregory McNamee.

Have students retell the story and vote what they think is stronger: the wind or the sun.

Another great read aloud when talking about the patterns of day and night is the book *Papa, Please get the Moon for Me by Eric Carle.* Students will learn about the moon and its phases and about how the reflection of the sunlight illuminates the moon.

Social Studies class

Connecting geography and climate and time of day

Locating where we live and the climate in our region. Another extension is looking back at patterns like day and night and how some parts of the world, it is day when it is nighttime in other locations.

Activity/Investigation

Science class

Divide the students into groups and have them investigate about the four seasons.

Have students fold a piece of paper into four sections. Then have the students list four types of weather (one per section) and draw a picture for each. Draw and label in each box a type of weather. Use the pictures that student drew to create a class season wheel. Play dress up with clothing that match with the weather pattern.

Mathematics class

Integrate Calendar Math with weather data. After a few weeks, have students read their weather chart and make a bar graph. Other ways to show how to keep track is to use tally marks to show the quantity or use the numeral next to each weather icon to summarize the total number of days recorded with that weather pattern. Use icons to record the daily weather on the weather graph and at the end of the week, ask questions like, "How many days this week was recorded as sunny? How many days did we have rain? Compare sunny and cloudy days? Which did we have more? Less? Comparing is one of the meanings for subtractions because we are looking for the difference between two quantities. Look and listen for students who may be subtracting or counting on to find the difference.

ELA class

Read aloud by *Too Hot? Too Cold? Keeping Body Temperature* by Caroline Arnold and Annie Patterson.

Watch a segment of a weather forecast on a news show. Let students be the "weather person" who looks out a window or goes outside to collect daily observations and then tells the class about the:

- Talk about the sky – sunny, mostly sunny, partly cloudy, and cloudy
- Talk about the precipitation – dry, rain, snow, hail, and sleet
- Talk about the temperature – hot, warm, cool, cold, and freezing
- Sing weather songs "Rain, Rain, Go Away," "Mr. Golden Sun," and "You Are My Sunshine")

Social Studies class

Show the map of our world and show where we live and how that explains our climate. This may be abstract for them so elicit some of their background knowledge by asking, "Have you visited family living very far from our neighborhood? What do you remember about their weather? Did anyone go to a really hot tropical place before on any trips? How about a really cold snowy place during winter?" This might trigger some conversation. If not, tell a story about your trips to very warm climates during your family vacations.

Explain

What components will the teacher explain/discuss/teach to students in this lesson?

Science class

How do people keep warm? How do animals stay warm in winter (fur, feathers)?

Show students artifacts like, fur coat, fake feathers, alligator boots, or purse to demonstrate the different textures of animals. Let the students use their five senses to describe the animal artifacts. Pick a few animals like a reptile and talk about them. For example, reptiles are cold-blooded and have scales to cover their skin. Create a three-column chart and label the columns fur, feather, and scales. Brainstorm what type of animals might have fur, feathers, and scales.

Mathematics class

Sorting and Categorizing Activity:
Make a T-chart of clothing for hot heather and for cold weather.

Clothing for hot weather	Clothing for cold weather

ELA class

Use the chart created with visuals in science class to be used as an opportunity to engage the students in share writing on a large poster to tell what the chart was about (see science class lesson).

Social Studies class

Look around the globe and talk about how people from different regions dress differently based on climate.

Extend/Apply Knowledge

What opportunities will students have to apply what they have learned through their work in this lesson explicitly, if any?

Science class

How do animals stay warm or dry?
   Draw a scene with a season and people and animals in the drawing showing how they stay warm.

Mathematics class

Summarize the weather graph and count and tally the total number of days in each category. Use that chart to be the "meteorologist" and tell about the pattern(s) we have had in our weather.

Table A.2.10 Weather Pattern Flip Book Rubric

Weather Pattern Flip Book RUBRIC

Objectives	Excellent	Good	Fair	Needs improvement
Student can identify and describe at least three different weather patterns (i.e. sunny, rainy, snowy, cloudy, and windy) with specific characteristics of the weather.	4 Student can identify more than three different weather patterns and describe the characteristics.	3 Student can identify at least three different weather patterns and describe the characteristics.	2 Student can identify at most two different weather patterns and describe the characteristic.	1 Student cannot identify different weather patterns and/or can name one but cannot describe the characteristics.
Drawing shows elements of weather (i.e. clouds in the sky, raindrops, and puddles)	4 Drawing shows elements of weather with lots of details.	3 Drawing shows some distinguishable elements of weather.	2 Drawing shows one sign of weather.	1 Drawing does not show any distinctive elements of weather.
Drawing shows people, animals or plants responding to weather (i.e. dressed for the weather with umbrella, raincoat, and rain boots)	4 Drawing shows people, animals, and or plants responding to weather with details.	3 Drawing shows people, animals, and or plants responding to weather.	2 Drawing may show either people or other things responding to weather.	1 There is no evidence that anyone or anything responding to weather.

Table A.2.11 Key Vocabulary – Lesson Two

Key Vocabulary	Definition
Adapt	To change in response to the environment or situation.
Adaptation	A change that a living thing goes through so it fits better with its environment.
Camouflage	Coloring or covering that makes animals, people, and objects look like their surroundings.
Environment	The natural world of the land, sea, and air.
Hibernate	An extended period of deep sleep that allows animals to survive winter extremes
Migrate	The seasonal movement of animals from one region to another.

ELA class

Use picture vocabulary cards to make match picture cards of weather patterns: clear, cloudy, cold, fair, fall, hot, rainy, spring, summer, sunny, temperature, warm, windy, and winter

Ask the students think back at the weather report, they have seen on TV.

Ask students whose job it is to give these reports and to research and make predictions of the weather (meteorologists).

Social Studies class

Use the map to show where we live and what our climate is in different regions of North America. Learn about how people specialize in a job, and a meteorologist is a specialized job that helps the community. Ask students, how the meteorologist helps them in their daily lives or when their family is planning an event.

Assessment

Performance Tasks

- Create a flip book with weather patterns and draw people, animals, and plants in the picture to show how they respond to the weather pattern.

Other Measures

- Mathematics class: Sorting clothing with weather pattern
- Sorting and Categorizing Activity
- Make a T-chart of clothing for hot weather and for cold weather.

Internet Resources

Journey North. (2014, September 2). *Fall 2014 teacher resources*. Retrieved from http://www.learner.org/jnorth/season/

Regents of the University of California Berkeley. (2000). *Eye on the sky*. Retrieved from http://cse.ssl.berkeley.edu/first/EyeontheSkyWeatherJournal/weather.asp

Special Education Technology British Columbia. (2006, August 14). *Sorting outfits for seasons*. Retrieved from http://www.setbc.org/pictureset/resource.aspx?id=280

TVOKids. (n.d.). *Seasons*. Retrieved from http://www.tvokids.com/games/sticksandseasons

TVOKids. (n.d.). *Dressing based on weather*. Retrieved from http://www.bbc.co.uk/wales/bobinogs/games/game.shtml?1

Books

Arnold, C., & Patterson, A. (2013). *Too hot? Too cold? Keeping body temperature just right*. Watertown, MA: Charlesbridge.

Carle, E. (1986). *Papa, please get the moon for me*. New York: Simon and Schuster.

Gibbons, G. (1996). *The reason for seasons*. New York: Holiday House.

McNamee, G. (2004). *The north wind and the sun and other fables by Aesop*. Einsiedeln: Daimon Verlag Press.

Lesson Plan: A Glance at Weeks 3–5

Kindergarten STEM Unit

Lesson Title: *Patterns in Our World and How It Impacts Living Things*

Lesson Summary

Introduce the PBL project with a field trip to the Petting Zoo and a book called *Beatrice's Goat* by Page McBrier

To launch the PBL project for this unit, students will learn about goats using the book called *Beatrice's Goat* by Page McBrier. The following lesson will launch a series of lessons where students can learn more about what animals need to survive and how they adapt to different seasons. This will prepare them for the PBL project in creating a year-long calendar for the petting zoo.

A petting zoo needs you to investigate how the patterns of the sky and the animals on Earth adapt to changes over one year and create a year-long calendar to demonstrate what you have observed throughout the year. Create a presentation for the petting zoo to explain to their customers the changes that animals experience over a year.

Essential Question(s)

What questions will guide student learning in this lesson?

- How do different animals, plant, and people adapt to the changing seasons?

Established Goals/Objectives

Students will understand (big ideas/key knowledge), know, and be able to do what (key skills)?

- Use observations to describe patterns of what plants and animals (including humans) need to survive. Use a model to represent the relationship between the needs of different plants and animals (including humans) and the places they live.
- Participate in shared research and writing projects.
- Add drawings or other visual displays to descriptions as desired to provide additional detail.

Time Required: Launching a three-week project (15 days)

Necessary Materials:

Beatrice's Goat by Page McBrier, books about animals and how they adapt to seasonal change, modeling clay, plastic toy animals, video camera, drawing paper, and crayons.

Teacher Background Information

What content does the teacher need to know about to deliver this lesson? Brief summary.

Teachers will need basic knowledge about goats and the changes they go through throughout the season.

Lesson Preparation

What will the teacher need to plan ahead of time for this lesson?

Teachers will need basic knowledge of farm animals or petting zoo animals: sheep, lamb, pony, rabbits, goat, pigs, and llamas. A good doe (a female goat) can produce from one-quarter to half a gallon (one to two quarts) of milk a day. Woolly and hairy animals should be sheared before the start of hot weather. Spring shearing allows sheep to have adequate wool growth to keep them cool in the summer and avoid sun burning and a full wool coat in the winter to keep them warm. Sheep and goats should not be sheared in extreme heat.

Learning Plan Components

Introductory Activity/Engagement

Science

Animals on the Farm and how they help humans

Learn about farm animals and how they help humans. Farm animals produce goods for humans. Animal scientists study about animals and learn ways to help animals grow strong and healthy. When animals grow well and stay healthy, a farmer can produce more meat, milk, or egg for human consumption.

Mathematics class

Billy Goat Math – Goat Milk?
How Much Milk Can a Goat Produce?

More people consume milk and milk products from goats worldwide than from any other animal. Goat milk is used for drinking, cooking, and baking. It is also used to make cheese, butter, ice cream, yogurt, candy, soap, and other body products. In addition to milk, dairy goats provide meat, leather, and fiber.

Measurement concepts:

- A good doe can produce from one-quarter to half a gallon of milk a day. Show how much milk that is using a milk carton.
- Compare your weight to a full-grown goat. Newborn kids average about two pounds at birth, but grow quickly. The average adult weight is 75 pounds. Compare with students' weights.

ELA class

Literature connection – *Beatrice's Goat* by Page McBrier

Read aloud the story, *Beatrice's Goat* by Page McBrier. Have students re-tell the story. This is a story about how a goat saves Beatrice and her family.

Social Studies class

Goods animals produce as natural resources

With the teacher's help, find the country that Beatrice is from on the map. Generate a list of goods that a goat can provide as a natural resource for humans.

Activity/Investigation

Science class

Learning about animal characteristics. Students will watch videos about the different animals and what the unique characteristic means to the animal and his life. Fur, color, size, eye, ear, nose size, teeth, etc. are important points to discuss. Questioning should encourage thinking skills. How would this animal get food, what kinds of food could he eat, where would he be able to survive, etc. Learn about baby animals on farms. Gather video resources and books on baby animals. Discuss how animal babies grow fast and research which animal grows the fastest: a calf, a chick, or a piglet?

Mathematics class

A day in the life of a goat.

Model this class math book after *Chimp Math* by Ann Nagda that tells a day in a life of a baby chimpanzee using time (calendar, clocks). Make a similar class book called *Goat Math* using the information about a goat's day.

ELA class

Designing a pen. Describe the plan for the designed pen and shed for your Billy goat for all four seasons. (Billy goats play in the yard in the Spring–Fall. Goats grow a thick, fuzzy undercoat of cashmere to keep them warm during the winter, so adults are usually fine in unheated goat barns in most of North America.

Social Studies class

Job specialization. Students learn about a basic economic idea about job specialization: Introduce students to people who specialize working with animals: Farmer, veterinarian, naturalist, biologists, and zoologists.

Explain

What components will the teacher explain/discuss/teach to students in this lesson?

Science class

- Animal diaries. Students will learn about how an animal lives and write a diary on an animal as shared writing.
- Students learn animal characteristics of horse that help human needs and what they need to survive. *Activity: Horses and Ponies: Providing transportation.*
- Students learn animal characteristics of chicken that help human needs and what they need to survive. *Activity: The life cycle of a chicken.*
- Students learn animal characteristics of sheep that help human needs and what they need to survive. *Activity: Learn about the wool that the sheep produces that warms people.*

Table A.2.12 Calendar of Events and Infomercial Rubrics

Calendar of Events at the Petting Zoo

Objectives	Excellent	Good	Fair	Needs Improvement
Student can identify and describe characteristics of all four seasons (spring, summer, fall, and winter)	4 Student can identify and describe characteristics of all four seasons and in order with details.	3 Student can identify and describe characteristics of all four seasons and in order.	2 Student can identify the seasons but may not describe as much details for some of the seasons.	1 Student cannot name the four seasons or describe the characteristics of the different season.
Student can show a sequence of numbers on the calendar.	4 Student can count from 1 to 12 and name the 12 months in the year fluently and without any prompt.	3 Student can count from 1 to 12 and name some of the 12 months in the year but with some prompting.	2 Student can count up some of the numbers from 1 to 12 for the 12 months in the year.	1 Student has difficulty counting from 1 to 12 for the 12 months in the year.
Student has illustrated four seasons to create a calendar of events poster.	4 Drawing shows elements of seasons with lots of details.	3 Drawing shows distinguishable features of the seasons.	2 Drawing shows one sign of the seasons but needs more details.	1 Drawing does not show any distinctive elements of the seasons.
Drawing shows how animals at the petting zoo respond to the changing seasons. (i.e. giving birth to young in spring or sleeping inside the shed in the winter).	4 Drawing shows animals responding to weather with details.	3 Drawing shows animals responding to the season.	2 Drawing may show an animal responding to the season but needs more details.	1 There is no evidence that of showing an animal responding to weather.

Infomercial about Visiting the Petting Zoo

Objective	Excellent	Good	Fair	Needs Improvement
Using their calendar event poster, the student can highlight an event that people should come to see.	4 Student is expressive with their oral language skill and has a dramatic flair.	3 Student is expressive with their oral language skill.	2 Student shares an event but is not very coherent.	1 Students need to develop oral presentation skills.
Student can talk about and explain how different animals respond to the seasonal change. (i.e. Baby farm animals born in spring time; sheep shearing in the spring time to shed the winter wool coat)	4 Student shares a lot of background knowledge about animals and how they respond to seasonal change.	3 Student can talk about and explain how different animals respond to the seasonal change.	2 Student can talk about how different animals respond to the seasonal change but could add more details.	1 Student has difficulty talking about how different animals respond to the seasonal change.

Mathematics class

What kind of animal is your favorite at the zoo? Read Tiger Math and learn about a baby tiger's life while learning to graph.

ELA class

- Read the book, *Is Your Mama, a Llama?*
- Matching Activity – parent and young.

Discuss how adult animals care for their babies.

Extend/Apply Knowledge

Science class

Create a calendar of events. Students use the pictures to make scene of all four seasons and what animals do in the spring, summer, fall, and winter and tell a story to the class.

Students learn about how animals not only prepare for winter, but use coats, fat deposition, to survive in the cold, hibernation, and migration.

Mathematics class

Calendar math. Students will use their knowledge of the sequence of the 12 months to create a calendar of events at the Petting Zoo.

ELA class

Work on the Presentation. Petting Zoo Infomercial Video. Read *Petting Zoo* by Gail Tuchman. Discuss the lives of the petting zoo animals. Students will create and present their calendar and make an infomercial for visitors about the calendar of events and highlights to look forward to at the petting zoo.

Social Studies class

Specializations. Learn about the specialized jobs that people have at the petting zoo or at a local farm to help with raising and caring for animals.

Assessment

Performance Tasks

- Create a Calendar of Events at the Petting Zoo: A petting zoo needs you to investigate how the patterns of the sky and the animals on

Earth adapt to changes over one year and create a year-long calendar to demonstrate what you have observed throughout the year. Create a presentation for the petting zoo to explain to their customers the changes that animals experience over a year.
- Create a Petting Zoo Infomercial Video: Students will present their calendar and make an infomercial for visitors about the calendar of events and highlights to look forward to at the petting zoo.

Extension

Building a pen and shed: Choose an animal and design and build a shelter for the animal.

Students will use the information they learned about the different animals to design their pen.

For example, goats are excellent at crawling through small gaps or climbing over fencing. How do you design the fencing on the pen so your goat does not escape? In addition, your goats will need a place to go in the winter and when it's raining. Goats with thick coats may be able to withstand colder temperatures.

Internet Resources

Charlotte Nature Museum. (2014). *What do animals do in autumn?* Retrieved from http://www.charlottenaturemuseum.org/blog/post/84/What-do-animals-do-in-autumn

PBS Kids. (2014). *Baby animals.* Retrieved from http://pbskids.org/dragonflytv/show/babyanimals.html

Scholastic. (2014). *Study jams: Animal adaptations.* Retrieved from http://studyjams.scholastic.com/studyjams/jams/science/animals/animal-adaptations.htm

Sheppard Software. (n.d.). *Animal classification: reproduction. (Baby animals!).* Retrieved from http://www.sheppardsoftware.com/content/animals/kidscorner/kc_classification_babies.htm

Smithsonian National Zoological Park. (n.d.). *Kid farm at the national zoo: Caring for goats.* Retrieved from http://nationalzoo.si.edu/Animals/Kids-Farm/InTheBarn/Goats/care.cfm

Books

Dunn, M. R. (2011). *Owls (Nocturnal animals).* Mankato, MN: Capstone Press.
Markle, S. (2013). *What if you had animal teeth?* New York: Scholastics Books.
Markle, S. (2014). *What if you had animal hair?* New York: Scholastics Books.
McBrier, P., & Lohstoetler, L. (2004). *Beatrice's goat.* New York: Aladdin Publishing.
Nagda, A., & Bickel, C (2000). *Tiger math.* New York: Holt and Company.
Nagda, A., & Bickel, C. (2002). *Chimp math.* New York: Holt and Company.

National Geographic. (2010). *National geographic wild animal atlas: Earth's astonishing animals and where they live.* Washington, DC: National Geographic Books.

Tuchman, G. (2013). *Scholastic discover more reader level 1: Petting zoo.* New York: Scholastic Books.

Whipple, L., & Carle, E. (1989). *Animals animals.* New York: Philomel Books.

Index

Note: **Bold** page numbers refer to tables and *italic* page numbers refer to figures.

active learning 22, 219, 220, 222
actuaries 56, 112
advocacy, STEM 228, 235–240, *236*, *237*, 243, 244, 250
aerodynamics 76, 315, 316, 317
aerospace engineers 71, 121
agriculture 16, 96, 116, 154
alternative energy 277, 278; *see also* renewable energy; solar energy; thermal energy
America COMPETES Act 18, 19
amusement parks (6) (12) 48, 104, 105, 106, **107**, 135, 164, **165**, 166
analytic rubrics 190
animals (K) 337, **334**, 340, 341, 346–347, **348–349**, 350–352; *see also* goats (K); petting zoo (K)
app creation **146**, 150, 171
aquariums/terrariums (3) 74
architects 112
Arizona STEM Network 247
art 45, 48, 71, 102, 134
assessments 177–200; backward design 178–179, *179*; benchmark 232, 234, 241; Bloom's Revised Cognitive Taxonomy and 180, **181**; DDDM and 193, **194**, 195–196, 199, *197–198*; diagnostic 177–178, 196; errors and 184, 192–193; formative 177, 196; integrated curriculum and 246; item analysis and 200; keywords 180, **181**; LOs and 179–181, *180*; matrices and 196, *197*, 198; multiple-choice 183–187, **194**; objective 183, 195, *197*; petting zoo (K) and 262; rubrics 189–193, **190**, **191**; self-constructed 180–184, 187–188, 195, **198**; smart phones (12) and 164, 166, **167**; state standards and 178–179, 181–183; STEM integration and **28**; STEM notebooks (K-2) and 48; STEMx Sustainability Compass and 240; summative 177, 196; teachers and 177; tools 183–189
astronaut 71
astronomers 55, 56, 328
atoms (11) 155, 194
at-risk students 206
audio engineers 69
authentic contexts 30, 31, 246
automotive x-challenge (7) 308–326; engineer it! 256, 302, 316, 323–326; fact or friction? **263**, 308–313; overview 321–326; ready, set race: the x-challenge **263**–264, 313–320; transportation/motorsports and **115**, 255

ball drop activity (7) 284, 285, 287
Beatrice's Goat (McBrier) 344–346
benchmark assessments 232, 234, 241
biodiversity 134, **146**, 147, 150
biology 83, 133
biomedical engineers 121
Bloom's Revised Cognitive Taxonomy 180, *180*
blow it up! (7) 279, 281, 283
bridges 126, 129
Burning Glass Technologies 231
Bush Administration 18
business of amusement parks, the (12) 163, **166**
business/industry partners 224, 227, 228; *see also* multi-sector partnerships

354 *Index*

car crashes (12) 163, **165**, 167, 168, 169, 286
car design (7) 256, 284, 299, 301, 302, 308, 312–314, 316, 321
carbon cycling 142
career awareness *see* STEM careers
carpenters 129
cause and effect: eighth grade and 122–123, **124**; eleventh grade and 154, 156, **157**; fifth grade and 93, **94**; first grade and 57–58, **58**; fourth grade and 84–85, **86**; grades K-2 and 45, **47**; grades 9-12 and 134; grades 6-8 and 102–103; grades 3-5 and 74; kindergarten and 48–49, **50**; ninth grade and 137–138, **138**; PBL and 22; second grade and 65, **66**; seventh grade and 114, **115**; sixth grade and 104, **106**; as STEM theme 6; tenth grade and 145, 147, **148**; third grade and 76, 77; twelfth grade and 164, **166**; *see also individual topics*
change over time—our schoolyard garden (2) **69**
changing environment, the (natural hazards) (2) 47, 52, 65, **66**
characteristics, effective STEM 244–248
chemical energy **275**, 279
chemistry 54, 125, 133–135, 154, **155**, 156, 163
Chimp Math (Nagda) 347
civil engineers 63, 90, 96, 112
class sizes 204–205, 244
Clever Crazes for Kids (website) 46, 81
climate 62, 64, 104, 340–341, 343
climate change 104, **165**, 168, 171
climate change mitigation (5) **93**, 99, **100**
climatologists 64, 81–83
coalitions, STEM 227–228, 241
coding a rainbow (1) 58–59, **60**
cognitive taxonomies 180, *180*, 184
collaboration 34, 103, 135, 284, 316, 323
collective impact model 235
collective participation, teacher 219–220, 222
college 3, 232, 233; *see also* postsecondary pipeline
communication 50, 103, 104, 107, 108, **109**, 134–135

compost (5) 74, 92, 96, 97, 98, **98**
computer scientist 64
conservation, energy 83, 153, 171, 255, 275; *see also* Law of Conservation of Energy
conservation, water (4) **84**, 89–90, **91**
conservation organizations 139, 147
construction materials (11) 154, **155**, 156, **158**
construction occupations 129
content knowledge 30–31, 214, 219, 221, 222
Core Conceptual Framework for Professional Development 219, 222
cost estimators 121
cost-benefit analyses 104, 108, 161–162
creating the next smart phone (12) 163, **167**
critical thinking 4, 35, 73, 103, 135, 163; *see also* 21st century skills
cross-cultural education 35, 208, 209, 215; *see also* cultural inclusivity; sociotransformative constructivism (sTc)
cultural inclusivity 30, 31, 207, 214
curriculum, integrated 246

data collection, STEM 193, **194**, 195–196, *197*, *198*, 199, 229, 241
database administrators 122
Data-Driven Decision Making (DDDM) 193, **194**, 195–196, *197*, *198*, 199; *see also* data collection, STEM
day and night (K) 338, 339
decision models **123**, 126
design engineers 114, 255, 265, 266
design justification 28, **29**
dialogic conversation 209, 211–213
differentiated instruction 177
drivers, Pre-K-12 STEM 241
Duncan, Arne 193

earth drillers 145
earth formation (9) 137–138
earthquakes 65, **65**, 85, 109, 122, 171
earth's spheres (9) 142
earth's systems (5) (9) 74, 92, 93, 145
earth/space sciences 21, 74, 145, 246
ecological sustainability 74
ecologists 56, 62, 328
economists 122
ecosystem preservation (3) 80, **81**

Index

ecosystems 75, 80, **81**, 83, 98, 116, 134, 149–150, **151**, 241–242; *see also* education ecosystems
ecosystems modeling (10) 134, 145, 147, 149–150, **151**
EDP (engineering design process) *see* engineering design process (EDP)
education ecosystems 229, 241, 249, 250
education pipeline (pre-K-12) 229, 232–234, 236, 239, 241; *see also* postsecondary pipeline
effective STEM program characteristics 244–248
eighth grade 122–131
elastic potential energy **263**, 275, 284–286, **285**, 292, **293**, 296–298
elastomers 292–299
electrical energy **275**
electricity 164, **165**
electromagnetic radiation 85, 134, 159, 164
eleventh grade 154–163
enablers 241, 251, *251*
endangered species 61–62, *63*
energy, renewable 74, 83, **146**
energy, solar 84–85, **84**, 87, 129
energy, sound **275**, 279, 282, 286
energy, thermal **123**, 129, **275**, 286, 292, **293**, 298, 309, 312
energy alternatives 277, 278; *see also* renewable energy; solar energy; thermal energy
energy carbon capture and storage 154
energy conservation 83, 153, 171, 255, 275; *see also* Law of Conservation of Energy
energy consumption **146**, 152, **165**, 168
energy conversions 183, 277, 299, 323
energy efficiency 154, 159
energy flow activity 276, 277
energy modeling 134
energy production 150–151
energy trading 154
energy transformations (7) **113**, 114, 255, 256, **262**, 273–274, **275**, 282
engineering design 5, 9, 33–34, 96, 108, 129, 156, 214
engineering habits of mind 9, 16, **28**, 34, 131; *see also* engineering thinking

engineering thinking 5, 27, 34, 38, 265; *see also* engineering habits of mind
engineers 54, 159; *see also individual types of engineers*
English Language Learners' (ELLs) 206, 207, 212
environment 103, 134, 137, 150–151, 153, **153**, 343; *see also* climate change; climate change mitigation (5); ecologists; ecosystem preservation (3); green economy sector
environmental engineering technicians 112
environmental engineers 63–64, 112
environmentalist 64
environmental management (10) 134, 145, **149**
environmental scientists 83, 111, 121
erosion and weathering management (9) 137
erosion modeling (4) 87, **88**
errors, assessment 183–185, 192–193

fact or friction? (7) **263**, 308–313
farmers 346
federal funding 18–21
field station mapping (4) 84–85, **86**
fifth grade 92–100
first grade 56–64
fission 155, 159
floods 65, 85, 99, 109
food 118, 121, 126, 139, 145, 147, 213
footprint reduction (3) 75, 80, **82**
force **285**, 292, **293**, 310
formal and informal STEM learning opportunities 232–233, 247–248
fossils 85, 126
fourth grade 83–92
Framework for 21st Century Learning (Partnership for 21st Century Learning) 12
Framework for K-12 Science Education, A (NRC) 36–37, 46, 246
Framework for STEM Integration in the Classroom (National Research Council) 26
freshman (college) remediation 234
friction 114, 255, 256, 263, **285**, 308–313
fusion 158–159
future transportation (3) 76, 77, **78**

Index

Galilei, Galileo **155**, 156
gateway courses 234
genetic disorders (7) 103, 113, **114**, 116, **118**
genetically modified organisms (GMOs) (7) 103, 113, **114**, 118–120, **120**, **135**
geographers 56, 145
geography 48, 52, 56, 62, 63, 90, 92, 93, 269, 306, 339
geology 56, 62, 64, 92
geoscientists 131, 145
geotechnical engineers 99
global bonds (12) 163, **165**, 168, **170**
Global Climate Change (GCC) 6, 64, 99, 104
global competitiveness 3, 14
global models (9) 134, 141, **141**
global warming 104, **105**; *see also* climate change
global water quality (6) 90, 103, 104, **105**, 108, **110**
GMOs (genetically modified organisms) (7) 103, 113, **114**, 118–120, **120**, **135**
goats (K) 344–347, 351
grade overviews: K-2 45–47; 9–12 133–134; 6-8 102; 3-5 73
graphic artists 99
gravitational potential energy (GPE) **263**, 275, 284, **285**, 288
gravity 312
green building rooftops (11) 154, **155**, 159, **161**
green construction 154
green economy sector 145, 150, 154
growth (1) 62

habitats, U.S. (K) **47**, **49**, 52, **53**
habitats—local and far away (1) **49**, 65
habitats—our changing environment (2) 65, **66**
Hamilton, L. 195
healthy living (10) 145, 147, **148**
hearing specialists 70
hibernation **336**, 350
high schools 38, 131, 133–135, 233, 234, 245; *see also* grade overviews
higher level thinking skills 195; *see also* engineering habits of mind; engineering thinking
historians 131
holistic rubrics 191–192, **191**

horizontal alignment STEM partnerships 236–238, *236*, *240*, 240–242, 247
horticulturalists 62
human impact, nature (9) 135, **137**, 142, 144, **144**
human impacts, climate (6) 104, **106**
hurricanes 108
hydrologists 99, 100
hydropower efficiency (4) 84, 87, **89**
hydrosphere 93, 96
hypotheses 212, 279

ice cream, sociotransformative STEM and 210–215
impact minimization (8) 122, **128**, **135**
Individual Professional Development Plans (IPDP) 223
inelastic collision 285–286
influence of the waves (1) 57–58, **58**
infographics 141, 142
informal and formal STEM learning opportunities 232–233, 247–248
information, media, and technology skills 35, **261**, **330**; *see also* 21st century skills
information and media literacy 35, 103, 135
infrastructures 126
infusion (engineering) 8–9, 18
innovation and progress: eighth grade and 123–125, **125**; eleventh grade and 155, 156, **158**; fifth grade and 92, 93, 94, **95**, **95**; first grade and 58–59, **60**; fourth grade and 84, 85, 87; grades K-2 and 45, **47**; grades 9-12 and 133; grades 6-8 and 102; grades 3-5 and 73; Kindergarten 49–51, **51**; ninth grade and **136**, 139, **140**; PBL and 22; second grade and 66–68, **67**; seventh grade and 114–116, **117**; sixth grade and 104, **105**, 106, **107**; as STEM theme 6; tenth grade and **146**, 147, **149**; third grade and 75, 76, 77, **78**; twelfth grade and 164, 166, **167**; *see also individual topics*
inquiry based instruction 246–247
Integrated STEM *see* STEM integration integrator, STEM 25
inter/multidisciplinary approach 4, 9, 27

Index 357

jobs/workforce 14, 229–235; *see also* global competitiveness; postsecondary pipeline
journalists 56, 90, 92

kindergarten 48–56; *see also* earth and sky patterns (K)
kinetic energy 190–191, **190–191**, 257, 258, **263**, 273, 274, 276–278, 285–286; *see also* potential energy
kinetic friction 309, 313; *see also* friction

land use planning 68
landslides 85, 86, 139
Law of Conservation of Energy 114, 255, 273, 274, 277, 278, 284–288
lead teachers 102, 134
learning and innovation skills 35, **260**, **330**; *see also* 21st century skills
learning centers 210
learning from our past (8) 103, 122, **127**, 135
learning objectives (LOs) 16, 26, 179–181, 183, 184, 186, 188, 195, 196, 199
let's get energetic! (7) **263**, 273–283
levees 74
life and career skills 35, **260**, **330**; *see also* 21st century skills
life in space/space travel (7) 114–116, **117**
life sciences 21, 56, 80, 154, 246
light 57–58, 115, 116, 277, 278, 280
light up my life (7) 278, 279, 282
lobbyists 131
local/regional STEM partners 231
logisticians 122

Maglev trains 74, 164
magnetism 164, **165**
making music (7) 279, 280, 282
manufacturing 213, 294, 316
mapping (4) 84–85, 104
mapping (engineering) 9
marine ecosystems 150
marketing managers 113
material engineers 70–71
material science and space (2) 156
materials design 134; *see also* construction materials (11)
materials matter (7) **263**, 284–291
materials science 156, **263**, 285

mathematical thinking and reasoning 33, 34–35
mathematicians 34, 56, 83
mathematics: actuaries and 112; *Benchmarks for Science Literacy* (AAAS) and 16; as career 154; eighth grade and **123–124**, 125, *125*, 126, **127**, **128**, 129, **130**; eleventh grade and **155**, 157, **158**, **160–162**; fifth grade and 92, 93, **94–95**, **97–98**, 99, **100**; first grade and 57, 58, **60–61**, 63; fourth grade and **86–89**, **91**; freshman (college) remediation and 234; grades 9-12 and 133; high-stakes testing and 220–221; historical curriculum initiatives and 15; international student scores and 233; jobs/workforce and 14; kindergarten and 48, 49, **49–51**, 327, **334–336**, 338, 341, 346; learning centers and 212–213; national assessments and 233; NECAP and 232; NGSS, K-2 and 45; ninth grade **138**, **140–141**, **143–144**; NOM 10; practices and 33, 34; *Project 2061: Science for All Americans* (AAAS) and 16; second grade and **65–67**, 69, **69–70**; seventh grade and 113, **113–115**, 116, **117–118**, 119, 120; sixth grade and 104, **105–107**, 108, **109–111**, 112; STEM integration and 25, 27, **28–29**; teacher content knowledge and 31; tenth grade and **146**, **148–149**, **151**, **152**, **153**; third grade and 75, 76, **77–79**, **81–82**, 83; twelfth grade and **165–167**, 168, **169–170**, 172; *see also* Common Core State Standards in English Language Arts (CCSS-ELA) and Mathematics (CCSS-M); transportation-motorsports (7)
matrices 196, *197*, *198*
matter 134, 156
mechanical energy 275, 277–279, 286, 296, 297
media literacy *see* information, media, and technology skills
medical sonographers 129
medicine 121–125, **125**
metacognition 209, 213
meteorologists 82–83, 343

Index

microbiologists 112, 121
middle schools 37; *see also* grade overviews
migration **336**, 350
mineral resources (11) 154, **155**, 159–162, **162**
modeling ecosystems (10) 134, 145, **146**, 147, 149–150, **151**
models, global 134, 141, **141**
modules, STEM instructional 38, 71
moon 327, 328, 338, 339
motion 48–49, **50**, 78
motorsports *see* transportation-motorsports (7)
multimedia 95, 137–138
multiple-choice items 183–187, **194**
multi-sector partnerships 229, 235–240, **236**, **237**
music 45, 48, 71, 102, 279
music box (7) 277, 279, 280, 282

nanotechnology 54
National Academy of Engineering (NAE) 8, 18, 28
National Academy of Science (NAS) 14
National Aeronautics and Space Administration (NASA) 19, 71, 115, 142
National Assessment of Educational Progress (NAEP) 233
National Association for the Education of Young Children (NAEYC) 45–47, 57, 71
National Commission on Teaching and America's Future 204
National Council of Teaching of Mathematics (NCTM) 178
National Oceanic and Atmospheric Administration (NOAA) 19
National Research Council (NRC) 10, 16, 21, 28, 33, 35–37
National Science and Technology Council (NSTC) 19
National Science and Technology Summit 19
National Science Education Standards (NSES) 16, 21
National Science Foundation (NSF) 14, 17, 247
National Science Teachers Association (NSTA) 203
National Weather Service 76

natural catastrophes (12) 163, **172**
natural environments, rebuilding (10) 135, 145, 150–151, 152, 153, **153**
natural hazards (2) 65
natural hazards (6) 103, 104, **105**, 108–110, **111**; *see also* natural catastrophes (12)
naturalists **335–336**, 347
nature, human impact on (9) 135, 137, 142, 144, **144**
nature of engineering (NOE) 10
nature of mathematics (NOM) 10
nature of science (NOS) 10
nature of technology (NOT) 10
nature patterns *see* world patterns/living things impact (K)
Newton, Issac 154
Newton's Third Law **257**
Next Generation Science Standards (NGSS) 46; California STEM Learning Network and 239; eighth grade and **124**, **125**, **127**, **128**, **130**; eleventh grade and **157**, **158**, 160–162; engineering design and 33–34; equipment, class size and 204; fifth grade and **94**, 95, **97**, **98**, **100**; first grade and **57**, **58**, **60**, **61**, **63**; fourth grade and **86–89**, **91**; grades K-2 and 45–47; grades 9-12 and 133; grades 6-8 and 102; grades 3-5 and 73; infusion and 9; kindergarten and **50**, **51**, **53**, **55**; LOs and 183; national and the 233; ninth grade and **138**, **140**, **141**, **143**, **144**; second grade and **66**, **67**, **69**, **70**; seventh grade and **115**, **117**, **118**, **119**, **120**; sixth grade and **106**, **107**, **109–111**; as standard 21–22, 32; standards integration and 17; STEM curriculum and 4; tenth grade and **148**, **149**, **151**, **152**, **153**; third grade and **77–79**, **81**, **82**; twelfth grade and **166**, **167**, **169**, **170**, **172**
ninth grade 135–145
No Child Left Behind Act 20, 202
noise reduction 159
NRC (National Research Council) 10, 16, 21, 28, 33, 35–37
nuclear energy **275**
nuclear engineers 163
nuclear field 163
nurses, registered 112

Index 359

objective assessments 183, 195, *197*;
see also multiple-choice items
occupations, STEM 231, 234; see also
STEM careers; *individual careers*
optimizing the human experience:
eighth grade and **123**, 129, **130**;
eleventh grade and **155**, 159–162,
162; fifth grade and **93**, 99, **100**;
fourth grade and **84**, 89–90, **91**;
grades K-2 and 45, **47**; grades
9-12 and 133; grades 6-8 and 102;
grades 3-5 and 73; kindergarten
and **49**, 52, **54**, **55**; ninth grade and
137, 142, **144**, *144*; PBL and 22;
seventh grade and **114**, 118–120,
120; sixth grade and **105**, 108–110,
111; as STEM theme 8; tenth grade
and **146**, 150–151, 152, 153, *153*;
third grade and 80, **82**; twelfth
grade and **165**, 171, **172**; see also
individual topics
our changing school environment (K)
52, **54**, **55**
our school yard garden (2) **65**, 68

Papa, please get the moon for me
(Carle) **334**, 339
parent involvement 205–206
partners, STEM 227–230, 236–238,
236, *237*; see also states
Partnership for 21st Century Skills
(P21) 12, 28, 35
patterns and the plant world (1)
59–60, **61**
patterns on the earth and in the sky
(K) see earth and sky patterns (K)
pedagogy 204, 207–208
performance assessments see
assessments
periodic tables **155**, 156
petting zoo (K) 49–50, 327–328,
330–332, *331*, 333, **335–336**,
344–345, **348–349**, 350–351
physical education 46
physical sciences 19, 21, 38, 56, 246
physics 48–49, **50**, 133, 154
plants and gardens: compost (5)
96 and; earth and sky patterns
(K) 328, **334** and; green building
rooftops (11) 154, **155**, 159, **161**;
and healthy living (10) 145; and
horticulturalists 62; our school
yard garden (2) **65**, 68; rainwater

analysis (5) and **92**, 93, 95–96, **97**,
101; window box gardens (1) **57**,
59, 60
plant world 59–60, **61**
plate tectonics 122, **123**
policies, STEM 229–244; data and
229–235; definition of 229; drivers/
enablers and 241; education
pipeline and 229, 232–234, 236,
239, 241; horizontal/vertical
alignment and 236–238, *236*,
240, 240–242, 247; multi-sector
partnerships and 229, 235–240,
236, *237*; professional development
and 19, 31, 37, 38, 203, 205, 206,
219, 222–225, 245, 248–250;
states and 245; sTc and 208;
transformative 229; workforce
needs and 230–232
policy makers 3, 208, 215, 229,
230, 245
political scientists 131
population density (7) **114**,
116–117, **119**
postsecondary education 231, 233
postsecondary pipeline 234, 235
potential energy **191**, 258, 263, 275,
277, 284–286, 292, 296; see also
elastic potential energy
precipitation **337**, 338
predict, observe, explain
(POE) 210
pre-K-12 pipeline 232
prior knowledge 30, 137, 138, 150,
177, 183, 210–212, 214
probability and statistics **146**
probeware 211, 213
problem-/project-based learning
(PBL): definition of 22; eighth
grade and 122;eleventh grade
and 154, 164; fifth grade and
93; fourth grade and 85; inquiry
based learning and 246; integrated
curriculum and 246; kindergarten
and 48; ninth grade and 135;
professional development and
220, 224–225; second grade and
65; seventh grade and 113; sixth
grade and 104; STEM integration
and 30; student thinking and 8;
tenth grade and 145; third grade
and 74; twelfth grade 164; see also
individual topics

360 Index

professional development 19, 31, 37, 38, 203, 205, 206, 219, 222–225, 245, 248–250
promotions managers 113
prototype design rubric (transportation-motorsports) (7) 317–320
P21 Framework (Partnership for 21st Century Skills) 35
public policy *see* policies, STEM; states
public-private partnerships 239, 240, 245; *see also* multi-sector partnerships
Punnett Squares 116

race day event 114, 255, 267, 314, 321, 323, 324
race engineers 266
racecars (7): fact or friction? and 263, 308–313; Internet resources and 270; materials matter and 263, 284–291; rubber band racers and 303, 305, 307–308; stretching it and 263, 292–301; *see also* automotive x-challenge (7); race day event
radioactivity (11) 154, 157–159, **160**
rainwater analysis (5) **92**, 93, 95–96, **97**, 101
reaction rates **155**, 156
ready, set race: the x-challenge (7) **263**–264, 313–320
Reason for Seasons, The (Gibbons) **334**
rebuilding the natural environment (10) 135, 145, 150–151, 153, **153**
recreational STEM (3) 79, **79**
recycling and waste reduction 154
reflexivity 209–210, 212, 213
registered nurses 112
remediation 234
renewable energy 74, 83, **146**; *see also* solar energy; thermal energy
represented world, the: eighth grade and 126, **127**; eleventh grade and 157–159, **160**; fifth grade and 95–96, **97**; first grade and 57, 59–60, **61**; fourth grade and 85–87, **88**; grades K-2 and 45, **47**; grades 9-12 and 133; grades 6-8 and 102; grades 3-5 and 73; ninth grade and 139, 141–142, **141**; PBL and 22;

second grade and 68, **69**; seventh grade and 116, **118**; sixth grade and 107–108, **109**; as STEM theme 7; tenth grade and 147, 149–150, **151**; third grade and 78–79, **79**; twelfth grade and 167–168, **169**; *see also* earth and sky patterns (K); *individual topics*
Rising Above the Gathering Storm (National Academy of Science) 14
rock formations 85
rocks/fossils 126
roll of physics in motion, the (roller coasters) (K) 48–49, **50**
rubber bands (7) 297, 299, 306
rubrics 189–193, **190**, **191**

sand energy/shakers (7) 276, 279, 280, 283
scavenger hunt for patterns (K) **334**, 338
school administration 208, 220, 224–225
school climate development 249
school nutritionists 147
schoolyard engineering (5) **92**, **93**, **94**
science: eighth grade and 122; elementary schools and 232; eleventh grade and **155**; fifth grade and **93**, 96; first grade and 62; fourth grade and 85; *Framework for K-12 Science Education, A* (NRC) and 36–37, 46, 246; grades K-2 and 45, 47; grades 9-12 and 133; grades 6-8 and 102; grades 3-5 and 73; high-stakes testing and 220–221; historical evolution 15–16; international student scores and 233; jobs/workforce and 14; kindergarten and 48, 49, **49**–51, 327, 334–336, 338, 341, 346; learning centers and 210; modules and 38; NAS 14; national assessments and 233; national science standards 32, 233; NECAP and 232; ninth grade and **138**, 140, **141**, 143, **144**; No Child Left Behind and 20, 202; NOS 10; NSF 14, 17, 247; NSTC 19; practices and 30; *Project 2061: Science for All Americans* (American Association for the Advancement of Science) (AAAS) 16; second

Index   361

grade and **65–67**, 69, **69–70**; seventh grade and **113**, 114, **115**, 116, **117–118, 120**; sixth grade and 104, **105–107**, 108, **109–111**, 112; STEM integration and 25, 27, **28–29**; teacher content knowledge and 31; tenth grade and **146, 148–149, 151, 153**; third grade and **75**, 76, **77–79, 81–82**, 83; transportation-motorsports (7) and 38, **113**, 255, 265, 268–269; twelfth grade and **165–167**, 168, **169–170, 172**; *see also* Next Generation Science Standards (NGSS)
scientific inquiry 33, 135
scientific method 266, 268
scientists 54–56
seasons (K) 51, 327–328, **331–332**, 333, **334**, 336, 337–339, **337**, 341, 344–345, **348–349**, 350
second grade 64–71
seismic activity 74, 84; *see also* earthquakes
seismologists 90
self-constructed assessments 80–184, 187–188, 195, **198**
semiconductor processors 131
seventh grade 113–122; *see also* transportation-motorsports (7)
shadows data (schoolyard engineering) (5) 92–93
sixth grade 103–113
smart phones (12) 150, 163, 164, 166, **167**
snow day friction! 310
snow-proof school challenge 270
social relevance **29**
social studies: eighth grade and **123**, 124, 125, 126, 129; eleventh grade and **155**; fifth grade and **92**, 93, **93**, 97, 99; first grade and **57**; fourth grade and **84**; grades 9-12 and 133; kindergarten and **49**; modules and 38; NAEYC and 45–47, 57, 71; ninth grade and **136–137**; No Child Left Behind and 202; second grade and **65**; seventh grade and **114**, 116; sixth grade and **105**, 109; STEM and 71; tenth grade and **146**; third grade and **75**; twelfth grade and **165**; *see also* transportation-motorsports (7)

sociotransformative constructivism (sTc) 208
software developers 131
soil erosion 84, 96, 99
solar energy **84**, 84, 85, 87, **123**, 129
sound 57, 58, 69, 116, 203, 279
sound energy **275**, 279, 282, 286
space 66–67, **67**
space life/travel (7) 113, 114, 115, 116, **117**
speed 114, 126, 128, **128**, 303
stakeholders 147, 229, 236, 243, 249; *see also* multi-sector partnerships
standardized testing 202–204
Standards for K-12 Engineering Education (National Academy of Engineering/NAE) 32, 36
Standards for Technological Literacy 32
start your engines (7) **263**, 265, 267, 268
state standards 78–179, 181–183
states 245; *see also individual states*
static friction 309
statisticians 99, 112
statistics and probablity **146**
sTc (socio transformative constructivism) 208
STEM: content areas and 4; definitions of 17–18; ELA and 4, 5, 32, 38; engineering and 4, 5, 8–9; federal government and 18–21; global competitiveness and 3, 14; historical evolution 15–16; inter/multidisciplinary approach and 4, 9, **27**; jobs/workforce and 14; mathematics and 4; nature of 9–10; NGSS and 21–22; NSTC goals and 19; practices 33–35; social studies and 5; teachers and 19, 20; teams, student and 5; technology, engineering infusion and 10; *see also* STEM careers; STEM integration; STEM Road Map Overviews; STEM themes; individual grades; individual topics
STEM careers: eighth grade and 129–131; eleventh grade and 163; fifth grade and 99, 100; first grade and 62–64; fourth grade and 90–92; future career explorations and 71, 72; grades K-2 and 45–48; Kindergarten 54–56; ninth grade

Index

and 145; seventh grade and 121–122; sixth grade and 111–113; STEM drivers and 241; tenth grade and 154; third grade and 80–83; twelfth grade and 171; websites and 145; *see also* individual careers
STEM Immersion Guide (Arizona) 247
STEM Innovations, LTD 249
STEM integration 25–35; characteristics of 26–27, **28–29**; content/context integration and 26; definition of 26; engineering and 4–5, 27–28; Framework for STEM Integration in the Classroom (National Research Council) 26; instruction practices 33–35; K-12 continuum 36–38; overview of 38; PBL and 35; pedagogical practices and 27–30, 35–36; teachers and 30–32; 21st century skills and 35
STEM notebooks/journals 48, 50, 327; *see also* design journals, transportation-motorsports (7)
STEM Road Map Curriculum Module Planning Template 12
STEM Road Map overviews: computational thinking 10–11; eighth grade and 122; eleventh grade 154; fifth grade 92; first grade 56–57; fourth grade 83–84; grades 9-12 133–135; grades K-2 47–48; grades 6-8 and 102–103; grades 3-5 73–74; kindergarten 48; ninth grade 135; second grade 64; seventh grade 113; sixth grade 103–104; tenth grade 145; third grade 74–75; twelfth grade 163–164; *see also* individual topics
STEM School start-up process 248–249
STEM themes 4–8; *see also* cause and effect; human experience optimization; innovation and progress; represented world, the; sustainable systems
stretching it (7) **263**, 292–301
student achievement 202, 203, 205, 206, 222, 233, 241, 244; *see also* assessments
sun, the (K) 51, 56, 93, 327, 328
sun's role, earth life (8) 122, 129, **130**, 135

survival and reproduction (10) 135, 145, 150, **152**
survival on earth-water (10) **152**
sustainable systems: eighth grade and 126, 128, **128**; eleventh grade and 159, **161**; fifth grade and 96, 97, 98, **98**; first grade and 61–62, **63**; fourth grade and 87, **89**; grades K-2 and 45; grades 9-12 and 133; grades 6-8 and 102; grades 3-5 and 73; kindergarten and 52, **53**; ninth grade and 142, **143**; PBL and 22; second grade and 68, **70**; seventh grade and 116–117, **119**; sixth grade and 108, **110**; as STEM theme 7–8; tenth grade and 150, **152**; third grade and 80, **81**; twelfth grade and 168, 169, **170**; *see also* individual topics
swing set makeover (3) 78–79, **79**
systems theory 142

teachers: change resistance and 207–208; educational leadership and 31; effective STEM and 222, 228–248; engineering and 8–9; high school 133; K-2 STEM Road Map themes and 45; lead teachers (9-12) and 102, 134; resources lack and 205; as STEM drivers 241; STEM integration and 30–32; STEM road map curriculum and 4, 133; STEM school start-up and 248–249; STEM training and 19, 20; teams of 225; *see also* assessments; Data-Driven Decision Making (DDDM); professional development
teams, community 249
teams, student 36, 38; *see also* individual topics
teams, teacher 225
technology: DDDM and **194**; early childhood education and 52; embedded 247; Framework for K-12 Science Education, A (Framework) (NRC) and 21, 245; genetic disorders (7) and 116; habitats-near and far (1) and 62; ITEA 18, 31; jobs/workforce and 14; learning centers and 210; NOT 10; practices and 33; STEM definition and 17; teacher content knowledge and 30

Index 363

technology literacy 35; *see also* 21st century skills
tectonic plates 122, **123**
temperature probes 211
temperatures 76, 96, 104, 150, 211, 275, 276, 277, 279, 281, 338
templates 12, 290
tenth grade 145–154
terrariums/aquariums 74
terrestrial ecosystems 150
Texas 228, 238, 240, 247, 249–251
thermal energy **123**, 129, **275**, 286, 292, **293**, 298, 309, 312
thermonuclear power 157
third grade 74–83
tires 292, 294–296, 306, 308, 310–313
Too hot? Too cold? Keeping body temperature just right (Arnold and Patterson) 339
topographers 90, 92
tornadoes 76
trains 74; *see also* Maglev trains
transformation stations 274, 276, 277, 278
transformative STEM policies 229
transportation: earth formation (9) and 137; footprint reduction (3) and 75, 80, **82**; grades 6-8 and 103; local departments of transportation 126; as STEM career (10) 154; third grade and 75; urban planners and 62–63, 90
transportation-motorsports (7) 255–326; assessment and 270; cause and effect and 114; content standards and **257–259**; engineer it! and 270–273; English/language arts and 277, 299, 306, 311; essential questions and 256; fact or friction? and 257, 308; goals, objectives 255–256; Internet resources 270; launch and 256; learning plan components and 267–268; lesson preparation and 267; let's get energetic! and 263; materials matter and **263**, 284–291; mathematics and 255–273, 284; module summary 38; NGSS and 256; outcomes and 262; overview of 114; ready, set race: the x-challenge and **263–264**, 313–320; rubber bands and 297, 299, 306;

snow-proof school challenge and 270; social studies and 255–273, 284; start your engines and **263**, 265, 267, 268; teacher background information and 265–267; timeline of **263–264**; 21st century skills and **260–261**; *see also* automotive x-challenge (7); design journals, transportation-motorsports (7)
tsunamis 109
twelfth grade 163–173
21st century skills: definition of 25, 31; eighth grade and **124, 125, 127, 128, 130**; eleventh grade and **155, 158, 159, 160, 162**; fifth grade and **94, 95, 97, 98, 100**; first grade and **60, 61, 62, 63**; fourth grade and **86, 87, 88, 89, 90**; grades K-2 and 46; grades 9-12 and 73, 135; grades 6-8 and 102, 103; grades 3-5 and 103; interdisciplinary themes and 260; kindergarten and **50, 51, 53, 55**, 260; ninth grade and **138, 140, 141, 143, 144**; Partnership for 21st Century Skills (P21) 35; second grade and **66, 69**; seventh grade and 38, 114, 116, **119, 120**, 260; sixth grade and **106, 107, 109, 110, 111**; tenth grade and **148, 149, 151,** 152, **153**; third grade and **77, 78, 79, 81, 82**; twelfth grade and **166, 167,** 170, 172
21st Century Skills Framework 4, 28, 32

urban planners 62–63, 90

velocity **303**, 303
Vernier probes 211–213, 215
vertical alignment STEM partnerships 236, 237, 238, 240, *240*
veterinarians 347
volcanoes 84, 85

water: alternative sources of 126; change over time—our schoolyard (2) and **65**, 68; global water quality (6) 103, 104, **105**, 108, **110**, 135; green building rooftops (11) and 154, **155**, 159, **161**; hydrologists 99, 100; hydropower efficiency (4) 84, **87**, **89**; rainwater analysis (5) and **92**, 93, 95–96, **97**, 101; soil

erosion (4) and 96; survival on earth-water **152**
water conservation (4) 84, **84**, 89–90
waterwind turbines **92**, 93, 94, 95
waves (1) **47**; *see also* influence of the waves (1)
weather (K) (3) 76, 77, 327–328, **329–332**, **334**, **335**, 337; *see also* climatologists; meteorologists; snow-proof school challenge
wetlands 147
Why Some Schools with Latino/a Children Beat the Odds and Others Don't (Waits et al.) 204
wind up flashlight (7) 277–280
window box gardens (1) **57**, 59, 60
Word Wall 212
workforce *see* jobs/workforce world patterns/living things impact (K) 319–21; *see also* goats (K); petting zoo (K)

X-challenge *see* automotive x-challenge (7)
X-challenge engineer it! (7) 314, 315, 324–326

zoologists 347

CPSIA information can be obtained
at www.ICGtesting.com
Printed in the USA
BVHW051832180522
637434BV00016B/151